WATERFRONT WORKERS OF NEW ORLEANS

Race, Class, and Politics, 1863–1923

ERIC ARNESEN

New York   Oxford

OXFORD UNIVERSITY PRESS

1991

Oxford University Press

Oxford   New York   Toronto
Delhi   Bombay   Calcutta   Madras   Karachi
Petaling Jaya   Singapore   Hong Kong   Tokyo
Nairobi   Dar es Salaam   Cape Town
Melbourne   Auckland

and associated companies in
Berlin   Ibadan

Copyright © 1991 by Eric Arnesen

Published by Oxford University Press, Inc.,
200 Madison Avenue, New York, New York 10016

Oxford is a registered trademark of Oxford University Press

Library of Congress Cataloging-in-Publication Data
Arnesen, Eric.
Waterfront workers of New Orleans :
race, class, and politics, 1863-1923 / Eric Arnesen.
p. cm.   Includes bibliographical references and index.
ISBN 0-19-505380-X
1. Stevedores—Louisiana—New Orleans—History.   I. Title.
HD8039.L82U63   1991
331.7'61387164'0976335—dc20
91-39686

1 3 5 7 9 8 6 4 2

Printed in the United States of America
on acid-free paper

For my parents, Alice and Jack

PREFACE

During the nineteenth century, American and foreign travelers often found New Orleans a delightful, exotic stop on their journeys; few failed to marvel at the riverfront, the center of the city's economic activity. What impressed them was the forest of masts and flags, the range of languages heard and nationalities seen, the quantities of freight accumulated, and the scale of business transacted along the miles of wharves. Many observers also took notice of the army of dock workers and the variety of labor it performed. Workers on or near the riverfront loaded and unloaded steamships, screwed cotton in their holds, transported bales and barrelled freight across the wharves to the cotton yards and warehouses, compressed that cotton at the presses in these yards, and worked on board the vessels that sailed up and down the Mississippi River. Most often arriving by river, travelers would be struck immediately by New Orleans' dependence upon commerce, and commerce's dependence upon human labor.

Today, the riverfront continues to attract visitors to New Orleans, but it bears scant resemblance to the nineteenth-century wharves. Far fewer (though much larger) ships are visible, as postwar containerization has revolutionized the longshore trade, drastically reducing the size of the labor force. The tourism industry, which in large part has supplanted commerce as New Orleans' economic base, promotes the local color of the French Quarter and the Mississippi River to conventioners and other visitors. Where nineteenth-century travelers encountered hundreds of ships and thousands of men, their late twentieth-century counterparts see restored steamboats ferrying ticket holders on scenic river cruises and tourists walking along a waterfront promenade, browsing in gift shops, and sampling local cuisine and fast food in the revitalized Riverwalk development. In packaging New Orleans' past, the modern tourism industry pays tribute to the vital role that commerce once played in New Orleans' economic history. Along the river, historical markers describe for the passer-by the character of the Mississippi River steamboat traffic, the importance of trade, and New Orleans' place as an immigration center for Irish, German, and other groups. But absent from the tourism industry's historical recollection is any reference to the immigrants or black migrants and their children who constituted the army of laborers along the riverfront and provided the essential human power to keep the cotton, sugar, and other goods flowing. Today, it is as easy to ignore the human dimension of New Orleans' commercial past as it was impossible for nineteenth- and early twentieth-century New Orleans to avoid it.

In the renaissance of both African-American and labor historiography since the 1960s, few studies have focused on black urban workers, black and white common laborers, and the attendant questions of class and race relations. Historians of the African-American experience have documented over the past decades important aspects of the urban black environment: the institutions that African-Americans built, the values they maintained, the racial and economic handicaps under which they operated, and the efforts they made to transcend racial proscriptions. Yet while these historians have produced excellent occupational outlines, they have paid little attention to the daily world of work experienced by black workers and their efforts to secure some measure of control and dignity in their working lives.

Following a distinct and parallel path, labor historians have transformed the way we conceptualize workers' worlds. Inspired by English historian Edward Thompson and U.S. historians Herbert Gutman, David Montgomery, and others, practitioners of the so-called new labor history have shifted the field's focus away from trade unions and institutions to the broader world of culture, community, politics, gender, and ethnicity. The earlier subjects of these studies were most often skilled artisans and craftsmen; more recently, factory operatives in northern and midwestern industries and communities have received attention. But gaps remain in the new portrait of American working people. Over two decades ago, Herbert Gutman regretted that the "absence of detailed knowledge of the 'local world' inhabited by white and Negro workers" meant that "little is known of the interaction between Negro and white workers, North and South, in particular communities." Suggesting the need to bridge the barriers between labor and African-American history, Gutman's own research on black union leader Richard Davis and the racial policies of the United Mine Workers represented an important start. But the project that he proposed—to shift attention from the policies of racial exclusion and discrimination by union internationals to the practices of local unions and communities—remains, with several notable exceptions, largely unaccomplished. Southern workers, blacks, and common laborers await fuller exploration.[1]

In examining one diverse group of workers—the 10,000 to 15,000 cotton screwmen, longshoremen, cotton and round freight teamsters, cotton yardmen, railroad freight handlers, and Mississippi River roustabouts—this book focuses primarily on the workplace and the labor movement that emerged along the riverfront. For both black and white dock workers, trade unions and interunion alliances produced a structure for mediating racial and occupational tensions, formulating strategy, and devising work rules to govern on-the-job behavior, limit exploitation, and increase wages and union control over work. In the years before 1880, racial and occupational divisions kept black and white dock workers fragmented and weak. The emergence of biracial union agreements and biracial and multitrade alliances in 1880, however, transformed racial and class experiences on the city's docks. In two periods—1880 to 1894 and 1901 to 1923—blacks and whites joined together in a series of alliances that enabled them to expand substantially their collective power over the conditions of labor.

The economic structure of the longshore trade and labor market lent an urgency to the formation of these alliances. As largely unskilled workers in often crowded trades, dock workers learned that employers could defeat efforts to improve working conditions and raise wages by manipulating racial or occupational divisions or by drawing upon a large pool of potential strike-breakers. Most of New Orleans' dock workers faced a stark choice: ally across racial and occupational lines or suffer the consequences of division. But this functional imperative only tells us that interracial and inter-trade collaboration at the workplace was the answer to the problem of a racially divided, unskilled labor force; it does not explain how those alliances formed and functioned, or how they could survive—even flourish at times—in a period marked by deteriorating race relations in other sectors of urban life.

During the 1880s and after the turn of the century, longshore workers sustained a movement that ran counter to the dominant trend of black subordination, exclusion, and segregation in the age of Jim Crow. *Waterfront Workers* argues that two key factors help explain the existence of powerful biracial union alliances in these periods. First, the strength of black unions was central in limiting white workers' ability to impose a racially exclusionist solution on the problems of competition and unemployment on the city's docks. To achieve their goals, white workers often had little choice but to secure black cooperation. Second, New Orleans workers benefited from a political culture that proved receptive to dock workers' efforts to build unions. Dependent upon a large white working-class electorate, the Democratic party machine, for all of its corruption, inefficiency, and institutional racism, avoided breaking strikes at critical moments and ignored workers' violations of the hardening racial codes.

As a social history of race and class relations in New Orleans during the sixty-year period following the Civil War, this book is a contribution to the effort to bridge the chasm between the fields of labor and African-American history. It focuses upon the institutions built by black and white workers to protect themselves and address their needs, the world of work, the relationship between work rules, union power, and race relations, and the ongoing struggles over workers' control and managerial power. But this book is neither a community study nor a comprehensive survey of race relations and working-class culture. In the past two decades, both African-American and labor historiography have stressed the importance of community, neighborhood, and family in understanding the experiences of white workers and black Americans. Such works have been invaluable in broadening our conception of working-class and black culture in the nineteenth and twentieth centuries. *Waterfront Workers* touches upon such issues peripherally only as an explanatory backdrop to the narrative account of other aspects of race and class relations on the city's docks. New Orleans, its working people, and their communities certainly deserve far more historical scrutiny than they have hitherto received, and I hope that this study suggests sources, topics, and questions that others pursuing the city's history will find useful.

Several theoretical propositions run through this book. First, while racism among whites was certainly ubiquitous during the sixty-odd years covered by this study, the forms that it assumed (embodied in institutions, attitudes, and behaviors) and its particular salience in certain periods demand careful historical identification. Racism was never static nor uniform. As social theorist Stuart Hall has argued, there "have been many significantly different racisms—each historically specific and articulated in a different way with the societies in which they appear," making it necessary to study racism both "sequentially" (the construction of racism over time) and "laterally" (racism's connections to other societal developments).[2] Snap-shot histories, focusing on single, extraordinary events such as interracial unity during a particular strike or racial violence during riots, often result in overly broad and misleading conclusions. It is only when relations between black and white workers are studied *over time,* and within the context of larger social, political, and economic developments, that the evolving dynamics of race and class relations can be understood.

Second, while this study necessarily addresses a problem that is common to both African-American and labor history—the dichotomy of race and class—those looking for a new formulation of the race/class problem will not find one in these pages. Historians often have pitted these terms against one another, some emphasizing that race has been more fundamental to southern history than class, others giving greater attention to those rare moments when class temporarily has transcended race. Where black and white workers have bridged the racial divide, their more or less common economic position—class—is said to have triumphed over race; where white workers have rejected interracial alliances, holding themselves aloof from blacks or excluding or even attacking them, race is said to have triumphed over class. An underlying theme of this work is that the race/class dichotomy, which poses the two terms as binary opposites, is not a useful one.[3] The problem itself cannot be understood by, or reduced to, a simple global formula. In so arguing, it is not my intention to diminish the centrality of racism in the study of working people or to embrace empiricism over theory (although I do believe that a good deal more empirical research into southern labor and urban African-American history is necessary before we can arrive at fresh theoretical insights into the interactions of race and class in American history. The paucity of research on black workers and working-class race relations should caution us that recent calls for historical synthesis in the field are quite premature).

Waterfront Workers suggests that a different set of approaches is needed to replace the race/class dichotomy and to guide our efforts in uncovering the social history of race and class relations. The opposition of often ill-defined and transhistorical formulations of race and class should be replaced by historically specific definitions of racial and class identities. The issue is not which category assumed dominance at a given historical moment. Rather, it is how particular groups of workers defined themselves; precisely how did notions of racial identity inform workers' understanding of their class identity, and vice versa? How did those definitions, for whites, contribute to the development of strategies of inclusion or exclusion? How did those definitions, for blacks, translate into strategies for survival in an often hostile urban economy? Black workers along the New Orleans waterfront saw themselves as both *black* workers and black *workers;* white workers similarly saw themselves as working men who were white, as whites who were workers. That is, longshore laborers exhibited complex white working-class and black working-class identities. In day-to-day life, the class and racial components of these identities were not in competition; so too, at the level of analysis, they are not separable. Workers' experiences were not determined by abstract categories of race and class but instead reflected their positions within both the class and racial hierarchies in urban southern society. At the workplace and in the union hall, their behavior, their goals, and their agendas seldom could be reduced to either labor or racial identity.

As all historians, their friends, and colleagues know, research and writing are highly individual endeavors only made possible through a collective process. Along the way numerous people have provided me with assistance. Librarians and archivists alerted me to collections, newspapers, and folders that otherwise might have escaped my attention. I would like to thank the staffs of the Amistad Research Center at Tulane University, the Louisiana Historical Collection, the Archives and Special Collections of the University of New Orleans, the Historical New Orleans Collection, the New Orleans Public Library, the Louisiana Collection and Manuscripts, Rare Books and University Archives at the Howard-Tilton Memorial Library at Tulane University, Xavier University, the Department of Archives and Manuscripts at Louisiana State University, Wayne State University's Archives of Labor and Urban Affairs at the Walter P. Reuther Library, the Moorland-Springarn Research Center at Howard University, the Library of Congress, the National Archives, Yale's Sterling Memorial Library, and Harvard's Widener Library and Littauer Library.

I owe debts of gratitude to numerous individuals and institutions. In New Orleans, Wayne Everard, Edward Haas, Lester Sullivan, and Robert Skinner pointed me in fruitful directions; Raymond Nussbaum also read portions of the book. Art Carpenter, Karen Bell, and David Wells gave advice at an early stage. Joseph Logsdon, of the University of New Orleans, shared with me his considerable knowledge of local history and sources, directing me to rich collections of nineteenth- and early twentieth-century materials. At the National Archives, Ken Hall and John VanDereedt navigated through the records of the U.S. Shipping Board, while James Cassedy provided guidance at the Washington National Records Center in Suitland. Stuart Kaufman and Katharine Vogel shared documents from the George Meany Memorial Archives. At Harvard University, Claire Brown of Littauer Library and Nathaniel Bunker of Widener Library located or procured a wide range of materials relating to New Orleans, labor, and African-American history. Maura Henry and John Donaghy provided research assistance, while Laura Johnson aided in the manuscript's preparation. Sheldon Meyer, Rachel Torr, Karen Wolny, and Leona Capeless of Oxford University Press provided encouragement and editorial help. Yale University's Afro-American Studies Program and History Department and Harvard University provided travel grants at various stages.

Over the years, the manuscript has benefited from the critical reading of many people. Leslie Frane, Dana Frank, Cecelia Bucki, Jeffrey Gould, Judith

Stein, Jim Stoddard, Laura Wilson, Noel Ignatiev, and Elizabeth Haiken read portions or all of *Waterfront Workers*. John Blassingame read an early draft and offered bibliographic assistance. Melvyn Dubofsky's comments on a conference paper drawn from one of the chapters led me to reevaluate several central ideas. I especially benefited from discussions of New Orleans' history with Arnold Hirsch, who also generously read the manuscript. Historical writing, of course, is enhanced by exchanging ideas and debating interpretation, and a number of friends and colleagues made this project more exciting. Karin Shapiro has offered commentary, criticism, and support since our days in Yale's Afro-American Studies Program. I have valued highly my ongoing discussions with Daniel Letwin on issues in African-American, southern, and labor history, learning from his questions and insights. Their encouragement and friendship helped sustained me throughout this endeavor. Two teachers at Yale University shaped this study in important ways. David Mongomery, who deepened my interest in labor and black history, directed the dissertation that this book is based upon. As everyone who has worked with him can testify, his knowledge, insight, and inspiration make studying history a pleasure. Emilia Viotti daCosta was an unsparing critic of this work. A committed teacher and engaged scholar, she has constantly challenged me and her other students to rethink our arguments, interpretations, and conclusions. I am especially grateful for her support and guidance. Finally, Sarah deLone patiently listened to me work out my ideas, read the manuscript at virtually every stage, and offered editorial, political, and historical advice. I thank her for this, and for her friendship.

Cambridge, Mass.   E. A.
June 1990

CONTENTS

Waterfront Workers of New Orleans

1

The Remaking of a Union State: Labor and Politics in the Early Reconstruction Era

"I have read of men being suddenly struck dumb by some violent concussion of air," a *New York Times* correspondent noted in an 1863 article about New Orleans, "but the loud thunder of this war seems to have acted in a directly opposite manner, and given speech to millions who had hitherto been tongue-tied by the resistless influence of Southern despots."[1] Over the course of four years of bloody fighting, the Civil War had ended slavery and had initiated profound transformations in southern life. But with the North's final victory, the shape of the South's new social order remained contested. While southern planters sought to minimize the changes, white urban workers, free people of color, and ex-slaves in New Orleans advanced their own, more radical programs for Reconstruction. In August 1865, the free black newspaper, the *New Orleans Tribune,* reflected upon the alignment of social forces in Louisiana's largest city. "The poor men of this city will never lend their aid or votes to restore the old order of things," it noted. "All men, the poor white laborer and mechanic, and the colored tiller of the soil, are now free. The feeling for the old slaveholders—whose rule was one of oppression and brutality—is no longer that of respect or awe; it is one of contempt and hatred."[2]

But if the war had "given speech" to millions of previously tongue-tied people, they hardly spoke with one voice at the outset of the postbellum era. Subsequent events demonstrated that common opposition to the planter class, urban employers, and the old regime provided little basis for political or economic collaboration between two major groups in a reconstructed New Orleans—white craft workers and blacks, free and ex-slave. From the start, sharp conflicts over racial and class issues marked the struggle for political control in New Orleans. While both white workers and blacks strongly opposed a restoration of the old order, they also raised distinct and contrasting political agendas at the start of Reconstruction and utilized different means to achieve their goals.

Black and white workers occupied separate material and ideological worlds at the war's end. At the level of daily experience, occupational hierarchies and a sharply segmented labor force structurally divided black and white

3

workers. At the level of ideology, white journeymen and mechanics viewed themselves as the producers of society's wealth and the foundation of republican government. Turning to politics, they sought to secure the dignity of white labor through the reduction of the working day. Although they applauded the destruction of slavery as a victory for free labor, at no point did they extend their hand to the city's black population. For their part, the free blacks of New Orleans sought civil rights, black suffrage, and protection of unskilled freed men and women; black workers subject to widespread poverty, state repression, and new forms of labor coercion required a great deal more than the eight-hour day to secure the dignity of *black* labor. The basis for an alliance between blacks and white mechanics appeared slim indeed. Essential to understanding the dynamics of race and class relations in the postbellum era, the analysis of those separate worlds of white mechanics and blacks, free and ex-slave, will lay the basis for the subsequent exploration of the experiences of black and white workers on the New Orleans waterfront.

I

New Orleans' short-lived participation in the Confederacy ended in late April 1862 when Union Admiral David Farragut's fleet lay siege to the city. On May 1, Major General Benjamin Butler's troops began their occupation of the Crescent City. Placing the city under martial law, federal military authorities appointed officials to administer municipal affairs. But an early restoration of civilian rule in federally controlled Louisiana was important to President Abraham Lincoln's plan for winning the war. He hoped that the consolidation of Union control in the state might undermine Confederate morale, providing an incentive for other Confederate states to rejoin the Union.[3]

To accomplish that goal, military occupation officials in New Orleans tapped several currents of unionist sentiment. First, they turned to a small group of loyal southern men to form the nucleus of a political force that could facilitate a successful transfer of power. For the most part, the leaders of this unionist movement were atypical of southern politicians. According to historian Ted Tunnell, many had been prewar Whigs who had not held political office; a majority were born or had lived in the North, in areas of the upper South where slavery was weak, or abroad; overall, they embraced a strong American nationalism. Taking advantage of the political vacuum created by the military occupation, in the summer of 1862 these men formed the Union Association, which later evolved into the Free State party.[4]

A second, larger source of support for the Union came from Irish and German immigrants who constituted roughly two-fifths of New Orleans' 174,491 residents. For economic reasons, many white immigrants felt little allegiance to the planter class. "The great mass of foreigners who come to our shores are labourers, and consequently come in competition with slave labour," noted the *Morehouse Advocate* in the early 1850s. "It is to their interest to abolish Slavery. . . . [They . . . entertain an utter abhorrence of being put on a [economic] level with blacks whether in the field or in the workshop." In some cases, ideological considerations were equally important. "These men come from nations where Slavery is not allowed, and they drink in abolition sentiments from their mothers' breasts," the paper concluded. Louisiana senator John Slidell similarly assessed the political proclivities of New Orleans immigrants. Although the foreign-born did not support Lincoln in the presidential election of 1860, as he feared they might, neither did they vote for John C. Breckinridge, the candidate of the Southern Democratic party and the choice of southern hardliners. "[H]ere in the city seven-eighths at least of the vote for [northern Democrat] Douglas were cast by the Irish and Germans, who are at heart abolitionist," Slidell conceded after the election.[5] The exaggeration of abolitionist sentiment notwithstanding, many immigrants opposed the peculiar institution and the political power that the planter class derived from it. Few urban German and Irish immigrants were slaveholders, and many of the Forty-Eighters, radical German refugees, professed a democratic ideology which was, at its core, antithetical to property ownership in human beings. One prewar politician, the Bavarian immigrant Michael Hahn, articulated those sentiments during the 1850s and 1860s, and found strong backing for both his anti-slavery and anti-black positions from the New Orleans immigrant working class.[6]

From the start of the federal military occupation, General Benjamin Butler and his successor, Nathaniel P. Banks, recognized the importance of New Orleans' large working class to the successful restoration of civil rule. In the decade before the war, white workers had emerged as a distinct political force, eventually dominating the tightly knit ward organizations of the Know-Nothing party. In 1858, the party elected the president of the city's Printers' Association to the mayor's office, and in 1860, it elected a former stevedore and union member, John Monroe, to the same office. While the fall of Confederate New Orleans shattered the party's hold over municipal affairs, Union military officials understood that the creation of a loyal city government required the backing of the city's working-class voters. Military officials secured that support by distributing relief aid, imposing price controls on food, and providing large numbers of public works jobs to unemployed whites.[7]

More than relief, patronage, and pro-labor policies lay behind the support of some New Orleans workers. As the pro-Union Free State Movement took shape in 1863, white craft workers in New Orleans established a Working Men's National Union League to voice their concerns about the political and military upheaval around them. Four themes dominated their analysis of the crisis and their sense of place in the new order. The first was the belief in the centrality of the mechanic class to the health and security of society. Drawing upon elements of a tradition that extended back to the era of the American Revolution, New Orleans mechanics repeatedly asserted the dignity of labor. The "[m]an who earns his living by the sweat of his brow," delegates to the League's first state convention in June 1863 unanimously affirmed in a classic statement combining a labor theory of value and artisanal republicanism, was the "very corner-stone of a Republican Government." Second, these working men were not only the Republic's cornerstone but the South's true, loyal, and unconditional Unionists opposed to the slaveholders' rebellion as well. As the newest recruits in "the old contest for man's natural rights," U.S. Army Major Davis Alton told a League crowd, working men were "carrying forward the grand work of redeeming mankind from the thraldom of tyrannical oppression, and a blighting, crushing aristocracy." To accomplish that task and to assume their rightful place in a reconstructed South, League members demanded suffrage for all white men age eighteen years and older and a new state constitution in which the "rights of white men will be secured." Third, the Working Men's League expressed a hatred of the slave system itself. Sharing the outlook of northern free-soil advocates, craft workers saw their political and economic opportunities eclipsed by slavery and slave owners. In an address before the League, U.S. Army Surgeon P. K. Smith argued that the "poor white man in the South . . . had been bound down to this negro-ocracy . . . and kept in such abject poverty by this tyrannical system of capital over free labor. . . ." The slaveholders' purpose was nothing less than to build up the slave oligarchy, form a new government with an aristocracy, and to "enslave or grind down the poor people, and deprive them of the right of suffrage." Another northern speaker, Major Davis Alton, told League members that the primary object of the rebellion was to "reduce poor white working men in the South to the same degraded condition of the negro slave." League members no doubt shared these sentiments. At a meeting on June 2, 1863, they unanimously resolved that "the question of slavery is a question of capital over free labor; and when we behold this bloody war, waged for the maintenance of this monopoly, which is contrary to the very best interests of a free people, we think it high time to sap and undermine the foundation of this absurd and tyrannical system."[8]

New Orleans mechanics' opposition toward the institution of slavery and the slaveholding class produced little understanding of or sympathy for the plight of slaves or free blacks. The League's vision of a reconstructed, free labor Louisiana had no place in it for blacks. Rather, the mechanics advocated the colonization solution—the removal of all persons of African descent from Louisiana.[9] In adopting such a stance, white workers revealed that they indeed shared general American racist beliefs, but it is likely that several sources specific to their social and class position also informed their outlook. The segmentation of the labor force by race created a deep gulf between white mechanics and blacks; white craft workers monopolized many of the more highly paid jobs, or worked apart from slave or free black artisans. Inhabiting institutionally separate worlds of work, white mechanics sought not to raise the level of all labor but instead to protect their monopoly from competition from blacks below them. (Unskilled whites, for their part, behaved in similar ways.) Important ideological factors also contributed to white labor's exclusionary position toward blacks. Artisanal republican beliefs held by New Orleans craft workers that linked economic and political independence, property ownership, virtue, and citizenship reflected, and reproduced, racial hierarchies. White farmers and craft workers saw themselves as society's producers; their independence, competence, and civic virtue made them uniquely qualified to uphold the nation's republican heritage. In this schema, how-ever,blacks possessed few of these valued attributes. Lacking independence and property, they constituted a group apart, incapable of exercising civic virtue and not qualified to participate in the nation's politics. Their dependent and degraded status allowed whites to perceive them, in Peter Rachleff's words,as a "negative reference group: the -antithesis of the 'independent' white worker."[10] Drawing no distinction between free blacks and slaves, these white artisans ignored the fact that at least some free blacks actually possessed property and exercised the economic independence whites prized so highly.

In late 1863, the *New York Times* observed approvingly that the Working Men's National Union League aimed at nothing less than to "entirely revolu-tionize the social condition of this State."[11] If it failed to accomplish that task, white craft workers appeared to gain considerable, if temporary, political strength. First, they elected Michael Hahn governor in the military-sponsored election of February 1864. Opposed on their right by conservatives who sought to stem the tide of social change and restore the old order, and on their left by radical Free State men who advocated black suffrage, the moderate Free State men accepted emancipation but opposed political and social equality for blacks. A staunch opponent of slavery, secession, and planter power, Hahn put forth positions closely resembling those of the Working Men's National Union League; he secured the organization's support, along with that of the Mechan-ics' Association, the Crescent City Butchers' Association, and the Workingmen of Louisiana, whose members marched for Hahn in pre-election parades.[12] Second, following Hahn's election, city mechanics and manual workers turned their attention to the drafting of a new state constitution, presenting a memorial signed by over 2000 "loyal men." Sensitive to the electoral strength of New Orleans' white working class, convention delegates adopted two of its key demands—a two-dollar daily minimum wage and the nine-hour day for all employees on public works projects. More importantly, the 1864 constitution broke the monopoly of the state's planter class by making the basis of represen-tation the total qualified electorate, not total population, a move that shifted electoral power to white farmers and urban workers.[13]

Despite their auspicious start, white mechanics quickly found their influ-ence diminished in the turbulent politics of early Reconstruction Louisiana. The ascendancy of moderate Unionist forces in the civilian government proved temporary. Governor Hahn's election to the U.S. Senate in early 1865 placed Lieutenant-Governor James Madison Wells, a unionist, planter-aristocrat from Rapides Parish, in charge of the state. Unlike his predeces-sor, Wells ignored the labor basis of his party and instead courted conserva-tives, appointing as mayor of New Orleans Dr. Hugh Kennedy, a "strong advocate of the rebellion, a man who favored oppression, who believed in elevating the aristocracy and degrading the laboring classes," in contempo-rary Emily Reed's words. Together, Wells and Kennedy offered patronage jobs to returning ex-Confederates, restored the contract system on public works, and cut the pay of public employees.[14]

White workers expressed outrage at the new administration's efforts to turn back the clock. The mayor's wage cuts for city laborers, in particular, provoked a major protest. Workers converged on Lafayette Square only to discover that Kennedy had locked them out. Gathering instead at City Hall, they condemned the "present improvised and irregular Government" for attempting "to overrule the Constitution of the State by repealing the labor ordinances, thus removing one of the supports and guarantees due to labor." Denouncing Kennedy's administration as a failure, the mass meeting demanded the resignation of the mayor—"that incompetent functionary"—and "all others concerned in the movement against the interests of labor." The city's blacks also viewed the political changes with alarm. Kennedy's appointment, argued the *Black Republican,* "was the beginning of a new rule of Copperheads and rebels, out of which, if it were possible, slavery would be re-established, and all the old wrongs of the slavocracy would be again fastened upon us. Slavery never had a stronger advocate than Dr. Kennedy, nor a more practical supporter than Gov. Wells."[15] Events in the fall of 1865 consolidated the triumph of the ex-Confederates. In early October, several conservative political factions united under the banner of the National Democratic party, whose platform called for white supremacy, the destruction of the 1864 constitution, compensation for emancipation, a general amnesty for Confederates, and full property restoration. As the party's nominee for governor, Wells won a clean sweep in the November 1865 election.[16] For the time, it appeared that despite the Union military victory, a counterrevolution was well under way.

II

By the fall of 1865, it was unclear where the white labor movement would fit into the political spectrum. The newly formed Republican party, whose platform called for civil and political rights for blacks, held little appeal to the working-class whites. Yet while the Democratic party's racist orientation may have won it votes among white workers, the dominance of economic conservatives and the old planter class in the party's leadership troubled labor activists. In the fluid political situation of 1865 and 1866, New Orleans' organized craft workers maintained a critical distance from mainstream Democratic politics. They set forth a vision of white labor's position in society, shared by other white workers, northern and southern, of a producer republic whose objectives could be realized through politics. That they failed to carve out or sustain a distinct ideological space within the political realm attests to not only the weakness of white working-class organization in this period but also to the resilience of machine politics, the persistence of racial divisions, and the common racial ground between craft workers and Democrats.

The eight-hour movement became the mechanics' vehicle for promoting their vision of white labor's place in American society. Across the nation, the ten-hour day and the six-day week had been standard for skilled artisans and unskilled laborers alike for roughly two decades. In the years immediately after the Civil War, the struggle for the eight-hour day engaged the energies of workingmen in dozens of American cities, and New Orleans was no exception.[17] Following the Democrats' electoral sweep in November 1865, some 2000 journeymen and "honest mechanics"—plumbers, coopers, house carpenters and joiners, ship carpenters, plasterers, journeymen tailors, journeymen shoemakers, and boiler makers—inaugurated a movement to reduce the length of the working day through legislative means. The "social, physical, intellectual and moral condition of the mechanical and laboring classes is not what it ought to be," contended the delegates to a mass meeting at the St. Charles Street Opera House held in mid-December. Complaints against the long hours of labor were many. Physical prostration and premature death resulted from constant labor without sufficient recreation; "want of time to obtain mental food" gave rise to mental inactivity; and excessive labor at the "expense of the higher natural wants of man" generated moral turpitude. Long hours also posed a dangerous political threat to the Republic. Under the ten-hour system, the delegates argued, "we are deprived, by want of time, from obtaining the necessary information and intelligence which as American citizens, entrusted with the sacred privilege of self-government and guardians of liberty, we ought of right and necessity to possess." The eight-hour system, which would allow workers time to cultivate their minds so they would not "be used by political tricksters as tools to effect their designs," would solve many of these problems.[18]

Workers debated the eight-hour system in the streets, meeting halls, and press. In March 1866, between 800 and 1000 workers in numerous mechanics' and benevolent societies paraded in a "grand procession" proudly displaying their banners down the city's main streets. Headed by the Ship Carpenters' and Caulkers' Association, which carried a full-rigged miniature steamship, the procession made its way to Lafayette Square. In a two-hour speech, labor radical Richard Trevellick, national president of the Ship Carpenters' and Caulkers' union, articulated a labor theory of value for his listeners: wealth "was but an accumulation of labor unconsumed." Drawing on arguments of workers across the nation, Trevellick offered a classic critique of overwork. Under the present system,

> . . . a laborer returns to his home jaded and care-worn. From the depression experienced, his wife and children are treated with harshness, and life has but little pleasure. The common dictates of humanity call for a remedy. Let the laborer enjoy a few hours for mental culture, and science and art will, in the end, be advanced and benefited . . . There were now three thousand drinking saloons in the city. The exhausted laborer resorted to them for momentary relief and exhilaration; instead of supporting these, with reasonable hours of leisure, reading rooms and libraries would spring up, and the laboring man would have an opportunity to turn aside for the moment and encourage a rational mental culture.

The solution was simple. The day, he concluded, should be divided into three parts: "eight hours for labor, eight for repose, and eight for recreation and mental culture" in order to "elevate the tone of the mechanic."[19]

While New Orleans craft workers issued a powerful indictment of the conditions of labor, they often found they lacked the power to change them. Opposition from employers was strong. Businessmen and newspaper editors paid careful attention to the "active proceedings of the various mechanical societies" for the eight-hour day in New Orleans and elsewhere in the postwar era, and responded with their own critique of labor radicalism and a defense of the status quo. Master mechanics in the building trades, for instance, asserted "an identical interest" between the mechanic and his employer; defended the "uncontrollable laws of supply and demand" which alone determined labor's value; and rejected their journeymen's effort to legislate wages as an act in "direct opposition to what experience teaches in political economy." Moreover, New Orleans employers objected strongly to the injection of class issues into a politics that they were framing around race. ("No time could be more unfortunately chosen than the present for the agitation of a question involving so radical a change as that contemplated," remarked Veritas, a New Orleans employer, "when the whole labor question of the South is disturbed by the 'new order' of things.") Setting forth arguments that would become familiar elements in the New South boosterism of subsequent decades, employers admonished that capital required encouragement, not challenges, before it would venture south, and encouragement necessitated the "absence of anything tending to destroy the relations hitherto existing between the workmen and his employer."[20]

While the eight-hour system dominated the agenda of white labor, workers formulated several strategies to accomplish their goals. Painters, iron molders, and several other groups in the "Trades Assembly" unilaterally declared the eight-hour day in effect on March 12, 1866; some employers acceded to the demand, most successfully resisted. Denouncing the strike as injurious to all concerned, building trades employers vowed to maintain a ten-hour day and to employ no one "who will not comply with said uniform system." In response to opposition from proprietors of city iron foundries, pattern makers, mechanics, moulders, and blacksmiths attempted to establish a cooperative foundry based upon the eight-hour system. These efforts, it seems, had little impact.[21] Instead, most organized workers turned to politics. Over the past year, Mayor Hugh Kennedy and the pro-Confederate legislature had repudiated the pro-labor articles in the 1864 constitution. In December 1865, a mass meeting of workers resolved to create eight-hour leagues in each district of the city and to elect one man to canvass the whole state and organize new leagues. Most important, they now pledged to vote for only those local and state candidates who supported the eight-hour system.[22]

Craft workers' showdown with the Democratic party came in May 1866 municipal elections.[23] A newly formed Union Workingmen's party pledged to passing eight-hour legislation courted labor's vote, drawing its small membership from the ranks of organized craft workers, old Free State men, and disaffected Democrats. For their part, Democratic party orators denounced the "aggrandizement of one class at the expense of another" and rejected the relevance of the eight-hour question to the election. The attempt to regulate

capital and labor, one party activist insisted, was a "species of insanity" comparable to an "attempt to beat down Gibraltar with your feet." Condemning
their opponents' factionalism, Democrats stressed the importance of rallying
around party standard bearers and supporting President Andrew Johnson's
Reconstruction policy.[24]

The Workingmen's party lost soundly to the Democrats in the May 1866
election. Democrats' appeals to party loyalty and racial solidarity and their
control of patronage proved stronger than the Workingmen's advocacy of
shortened hours. But the eight-hour day notwithstanding, Workingmen and
Democrats had a great deal in common. Both parties called for national unity
and support for President Johnson. "The only hope for the American constitution and the principles of republican government," one speaker told a rally of
the Workingmen, "rested in the success of President Johnson's policy." As one
party activist put it, it was wrong for the National Democracy to call the eight-
hour men "Black Republicans for daring to assert their rights," because the
Workingmen "were the best Democrats."[25]

If not the best Democrats, then at least good Democrats. Over the course
of the following two years, the social condition of the state had indeed been
revolutionized, but not in the way predicted by the *New York Times* correspondent or hoped for by the original Workingmen's League or its descendants.
The small white labor movement had suffered a sharp decline by 1868, as
economic depression wiped out many local craft associations. On the political
level, Military Reconstruction temporarily broke the electoral hold of conservative Democrats by enfranchising black voters, enabling the Republican
party to take control of the state government. White craft workers, however,
made no alliance with the new political force. Although a new Workingmen's
party ran unsuccessfully in the municipal and state elections in April 1868, its
small number of supporters amongst white workers were back in the Democratic fold by the summer.[26]

The correspondence of one New Orleans mechanic, P. J. K., to the
Chicago-based *Workingmen's Advocate* in 1868 suggests the contours of white
craft workers' thinking toward the political changes that were taking place.
Particularly revealing was his commentary on the subject of race and black
political participation. The new state constitution, drafted by black and white
Republican delegates, guaranteed blacks the right to enter and associate "in
all public places." "Then the good time will come," P. J. K. remarked sarcastically, "and our city and state will become an Eldorado for these new citizens
and supposed hell for all others." Particularly disturbing to the correspondent
were the city Republican's educational policies promoting integrated schools,
measures which he believed would produce a "bleaching [of] the incoming
generation."[27] P. J. K. was not alone in his negative assessment of blacks. At a
meeting of a newly formed Workingmen's Union in March 1868, white craft
workers divided sharply over the presence of one Mr. Rogers, a black. No
sooner had Rogers "taken the stand than most of the audience left the hall,"
P. J. K. reported. "After the excitement had subsided, Rogers addressed those
remaining, in an eloquent speech, on the subject of a Labor Union."[28]

Most of P. J. K.'s complaints focused on the Republican party and its "reign of corruption" which was making "our once glorious country . . . a laughing stock of the civilized world. A good many of our best citizens would like to hear the news that Washington, the city of everlasting evil, had been wiped from the face of the earth." On the subject of state politics, P. J. K. condemned what he saw as the "evils of anarchy" which hovered over Louisiana as a result of the rule of "carpet-baggers and their new-made voters, who are worse than a nuisance to all good citizens." P. J. K. found the ratification of the new constitution that spring disheartening. But by late summer 1868, he reported that preparations for the November presidential election had breathed new life into the Democratic party, whose "platform and the candidates give universal satisfaction, especially to the working classes, who look upon said platform as the production of working men, to a certain extent, at least." Among the numerous Democratic clubs that formed were several workingmen's associations that boasted a membership of several hundred in the Third and Tenth wards of the city. A "new era is apparently approaching," the mechanic-correspondent noted approvingly. "The movement of the Democratic party is stronger than it ever was before."[29]

Independent political action proved challenging and difficult for New Orleans' small labor movement in the years of the Civil War and Reconstruction. Whether as supporters of the Free State party, the Conservative Democrats, or autonomous labor parties, many craft workers viewed themselves as a distinct element in the body politic with a distinct labor agenda. But white mechanics had failed to implement their proposals or sustain an independent electoral presence in the postwar era. By 1865, unrepentant ex-Confederates firmly controlled the municipal government, and a new Democratic machine, through its powerful neighborhood networks and ward committees, secured white workers' electoral support through the careful and at times liberal dispensation of patronage jobs.[30] The ascendancy of the Republican party by late 1867 did little to improve white labor's political standing. Instead, white craftsmen rejoined the ranks of the Democrats, whom they viewed as a bulwark against corruption and black government. For the remainder of the Reconstruction era, racial issues and opposition to the Republican government defined the content of politics for most New Orleans whites.

III

If white mechanics found it difficult to exert influence and improve conditions in the early Reconstruction years, blacks confronted even greater obstacles in their efforts to secure civil, political, and economic rights. After all, the Civil War that destroyed slavery hardly resolved the vital question of what position black Americans would assume in the postbellum era. In New Orleans, as elsewhere, blacks found that their efforts to overturn the inherited legacies of racial discrimination, exclusion, and oppression and to assume the rights of full citizens and wage workers seriously challenged from many quarters.

Northern military officials, southern Democratic politicians, and urban white wage earners all attempted, in different ways, to circumscribe the options available to free blacks and former slaves. During the Civil War and Reconstruction eras, the black struggle to define the content of freedom took place on many fronts.

While the New Orleans black community was united by common racial oppression in a period of great political, social, and economic upheaval, it was divided along class lines in ways which had important implications for the character of various struggles for a better life in the postwar South. Those divisions reached well back into the antebellum era. The *New Orleans Tribune,* the voice of the city's black elite, explained in December 1864 that the city's black population "has a twofold origin. There is an old population, with a history and momentos [*sic*] of their own, warmed by patriotism, partaking of the feelings and education of the whites. The only social condition known to these men is that of freedom. Their rights and status [have] been reserved and recognized." In contrast to these free men of color, or *gens de couleur,* as they were called, was another category of blacks. This was the "population of freedmen, but recently liberated from the shackles of bondage. All is to be done yet for them."[31] The ex-slaves had more cultural and spiritual resources to draw on than the free blacks acknowledged, and the status of all free blacks in the antebellum period was not as secure as the *Tribune* indicated. But the passage accurately described a fundamental division in the New Orleans black community during the Civil War and Reconstruction.

Chattel slavery was in decline in New Orleans, as it was in many southern urban centers, in the decade before the Civil War. During the 1850s, a cotton boom increased the demand for slaves and intensified their labor regimen on rural plantations; some urban slaveholders found it more profitable to sell their human property and employ free, immigrant labor instead. From a high of 23,448 slaves in 1840, the New Orleans slave population fell to 17,011 in 1850 and to 14,484 in 1860. New Orleans slaves encountered a far greater diversity of occupational opportunities and living arrangements than did their rural counterparts on the state's sugar and cotton plantations. The majority worked as house servants, porters, and unskilled laborers, while a smaller number found employment as skilled artisans. A few businesses—such as foundries, gas works, cotton presses, and a canal company—owned large numbers of slaves. Most urban slaveholders, however, owned only a few slaves, whom they frequently hired out on a temporary basis. One free black visitor to New Orleans, James P. Thomas, commented on the process of self-hire in the 1850s: "The owners of many [slaves] allowed those people to go out, do the Job, and return to give [the] master his portion," he noted in his autobiography.[32] The cultural life of urban blacks also compared favorably to those of agricultural bondsmen and women. Many exercised a considerable degree of independence in their personal and work lives, as white efforts at social control failed to stem the growth of a relatively autonomous black social and cultural life. Although slaves and free blacks faced constant surveillance and increased legal restrictions during the 1850s, they nevertheless created associations and institutions of their own. In particular, free blacks built and sustained their own churches, schools, businesses, and charitable and benevolent societies—organizations that could command considerable financial resources.[33]

The Civil War generated a vast influx of freed rural slaves into New Orleans' established black community. When Union troops reached Louisiana, thousands of slaves fled their plantations for safety and freedom behind Union lines; as well, many found their way to urban New Orleans, which had been under Union military occupation since mid-1862. Alarmed by the presence of so many hungry, impoverished, and unemployed refugees, General Butler's forces devised makeshift health and housing schemes, administering rations to some 10,000 blacks by September 1862. (Whites, too, often relied on the Union army to provide for them; in that same month, General Butler issued $50,000 worth of food to impoverished whites in the city.) In Union-controlled sections of the state, military officials relied upon vagrancy laws to force blacks back to work. In exchange for wages, clothes, and food, thousands of blacks built fortifications and labored in the Quartermaster Corps and on the levees. Federal policy in New Orleans was even more drastic. In an effort to reduce the number of people dependent on federal relief and to reconstruct an economically viable system of large-scale agriculture, General Butler and his subordinates ordered large numbers of "idle" blacks out of the city and back to the plantations. Despite the Union military's efforts, many freed men and women managed to remain in New Orleans. In the decade after 1860, the city's black population doubled to 50,456, increasing from 14.6 percent of the total population in 1860 to 26 percent by 1870.[34]

Visitors to New Orleans before and after the Civil War were struck by the distinctiveness of the city's free black population, the largest in the Deep South. "I very much doubt whether there is another city in the United States where so large a colored population exist who are so prosperous and well-educated as in New Orleans," a northern journalist concluded in July 1866. Boasting a freedom that could often be traced back several generations or more, many of these often light-colored *gens de couleur* were highly educated and cultured, spoke French as their first language, owned property, engaged in commercial activities, and had traveled widely in Europe.[35] Such differences in culture and status constituted a barrier between these free blacks and the far larger number of slaves. The *gens de couleur,* one Louisiana politician told Alexis de Tocqueville in 1831, "always make common cause with the whites against the blacks." A small number even owned slaves. This privileged position for free blacks has led some scholars to conclude that the New Orleans racial caste system resembled that of the West Indies more than it did the rest of the Old South.[36]

Yet these black Creoles represented only a small percentage of New Orleans' free black population. Most free blacks stood well below this refined elite. Many men found employment as artisans or as skilled or unskilled laborers in the building trades. Out of a sample of 1,792, an 1850 census of free black men's jobs ranked the top ten as carpenters, masons, laborers, cigar makers, shoemakers, tailors, merchants, clerks, mechanics, and coopers. While New Orleans offered a wide range of jobs to free black men, the occupational choices for women were far more restricted. Roughly 60 percent of the city's free blacks were women, and the overwhelming majority of them, as well as female slaves, worked as domestic servants. Moreover, impressionistic evidence suggests that Irish and German immigrants offered slaves and free blacks serious competition for employment as cartmen, porters, hotel porters and waiters, common laborers, and as skilled mechanics by the 1850s. Thus while a small number of free blacks were considerably well-off, most struggled to maintain a slim margin of comfort in the prewar decade.[37]

Their precarious position in a slave society led the *gens de couleur* to respond cautiously to the outbreak of the Civil War. In January 1861, whether from patriotism, expediency, or coercion, some New Orleans free blacks declared their allegiance to the government of Louisiana and petitioned the state's Confederate governor to organize black military companies to protect the city. The following year, New Orleans free blacks formed the Native Guards of the Louisiana Militia, at considerable expense to themselves. Black military action was not without precedent. Free black militias had existed well before the United States purchased the Louisiana Territory from the French; they had fought some four decades before under Andrew Jackson against the British in the so-called Second War for Independence and even had participated in suppressing a slave uprising in 1811. But despite their declared loyalty to Confederate Louisiana, the Native Guards refused to evacuate the city during the Union invasion, and black militiamen offered their services to the northern occupiers. Military necessity led Union officials to accept that offer in August 1862; early the next year, General Nathaniel P. Banks permitted the formation of the Corps d'Afrique, a new unit of 5000 blacks. As many as 24,000 Louisiana free blacks and former slaves may have served in the Union military by 1865. These soldiers were frequently the object of derision and racist remarks by white troops and officers, and the high command replaced black officers with whites in 1863 and 1864.[38]

However humiliating the experience could be at times, the army provided a crucial training ground for future black political leaders. Indeed, the Civil War and Reconstruction tremendously expanded the black elite's opportunities to exercise political and social leadership. Historian David Rankin's collective biography of Louisiana's black leadership shows that the majority of postbellum black politicians were free-born mulattoes who lived in the Creole section of New Orleans, on the downtown side of Canal Street. Financially more secure than the majority of the state's black population (including other free blacks), many were businessmen or skilled tradesmen, relied upon a single occupation for their livelihood, had some education, and had served in the Union army following the federal occupation of New Orleans. The *Daily Picayune,* a conservative white newspaper, described them as "a sober, industrious and moral class, far advanced in education and civilization." Jean-Charles Houzeau, the white Belgian-born editor of the free black paper, the *New Orleans Tribune,* concurred with that assessment: "Here was a core of men of African race, who in their intelligence, sense of rectitude, commercial talents, and acquired wealth, held a peculiar place, a unique place, in the southern states. Here was a sort of elite; here was the vanguard of the African population of the United States."[39]

This free black elite struggled continuously for political rights and played an active role in the restoration of civil rule in wartime Louisiana. When New Orleans fell to Union troops, free men of color called for the enfranchisement of those who had been "born free before the rebellion," basing their position on classic republican and nationalist arguments. As owners of real estate, as men of commerce, as artisans in the various trades, as loyal citizens who "ardently desire[d] the maintenance of National unity," these men sought "the privileges and immunities belonging to the condition of citizens of the United States." As the Free State Movement slowly registered new white voters in the second half of 1863, free blacks and a small number of white radicals held their own pro-Union meetings, calling in vain on the military governor to register free blacks.[40] Although the black elite called for universal emancipation, it stopped short of demanding the enfranchisement of former slaves. But, as Joseph Logsdon has recently pointed out, this was not the result of free black conservatism but of tactical considerations urged upon Creole radicals by their white allies; universal suffrage appeared too radical even for many northern Republican legislators. In the context of evolving federal wartime goals, the abolition of slavery and the right of suffrage were wholly distinct issues.[41]

Yet by the end of the war, the black elite's conception of itself, its relationship to the former slaves, and its sense of political possibilities had undergone profound changes. If free men of color had once "held themselves aloof from the slaves," one free black told Whitelaw Reid in 1865, now "we see that our future is indissolubly bound up with that of the negro race in this country; and we have resolved to make common cause, and rise or fall with them. We have no rights which we can reckon safe while the same are denied to the field-hands on the sugar plantations."[42] Despite the persistence of a group consciousness among some free blacks, the National Equal Rights League allied black New Orleans' "best men" and the ex-slaves. Unity between free and freed blacks was essential, noted a free black speaker at the League's founding convention in January 1865, because both groups were "equally rejected and deprived of their rights, [and] cannot be well estranged from one another." The *Tribune,* the official journal of the League, described the last day of the convention:

> There were seated side by side the rich and the poor, the literate and the educated man, and the city laborer, hardly released from bondage, distinguished only by the natural gifts of the mind. There, the rich landowner, the opulent tradesman, seconded motions offered by humble mechanics and freedmen. Ministers of the gospel, officers and soldiers of the U.S. army, men who handle the sword or the pen, merchants and clerks—all the classes of society were represented, and united in a common thought: the actual liberation from social and political bondage.[43]

This portrait of equality and unity between freedmen and freemen, how-ever, obscured the enduring reality of class divisions within the black commu-nity. Some free blacks, continuing to see themselves as superior, objected to "being placed in the same class with the freedmen just released from bond-age," one northern journalist noted in 1866.[44] Paternalism infused the atti-tudes of even those who allied with the former slaves. "Unless the people of refinement and education act for the benighted ones," declared Captain James H. Ingraham, vice president of the state National Equal Rights League, "the ignorant will be trodden down under foot . . . Our duty is to do some-thing for the new freedmen, to educate, to organize them." The *Tribune* expressed a similar sentiment in late 1864: The newly emancipated would find in the free people of color

> friends ready to guide them, to spread upon them the light of knowledge, and teach them their duties as well as their rights . . . The emancipated shall really be free only when we will see him associated with the educated and intelligent, who can better appreciate the value of freedom.[45]

The black elite promoted the notion that it was best qualified to lead the race in the troubled waters of politics, no doubt believing that advancement would come from the leadership of "our best men . . . men of intellect and education" (as Ingraham put it) and from "these natural representatives, these enlightened spokesmen for the black and colored population" (in Houzeau's words). Before and after the onset of Military Reconstruction in mid-1867, the *Tribune* urged Louisiana blacks to chose the "most able, sin-cere, and respectable men from among our people." The free men of color expected to benefit from the relationship as well: ". . . at the same time, the freemen will find in the recently liberated slaves a mass to uphold them," the *Tribune* editorialized in December 1864, "and with this mass behind them they will command the respect always bestowed to numbers and strength."[46]

As leaders of the black community, New Orleans' black elite defended the freed people from white criticism. In response to planters' constant com-plaints that the "Negro won't work" without compulsion, the *Tribune* argued that although blacks *did* work, a great change, unrecognized by white employ-ers, had taken place: In lieu of the lash, the *Tribune* advised white employers to try "the potency of cash."

> The blacks are no longer required to rise at four and work, work, work all day, till it is too dark to see; and then get up frequently during the night to wait upon the caprices of an indolent master or mistress to whom surfeiting forbids sleep . . . Try the potency of cash; and if there is not enough of that species of persuasion at your command, why, roll up your sleeves, and [be] at it yourself.

The economic vision of the free black elite combined the ideology of free labor with an emphasis upon economic and moral uplift, racial cooperation, and education. The ex-slaves would achieve success in the post-emancipation era, it believed, only through "diligence, industry, frugality, self-denial, sobriety, and faithful and honest toil."[47]

But as recent studies have amply shown, cash was hardly at the center of the post-emancipation labor system. The contract labor system on plantations in federally controlled Louisiana fell far short of a true free labor regime. Although it did away with legal slavery, ensured that blacks would receive wages, and extended to the ex-slaves some limited rights—to cultivate garden plots, maintain family units, and choose employers—the contract system offered scant opportunity or freedom to black agricultural workers. The military oversaw the system's implementation, enforced labor discipline, and required freedmen and women to sign annual contracts that severely restricted their mobility and ability to exercise their supposed market power. The Black Codes, adopted by the newly elected conservative state legislature late in 1865, were equally repressive. Responding to the prospect of continued rural labor shortages caused by increased black mobility and planters' inability to adjust to emancipation, the legislature joined other southern governments in constructing a coercive labor system designed to limit the rights of the freed men and women.[48]

Thus, by 1865, free labor was, as the *Tribune* put it, "not yet an accomplished fact"; the contract system constituted little more than "mitigated bondage" for "mock-freedmen." The *Tribune* advocated in their place both the sale or renting of abandoned lands to the freedmen and "good wages—paid as fast as the labor is done." Only a free market in labor could guarantee the ex-slaves' future success, for the "very nature of things will finally bring the plantation labor to be regulated without regulations—by the natural laws which govern all free relations between men."[49] Despite these admonitions, a free labor economy did not materialize in most of the agricultural South during Reconstruction, still less in the years that followed. The planters' inability to pay cash wages and the freed people's insistence upon autonomy in the fields and payment in shares of the crop locked the rural black peasantry into a very different economic order.

A somewhat different process unfolded in southern urban centers. In cities like New Orleans, the Freedmen's Bureau's contract system hardly constrained the freedmen and women. No longer the property of white slaveholders, they sold their labor power directly on the market for wages. While many urban slaves had had experience with wage-earning in the antebellum era through hiring-out practices, the transition from a slave to free labor system marked a significant shift in the relationship between black workers and white employers. In theory, black common laborers could exercise for the first time their power in the marketplace. In practice, whites who only reluctantly had made the transition from slave owner to employer of wage workers had little intention of purchasing the loyalty of their new employees with the "potency of cash." But those employers no longer possessed the same legal power to enforce labor discipline that they had enjoyed during the antebellum period. That task now fell to the state. Through 1865, coercion and harassment characterized police behavior toward the black community. The city's "rebel police"—composed of ex-Confederate soldiers—not only shut down churches and broke up meetings on a regular basis, but "set loose, like blood-hounds, on the colored laborers," arresting dock workers and other laborers on their way to work. Charged with vagrancy, some were forced to work on neighboring plantations.[50] Not only Confederate sympathizers but later Freedmen's Bureau officials and Republican party politicians would set sharp limits on the range of permissible behavior by urban black workers.

Still, urban life and labor control were different from those on rural plantations. In December 1865, freedmen who worked along the river had occasion to test their new power as free laborers. The ensuing events revealed to them that an agenda of their own would attract little support from either white employers, white workers, free black political leaders, or federal officials in the post-emancipation era.

IV

The war brought substantial changes to the New Orleans waterfront. Even before the Union occupation, the federal blockade of the port proved devastating to commerce. "The cotton ships disappeared; the steamboats were laid away in convenient bayous, or departed up the river to return no more," northerner James Parton noted of the blockade in 1864:

> The cotton mountains [on the wharves] vanished; the sugar acres were cleared. The cheerful song of the negroes was seldom heard, and the grass grew on the vacant levee. The commerce of the city was dead; and the forces hitherto expended in peaceful and victorious industry, were wholly given to waging war.

Where "formerly all was life, bustle, and animation," New Orleans resident Marion Southwood noted in her 1862 diary, "nothing is doing. . . . The place looks as if it had been swept by a plague, such is its bare and deserted appearance."[51]

The war years witnessed a racial recomposition of the waterfront labor force as well. While Irish and German immigrants did not displace blacks from dock work during the 1840s and 1850s, as some historians have maintained, they did dominate certain sectors—such as cotton screwing and cotton yard work—and competed with free blacks and slaves for the loading and unloading of cargo.[52] Secession rapidly changed matters. Enlistments—produced by unemployment, economic distress, and prohibitions against northward migration—and conscription accounted for the "absence of any whites of the laboring classes" along the levee that struck English newspaper correspondent William Howard Russell in late May 1861. When limited commercial activity resumed under federal auspices in 1863, employers turned almost exclusively to blacks. "There is one feature of labor here which must have attracted the attention of all who frequent our levee," the *Daily Picayune* noted in July 1865. "The loading and unloading of steamboats, which was once done chiefly by white labor, is now altogether in the hands of negroes." The unavailability of white workers only partially explained the change: ". . . the chief reason is that black labor is cheaper than white. At all events, whether from mutual repulsion, or from undercharging, or from whatever cause it might arise, there is not a white laborer to be seen along the levee." Only the watchmen on the docks were white.[53]

Both the threat of black economic competition and "moral revulsion" had long kept southern—and northern—white workers at odds with their black counterparts, free and slave. White mechanics claimed that blacks' acceptance of lower wages posed a real threat to white livelihoods and standards of living. On numerous occasions in numerous cities, white artisans and laborers sought to restrict black employment by law. Workingmen's associations petitioned city councils and state legislatures to eliminate free black and slave laborers from particular crafts or restrict them to designated trades. In New Orleans, for example, 200 to 300 white mechanics, or unemployed workers, protested the employment of "slaves in the mechanical arts" in 1835. Their action met resistance from the local militia and city police. In the 1850s, Frederick Law Olmsted observed "a great antipathy among the draymen and rivermen of New Orleans (who are almost to a man foreigners) to the participation of slaves in these branches of industry"; during that decade, white river pilots, cotton yardmen, and screwmen had successfuly excluded black workers from their ranks. For the most part, however, racial competition for waterfront employment in the antebellum era rarely erupted into large-scale racial violence, in New Orleans or elsewhere. The 1850s, of course, had been a decade of considerable commercial prosperity for New Orleans, and "antipathy" notwithstanding, it is likely that favorable economic conditions reduced racial competition for available work. But upon occasion, river and ocean fronts witnessed serious clashes. During the Civil War, Irish dockworkers in Cincinnati and New York violently attacked black longshoremen; in Baltimore, returning Confederate ship caulkers struck (without violence) in late 1865 to eliminate black caulkers from their trade.[54] Where blacks and whites labored in close proximity and in the same trades, tolerance and hostility constituted two ends of the behavioral spectrum.

How, then, would southern white artisans and laborers react to black competition—real or potential—at the end of the Civil War in 1865? The slaveholders' rebellion had shattered the South's economic prosperity, generating widespread unemployment and poverty among whites as well as blacks. Given general economic uncertainty, the wartime recomposition of local labor forces, and glutted urban labor markets, white and black workers in certain trades encountered one another in circumstances not conducive to promoting good will. But emancipation, northern political sensibilities, and the new logic of the marketplace made legal proscriptions on the basis of race (as well as the colonization proposal) an untenable long-term strategy for white workers. In subsequent years, whites relied upon a variety of approaches. Some, like the Baltimore caulkers, struck to exclude blacks from their trade; other skilled workers maintained their monopoly of certain jobs by refusing to work with blacks or for any employer hiring blacks. Numerous unions similarly barred blacks from membership or kept blacks out of apprenticeship programs. But not all situations lent themselves to completely exclusionary tactics. In certain trades and in certain moments, circumstances compelled white workers to experiment with new methods for dealing with blacks, whom they viewed as competitors. From late December 1865 to early January 1866, the New Orleans waterfront became one such testing ground, exposing for the first time, though not the last, both the possibilities and limits of interracial labor collaboration.

At the center of the city's first postbellum labor strike were the white cotton screwmen. Established in 1850, the Screwmen's Benevolent Association was one of New Orleans' oldest labor unions and the first to organize on the docks. Its all-white membership consisted primarily of Irish and German immigrants. Successful strikes in 1850 and 1854 established the screwmen as the highest paid daily wage workers on the waterfront, with each gang member receiving $3 and each union foreman $4; at the end of the war, wages had risen to $4 and $5 respectively. The screwmen's privileged position on the docks derived from the nature of the work performed. Not only strength but a good deal of skill was involved in the tight packing of cotton bales (with large jack-screws) into the holds of ships. While enlistment in the Confederate military significantly reduced their membership and almost destroyed their organization—they later condemned the contracting stevedores, their employers, for leading them astray and inducing them to join the Confederate army by "their false promises"—their skill and strength made the screwmen indispensable to the resumption of profitable cotton loading in the postwar era. With demobilization of the Confederate army at the war's end, the screwmen's union promptly resumed its position as the strongest labor organization on the waterfront. On December 20, 1865, white screwmen struck to raise wages to $6 per gang member and $7 per foreman for a nine-hour day. Their reasons for striking were straightforward: irregularity of employment, a constant problem on the docks over the next six decades, permitted only two or three days of work per week in the commercial season for union members. Wages were so low, they explained in a public statement, that "we are compelled to strike for higher wages, and nothing but dire necessity forces us to assume this position."[55]

The strikers' tactics included both intimidation and persuasion. In some cases, screwmen boarded ships and threatened those working under the old tariff; in other instances, they expanded their ranks through example. A procession along the levee by several hundred enthusiastic white and black strikers on December 21 effectively halted most work on docked steamships. "As ship after ship was passed," the *Tribune* observed, "the men at work on board were induced to join." The screwmen's strike and the subsequent walkout of other groups of "levee-men" brought the port's commerce to an immediate standstill, "causing more or less commotion among shippers, unsettling freights, in fact making a complete chaos among shippers and presses."[56]

Stevedores—the contracting middlemen who employed dock labor on behalf of ship captains and agents—and shipmasters wasted little time mounting a defense. At numerous meetings, employers denounced their workers for striking without notice "at a time when almost every ship has hundreds of bales of cotton and other valuable property exposed to the elements and depredations of petty thieves." Vowing to resist the strikers' "exorbitant prices," over fifty masters and stevedoring firms pledged to "exert every means in our power to replace them with negroes, who will be guaranteed protection in their labor by the Provost Marshal." They similarly resolved to employ no strikers "under any circumstances, if suitable men could be obtained in their stead."[57]

Indeed, as the captains' and stevedores' resolutions made clear, many believed that the issue of black competition would determine the outcome of the screwmen's strike. To some contemporaries and subsequent historians, this situation produced one of the first postwar examples of interracial cooperation in labor's ranks. "I thought it an indication of progress," a New Orleans politician commented to traveler J.T. Trowbridge, "when the white laborers and negroes on the levees the other day made a strike for higher wages."[58] But "white laborers and negroes" did not make a strike together, and the allusion to "solidarity" is contradicted by numerous black and white city press accounts of racial violence and intimidation. Armed with "shilleleys" (wooden cudgels), the screwmen inaugurated their strike by marching along the levee, "knocking down every black man who would have gladly worked for less than that high tariff," the black *Tribune* reported. On the first day, white strikers drove off a number of freedmen working along the docks at the lower rate with threats and assaults, a scene that was repeated throughout the strike. A group of six blacks arrested on charges of inciting a riot protested that a group of white strikers had ordered them to quit work and that they had been "most miserably cozzened" by the screwmen. Solidarity, in this case, was not even a marriage of convenience. And yet white workers relied upon persuasion as well as threats to secure black support. As soon as they heard that the shipmasters intended to request federal military protection for black strikebreakers, the screwmen, noted the *Tribune*, "out of mere policy extended to the dusky cotton rollers the hand of the fellowship, and commenced coaxing them, the better to use them as their tools." The *Daily True Delta* offered a similar, if racist, assessment: The determined screwmen "freely fraternize with darkies, probably for the first time in their lives, in order to keep them from going to work."[59] At this moment in time, fear that employers would "substitute freedmen for the whites who treated them with so little consideration" was a poor cornerstone upon which to build an interracial labor movement.

Whether under threat of violence or by conviction, some black workers did play an active role in the strike. Levee cotton rollers—presumably black—immediately "caught the contagion" and demanded $6 per day. On December 21, several hundred strikers marched along the levee; Trowbridge, the *Tribune,* and the *Delta* all noted the participation of a large number of freedmen, who followed up the rear of the procession.[60] More revealing, perhaps, were the freedmen's efforts to secure wage increases for themselves. Blacks employed on the Atlantic and Mississippi Steamship Company's wharf, who received 25 cents an hour, struck for a wage increase on December 24. Days before, the "trouble" had spread to the Mississippi River steamboats. With the arrival from Vicksburg of the *Fashion*, gangs of unloaders gathered around its plank and demanded a doubling of wages to 50 cents a bale. The dispersal of the crowd by the police left the boat's captain with no labor force, and he put his own crew to work unloading cargo. A similar scene on December 27 ended in a small "riot." A group of black strikers forced a gang of black workers to stop discharging the freight of the steamboat *Prima Donna* until the captain raised their hourly wage from 25 to 50 cents. When the boat's officer ordered the strikers to leave, they refused and attacked the steamship, its officer, and workers with bricks and stones. The arrival of one Lieutenant Rayne and a dozen officers of the Second District Police met similar resistance, as strikers stoned the law enforcement officials. With revolvers drawn, the police charged the crowd, and all but eighteen of the strikers managed to escape down city streets.[61]

Local police and the federal military authorities stationed in New Orleans provided shipmasters and contracting stevedores with much of the support they demanded. Military officials endorsed the employers' stance, but preferred to rely on civil authorities for day-to-day protection of laborers. At the outset, Major A. M. Jackson, Provost Marshal for New Orleans, issued a statement affirming employers' liberty to employ any worker—black or white—they chose. In this situation, his defense of free labor amounted to a defense of strikebreaking. Arguing that "all persons white or black are equally entitled to the privilege of honest labor," he made clear that military officials would suppress any interference with black laborers "properly employed" as an attempt to incite riot. Police did target a small number of whites thought to be "interfering with black laborers" (that is, inducing them to join the strike); but more often than not, police took black strikers into custody. For example, police allowed several hundred white strikers to parade along the levee for several hours and then arrested some six black laborers who had joined the waterfront procession (under duress, the blacks later insisted). In another incident, when black strikers gathered at various steamboats to prevent non-striking workers from unloading cargos, police dispersed the crowds and arrested the alleged ringleaders.[62]

The conduct and outcome of the strike underscored both the sharp limits of interracial collaboration, as well as the distinct positions blacks and whites occupied in the waterfront's occupational hierarchy. Under police protection, ship owners and captains easily replaced unskilled black strikers. Consequently, black laborers slowly returned to work. On December 28, for example, the Atlantic and Mississippi Steamship Company's black strikers surrendered and resumed work at their original wage of $2.50 a day. The success of the skilled white screwmen sharply contrasted with the black workers' defeat. Despite their desire to replace white labor with freedmen, stevedores recognized immediately that they "would have trouble in teaching the blacks how to screw in a load of cotton." Not for the last time in the history of waterfront labor conflict, employers remained entirely dependent on the port's most valuable workers. Shortly after the new year, ship captains, unable to afford idle vessels any longer, capitulated. The screwmen, "being firm, and being masters of the situation," finally won their wage demands of $6 for gang members and $7 for foremen per day.[63]

New Orleans' black elite drew some clear lessons from this experiment in interracial labor activity, taking the opportunity to lecture the freedmen on the philosophy of free labor. Although the *Tribune,* the voice of the free black elite, officially declared that both contracting stevedores and laborers were wrong, it regretted that "a respectable proportion of [the strikers] were freedmen," consistently disparaged the dock workers' cause, and advertised the folly of strikes. White workers—who had "spent a good deal of money in grog-shops, spoiled their temper, worn out their shoes to parade the streets, and . . . are no better off than before"—came in for special criticism. Whites had secured the participation of blacks largely through threats of violence; moreover, those whites who felt compelled to court (as opposed to coerce) black support did so not out of commitment to equal rights but rather as a method to prevent strikebreaking through control of the labor supply. The *Tribune* found this goal highly suspect. If "that public manifestation of white and black levee-men tend to remove among the former their prejudices against color," it would nevertheless be a false victory, an "unlooked for result of a movement which we cannot help to deprecate in itself, as unjust." According to the black paper, a monopoly of labor by unions, even if it promoted interracial cooperation, remained a harmful violation of the marketplace laws.[64]

The black elite's criticism of white union members extended beyond the immediate strike, deriving in part from the double standard permitting white workers to organize for their own advancement but depriving black workers of the same right. The *Tribune* took notice when New Orleans white workingmen met in mid-December to demand the legal enactment of the eight-hour day. That goal was a "move in the right direction," based upon "the broad platform of Republicanism." It had the potential to bring together black and white laborers, whose "interests are one and the same . . . and the prejudices of caste will eventually die away." But that day, apparently, was far off. The free black press could not help but bitterly complain: "Had the freedmen held such a meeting with the same object, what a fuss in the press, in the legislative halls, everywhere! 'Those Africans, those sons of Ham are so lazy!' Such would have been the mildest comments on their movements on that direction."[65]

The *Tribune*'s editors perhaps worried most about the consequences of black participation in the waterfront strike for the black strikers themselves. On different occasions and for different reasons, both employers and white workers had condemned the freedmen on the docks. "Poor negroes!" the *Tribune* complained. "[A]bused when suspected of being unwilling to work, and mauled when ready to labor! is there on earth a more miserable creature than they?" Possibly some free blacks welcomed the strike as an opportunity to demonstrate the reliability of free labor to shipmasters who appeared determined to replace whites with blacks. According to one report, several prominent blacks and their "influential political friends" urged freedmen to break the strike, that is, to "take every offer of work that is made, arguing that if the blacks once get firmly installed in the places vacated by the strikers, it will be almost impossible to dislodge them in the future, and no such opportunity may ever again present itself." But there is no evidence that this was a calculated strategy pursued by black leaders. Instead, black strike participation may have jeopardized the very possibility of a racial recomposition of work on the docks. The *Tribune* took pains to blame whites for the trouble, complaining that it was unjust that the city's white newspapers "should throw all the odium of the Levee laborers' strike on the shoulders of the black man" when it was the white workers who had "compelled by threats" the freedmen to join them. One observer of a labor "riot" on December 27 drew a sharp and ominous conclusion. Since the black steamboat strikers demanded such an "exorbitant price," the shipmasters should "all agree henceforth to employ none but white laborers" with the consequence that "very soon the colored ones would be compelled to seek other kinds of employment and quit the Levee."[66] In a racially divided labor force, such threats to play one group off another could work against black as well as white workers.

V

In the immediate aftermath of the Civil War, New Orleans blacks focused far more energy on politics than they did on workplace organizing. In September 1865 a newly formed Republican party drew together white radicals, the black elite, and freedmen. Its platform called for complete emancipation, the end to all forms of serfdom and forced labor, and the extension of "complete civil, social, legal, and political equality to the black population, with universal suffrage as cornerstone and the safeguard to that equality."[67] If initially without formal influence at the electoral level, the new party found that Louisiana politics would soon produce strange bedfellows. The Democratic party's 1865 political counterrevolution convinced leaders of the fragmented Free State Movement that political salvation required enfranchising blacks. Even conservative Governor Wells came around to that position, as the elections of March and May 1866 made a mockery of his policy of conciliation. Fearing the power of the "intolerant and rebellious spirit" in the legislature and city offices, Wells allied with Republicans and recalled the 1864 constitutional convention in order to grant blacks the vote. Black enfranchisement was the only thing that could break the ex-Confederate monopoly on political power in the state and, as Wells later testified, "place the loyal people of the south in a majority." But the attempt to reconvene the convention resulted in the New Orleans Riot of July 1866, which left 46 blacks dead and 60 severely wounded. "This riotous attack upon the convention, with its terrible results of massacre and murder, was not an accident," a Congressional Select Committee later concluded. It "was the determined purposes of the mayor of the city of New Orleans to break up this convention by armed force." This "prearranged and preconcerted" massacre, together with another anti-black riot in Memphis, demonstrated to many northern legislators that President Johnson's Reconstruction program was inadequate to secure racial peace. More substantial, radical measures would be required.[68]

In the spring of 1867, Congress passed the first Reconstruction Acts and placed the former Confederate states under military rule—inaugurating, in the words of the *Tribune,* a "new era in the history of our city and State." Under Military Reconstruction, General Philip Sheridan, commander of the Fifth Military District—which included Louisiana—removed the architects and beneficiaries of the Democratic counterrevolution from office, disfranchised many Confederate leaders, authorized the registration of black voters, and initiated the process of writing a new state constitution. The new voter registration drive created a black electoral majority in the state, enabling blacks to choose many of the delegates to the new constitutional convention that convened in November. Overwhelmingly dominated by Republicans, the convention adopted much of the black delegates' agenda for civil equality. After three months of deliberations, the convention had opened public schools to black and white children on an equal basis, barred segregation in state institutions of higher education, and, in Article 13 of the state's first Bill of Rights, guaranteed "equal rights and privileges to every citizen"—black and white—on public transportation and in licensed businesses and public resorts. In April 1868, Louisiana voters ratified the constitution by a vote of 66,000 to 49,800, with blacks heavily supporting the document and whites largely opposing it.[69]

Important as it was, the civil rights program had limited appeal for a large section of the Republican party's black constituency, for whom the struggle for existence took priority over matters of principle. Few blacks tested specific statutes by directly challenging segregation; only those who had the financial resources and knowledge required to pursue legal cases through the judicial system could. Black laborers, most of whom could hardly afford to patronize the city's fancier theaters or coffeehouses, inhabited a social world characterized not by sustained interracial contact but by racial separation. For them, legal desegregation was not a realistic goal, the *Picayune* noted, because "the colored race having only a slow course of redress through the courts—[have rendered] themselves liable to be not only summarily ejected from but severely injured in any place of resort which they desire to enter on equal terms with the whites."[70] The *New Orleans Tribune* also recognized the limited appeal that civil rights laws held for black workers: the "poor laborer, coming home from his hard work, with his soiled clothes, having gained hardly enough to live, will certainly not pretend to come and pay the price charged by the establishments frequented principally by the whites."[71] In theory, of course, the civil rights crusade carried on by middle-class black legislators benefited all blacks; in practice, those hard-won achievements had little impact on the lives of most freed men and women.

If civil rights issues failed to capture the popular imagination, black workers actively organized around other issues directly concerning them. More than anything else, freedmen focused their hopes on politics. The state Republican party, formed in September 1865, brought together ex-slaves, free men of color, and a small number of northern and southern white Republicans into an unstable alliance. Although led by the black elite and a few white radicals, the party's base rested squarely on the former slaves. New Orleans black workers formed ward clubs, held marches, and attended meetings. In October 1865, for example, black dock workers in the First District Emancipation Club marched with black Union soldiers to contribute $1 per person to the Republican party's central committee. It was Military Reconstruction, however, that provided the context for widespread black participation in political affairs. "The new order of things has a powerful effect upon those who were slaves before the war," the *Picayune* noted in May 1867. Ordinarily, "the negroes are a quiet people, yielding obedience to the law and respecting its authority," added the *New Orleans Crescent* in mid-May. "But no one can have failed to perceive that they have, of late, manifested a different disposition." City blacks responded enthusiastically to the unprecedented registration of black voters and attended in large numbers the Republican-sponsored mass meetings and marches.[72] On May 11, for example, "throngs of loyal men, marching in solid phalanx under the glorious Banner of Freedom," and organized into numerous Republican clubs, filed into Lafayette Square for a major rally. The 4000 present—of whom four-fifths were black—heard a political address by radical Pennsylvania congressman William D. ("Pig Iron") Kelly; the following week, another large crowd convened to hear Massachusetts senator Henry Wilson advise blacks to "insist upon having every right that the proudest white citizen may enjoy, and never to compromise this point."[73]

By May 1867 the new sense of possibility that prompted the widespread politicization of the black community had created considerable concern in white circles. Across the South, planters complained that former slaves were obsessed with politics. The *Picayune* noted bitterly that black laborers, "who were industriously employed in securing a livelihood" before, have "neglected their occupations and entered the political arena." General Joseph A. Mower, the head of the local Freedmen's Bureau (and later a Republican party official in Louisiana), offered a similar perspective: "The great change in the status of the colored man caused by the practical enforcement of the Military Bill," he wrote in his May 1867 report, "has created a great deal of excitement among the freedmen of this state and in some few instances led them into errors." The source of the problem, Mower believed, was that the former slaves had taken politics too far:

> The City of New Orleans has been disturbed during the past month by these errors. Politicians have addressed the freedmen in language which they cannot understand, and their harangues have tended to excite them to seek justice themselves, instead of endeavoring to do so through the Civil Courts, or the Military Authorities. They were so constantly told that they were the

equals of the white man and urged to obtain their just rights, that it is remarkable an open outbreak between the races had not occurred.[74]

His low opinion of the freedmen notwithstanding, Mower was correct in noting that former slaves did not wait for the impending election or rely upon military authorities; instead, they acted forcefully to achieve equality and "seek justice themselves." Their first target was the segregated streetcar system, in which blacks could ride only in specially designated "star cars." Free blacks had attacked unsuccessfully the star car system shortly after the arrival of Union troops, and their protests continued sporadically for the duration of the war. But in the aftermath of the New Orleans Riot of 1866 and the passage of Military Reconstruction Acts, the issue acquired a renewed importance. Beginning in late April 1867, a number of blacks boarded white cars, only to be forcibly removed by conductors or arrested by police. A crowd of black men and boys harassed and attacked passing white cars on May 4 and 5; on May 6 several black women boarded white cars and refused to leave when the driver ordered them off; that afternoon, blacks on Rampart Street, "without using any violence," attempted to "take possession of the cars." On one occasion, a group of blacks refused to board a car they had signaled when they realized that it was a star car (prompting the *Picayune* to recommend that the streetcar companies put stars on all their cars!). And fights between blacks and whites over the issue broke out throughout the city. Mayor Edward Heath, a moderate Republican recently appointed by Fifth Military District commander General Philip Sheridan, successfully entreated a crowd of angry blacks to disperse, despite jeers from some. The blacks' strategy was partially successful, as the streetcar company quickly surrendered and agreed to dismantle the segregated system. But sporadic violence against black riders continued. Whether due to intimidation or habit, observers noted little change in car ridership patterns, despite the first major victory of New Orleans blacks.[75]

It was at this politically charged moment that black dock workers struck for a second time. While black elites were challenging the inequities of segregation in educational facilities (the star-school system, they called it), the police force, voter registration staff, and in public restaurants and stores, black longshore workers targeted not equal rights but the system of employment contracting on the docks. Few steamboat captains hired their own labor to load and discharge cargo on the city's docks; rather, they employed white and black middlemen to secure the necessary workers. These contractors hired their own gangs, supervised the loading and unloading of vessels, and paid the workers directly from the sum they received from the steamboatmen. In the early Reconstruction years, disputes between black contractors and black workers were common. Longshoremen complained that contractors regularly denied them their "just dues" by pocketing much of the job's proceeds themselves. Employers, for their part, accused workers of presenting "false claims" for work they had not performed, of "playing possum" and "clamoring for pay." Racial solidarity did not negate the simple fact that black workers were still workers and black employers were still employers. Responding to what they considered to be a serious breach of contract, black dock workers in May 1867 took the matter into their own hands.[76]

On the morning of May 16, as many as 500 black laborers gathered on the levee at the foot of Girod Street to call for an end to the abuses of the contract system. Claiming that they had been paid only half the amount justly due them, the strikers demanded a wage increase from $3 to $4 a day and "announced their determination to prevent the further discharge of any boat, as well as their intention to seize the contractors, both white and black, under whom they have been working." They immediately carried out their threat by blockading two steamers, the *Henry Atkins* and the *Olive Branch,* and forcing dock hands to quit work. Proceeding to the *Clara Dolsen,* they spotted Moses Shepherd, a particularly notorious black contractor, and attempted to "lynch him." Shepherd and a policemen failed to restrain the strikers, who threatened to damage the vessel where Shepherd had taken refuge unless the contractor surrendered. Two policemen rescued Shepherd, but as they made their escape, strikers pelted them with stones and brickbats, injuring one officer in a "hot pursuit" through the streets near the river. At the same time, a separate group of strikers halted work on other ships.[77]

City officials viewed the situation as sufficiently dangerous to warrant serious intervention. Mayor Heath and his chief of police immediately visited the levee to stop what they considered to be a riot. Speaking from an elevated, makeshift platform, Heath counseled the crowd of black strikers to disperse quietly, return to work, and let the law address their grievances. Most strikers rejected his words of caution, declaring "openly that the Mayor's speech was not the thing to suit them." According to the *Times,* some strikers cried out, " 'Get down you d——d rebel son of a b——h,' and other expletives to that effect." Another insisted that "We've got enough of that; give us the other side. We are going to have things our way, now." Unable to control the crowd, the mayor retired quickly. A black speaker who mounted the stage next defended their actions and received "three cheers" from the crowd.[78]

Ignoring the mayor's entreaties, the strikers prevented other contractors from resuming work. Another group of 150 to 200 blacks, armed with clubs and stones, marched toward the New Basin to confront contractors there. Strikers shouted and kicked up a "pretty lively fuss as they marched along to the great terror of the inhabitants upon the street," the *Picayune* observed. This time, a federal army force succeeded where the persuasion of local politicians failed. General Joseph A. Mower ordered the strikers to halt as they paraded by his headquarters. "If you feel yourself wronged you must apply to the proper authorities for redress, and you shall have it; but if you take this thing into your own hands, you may lose what rights you already possess," he stated testily. As a cavalry unit blocked their path, the general warned strikers of "consequences . . . most fatal" ("a little grape and canister") if they continued their riot. The crowd dispersed. At Mayor Heath's request, federal troops stood "in readiness at a moment's warning," and a squad of U.S. infantry and the city police patrolled the levee, breaking up groups of black strikers who had gathered along the river and arresting the strikers' alleged ringleaders.[79]

It was, however, a long day, and it ended as it had begun—with an attack on contractor Moses Shepherd. Late in the afternoon, 150 laborers—mostly black but with a few white participants—marched from the waterfront to Lafayette Square. These men, who had unloaded the *Clara Dolson* for Shepherd after the earlier labor turbulence had died down, were "clamorous for their pay." But Shepherd claimed to have received only $160 from the firm for the job, an amount insufficient for him to pay his large crew. Upon the arrival of this new crowd, a lieutenant in command of a First U.S. Infantry squad, then in the Square, ordered his men to "load [their guns] at will." An "old levee policeman" named Peter Meyers intervened, and arranged for the military authorities to take Shepherd's money and divide it equitably among the laborers.[80]

Black longshoremen continued their efforts against black contractors on the following day, May 17. Early that morning, several hundred gathered at the head of Girod Street, where a large number of men, employed by a black contractor for $3 a day, were discharging barges of the Mississippi and Atlantic Steamship Company. The strikers removed the gangways from the barges and declared that since the "contractors were swindlers . . . no one should go aboard and work unless employed and paid by the Company" directly. Again, police put an end to the crowd's activity, arresting two men who appeared to be leaders, and forcefully dispersed the crowd. At the same time, a police officer reported that he had observed a gang of about 30 or 40 black strikers who "went along the Levee at an early hour . . . threatening to kill every man who was working for two dollars per day, if he did not quit." It was very plain, the *New Orleans Tribune* observed, "that the workmen are anxious to get rid of the contract system."[81]

If black dock workers failed to secure their wage increase, they did succeed in imposing a greater degree of order upon the chaotic system of employment and payment. On May 18, Major General Mower intervened on the workers' behalf and issued an order designed to remedy some of their complaints. "Steamboatmen!" his order declared, "Pay Your Hands Yourself!" He stipulated that all owners, captains, and agents of steamships, steamboats, and barges would be liable "and held responsible for wages due freemen for labor performed for loading or unloading the same, whether the labor has been performed under the immediate supervision of the officers of the vessels, or of a contractor employed by them." For their part, steamboat captains agreed to abandon the contract system. In its place, they distributed labor-tickets to prospective laborers, black and white, and commenced work only when a sufficient number of men were engaged. Although a somewhat more responsible system of stevedore contracting would soon emerge, Mower had rectified the system's worst abuses by making shipping firms ultimately responsible for paying the agreed-upon wages.[82]

Black workers' independent, collective activity troubled many in other quarters. Predictably, the city's white press issued a scathing critique of the strikers as well as their black employers. Riotous demonstrations were the "fruit of the lawless and vindictive spirit which artful and designing persons [namely, northern Republicans] have aroused among the freedmen," the *Crescent* editorialized. While also condemning the strikers, the *Times* sarcastically pointed out that the workers' grievances had received no "consideration from their professing champions and special friends" and that the "real and tangible wrong and oppression" of the laborers came from "men of their own color and race—which defrauds them of the rewards of their labor." In the view of the *Picayune,* the workers were "ignorant blacks" imposed upon by their "more knowing and unscrupulous colored brethren." These contractors, "with an active speculative faculty," had undertaken work at "ruinous low figures" because of rivalry, competition, and ignorance; to save themselves, they swindled their employees. But more important, the *Picayune* remarked, were "recent political changes in the status of whites and blacks."

> Ignorant as many of the colored laborers are, having passed the greater part of their lives on remote plantations, and with but little experience in the ways of the world, left to shift for themselves, they become the tools of designing men, and know of no redress but that offered by brute force. Encouraged by recent changes, their passions are easily aroused, and lawlessness becomes the order of the day.

Conservative whites saw the conflict between black workers and their black employers as just one more example of the tragedy that had befallen the South. But the *Picayune* extended the lessons it drew from the strike to the political sphere as well: "If such a want of capacity is manifested by this class in managing their own private affairs," the paper concluded, "how absurd, how indiscreet and reckless it is to almost force upon the government of our whole social fabric, and the control of cities and States."[83] Indeed, many Louisiana whites concurred with the general point; the Democratic party succeeded in winning most whites—workers included—to its banner in opposition to the "black Republicans," who maintained power through the presence of the U.S. Army.

Even the freedmen's allies expressed concern. The black elite was enthusiastic about equal rights and struggles for desegregation but wary of labor conflicts and disorder. The *Tribune* argued that Mayor Heath's recent declaration upholding segregation at the merchants' counters had emboldened the rebels; given the persistence of white violence, blacks would do better to let any provocation pass unnoticed. Thus, the paper advocated moderation and prudence, counseling the freedmen on the docks to "not jeopardize the future by rushing into some unreasonable excitement." At the same time, it concluded, the waterfront riot—such as it was—had been blown out of proportion. "In Philadelphia or Chicago such an event would have been looked upon as a job for a single policemen."[84]

Others "friends" took a harsher view. Major General Mower, who had reformed the contracting system, also threatened to suppress any riot with military force. In addition, he had just issued a notice to the "Freedmen of New

Orleans" in response to the successful campaign to desegregate streetcars and a few unsuccessful efforts to desegregate private shops and restaurants. The document's warning applied even more to the strikers, and the press reprinted the notice often during the strike. It was his duty, Mower told the former slaves, "to address and advise you . . . [as] your friend." Speaking for the government and the Republican party of the North, Mower stressed the limits of federal intervention, invoking the virtues of patience and moderation:

> You have been brought from a condition of slavery to freedom by the Government of the United States. You have been given the right of partici-pating in the election of State and Federal officers. You have the same political status as loyal white citizens. The United States Government, if necessary, will protect you through the military, but you *will not be protected in wrong doing.* Your enemies are watching you, and will be only too de-lighted to see you commit some excess or outrage. The eyes of your friends in the North, who fought for your freedom, are also fixed upon you, and I counsel you not to disappoint those who shed their blood for you. Commit no excess. Be patient. Submit to lawful authority, and by so doing you will gain friends. If you are turbulent, if you are disrespectful to lawful authority, if you are the guilty cause of bloodshed, the people of the whole North will turn against you, and you may lose those rights which it cost the country so much to gain for you.

Those rights clearly ended at equality before the law. Once political equality existed, order and submission to lawful authority became the priorities of the day.[85]

In the years immediately following the downfall of Confederate New Or-leans, neither white workers nor blacks managed to implement their visions of a postwar society. White workers who saw themselves as the "bone and sinew" of society offered a political challenge to the conservative Democrats in 1865 and 1866; with failure, they returned to the fold of the stronger party that ignored their specific goals but offered patronage and racial solidarity. Black elites' advocacy of full civil rights encountered hostility from white Republicans as well as Democrats and a degree of apathy from black workers. For their part, former slaves found that, despite their electoral support for the Republicans, neither party leaders nor the black elite would endorse the economic demands and tactics that transcended the Republicans' agenda of civil and political equality and violated their conception of a free market political economy.

But neither did white mechanics and blacks make common cause in the political or economic sphere. At the level of ideology and economics, the gulf between the groups was simply too great. Reinforcing the legacy of decades of racial antipathy was the white workers' version of artisanal republicanism. Defining the labor of white workers as the source of society's wealth, white mechanics and artisans believed that former slaves possessed few of the attri-butes, such as independence or skill, necessary to include them in the defini-tion of producers. Moreover, white workers guarded their privileged position in the city's employment hierarchy: black unskilled workers constituted a potential threat to their security, not potential allies in a larger struggle. In essence, given their immediate economic and political experiences and outlook, white workers neither understood nor sympathized with the specific needs of the former slaves. If both groups remained hostile to the planter class and welcomed the remaking of a Union state, blacks and white workers adhered to different agendas and occupied different ideological and political worlds.

2

"Raising an Arm in Defense of their Cause": The Emergence of the Labor Question

A large, interracial crowd gathered at a black church near the New Orleans levee on September 11, 1881. The previous day, union teamster James Hawkins became the "first victim of the war between labor and capital" when Police Sergeant Thomas Reynolds shot and killed him in a scuffle on Tchoupitoulas Street. Hawkins' funeral brought together an assemblage that would have been inconceivable only a few years before—approximately 2000 to 3000 people, "white and black, male and female, adults and children." Numerous black fraternal and labor organizations turned out their members to honor their murdered brother. Five hundred members of Hawkins' own Teamsters' and Loaders' Union and 400 black cotton yardmen, each headed by a brass band, led the several block-long procession to the black Lafayette Cemetery No. 2; three black longshoremen served as pallbearers. Marching behind the black mourners, in an unprecedented show of recognition, were 800 white cotton yardmen and screwmen. The crowd's "subdued tone of voice," one newspaper noted, was "far more impressive and ominous than loud threats of vengeance or wails of grief."[1]

Hawkins' death occurred on the tenth day of a waterfront cotton workers' strike which had united—however tenuously—most of the 10,000 laborers in the numerous waterfront trades for the first time in New Orleans' history. "Thousands of men had banded themselves together, white and colored . . . ," the black newspaper *The Weekly Louisianian* reported, in pursuit of higher wages. Riverfront workers had laid the groundwork for such cooperation the previous December when they established the Cotton Men's Executive Council. Representatives from thirteen black and white waterfront unions attended the Council's founding conference, uniting riverfront cotton laborers into a single organization vested with considerable decision-making power. The 1881 strike and the community support it received sprang from this new departure in working-class race relations in the port of New Orleans. Racial and occupational divisions would continue to plague the waterfront Council, its constituent unions, and its twentieth-century successor, the Dock and Cotton Council. Nonetheless, the alliance

34

between blacks and whites enabled dock workers to ameliorate partially the divisive effects of a racially segmented labor force.[2]

Not just race relations but also class relations were affected by the new riverfront labor movement. Longshore workers had exercised little power and won few gains during the Reconstruction and depression years; sporadic collective action on the part of small, isolated, and primarily black longshore unions constituted only minor irritations to steamboat captains, stevedores, and other contractors, who regularly called upon friendly municipal authorities to arrest union leaders, protect strikebreakers, and enforce "law and order." But widespread unionization during and after 1879, culminating with the establishment of the Cotton Men's Council in late 1880, dramatically shifted the balance of power toward workers. The mere existence of this riverfront labor federation, backed by important white Democratic politicians in city government, made police repression less certain and less effective. Beginning in 1880, waterfront unions won wage increases and greater control over the labor process itself. The institutionalization of a framework for expressing interracial and intertrade solidarity, including mechanisms for resolving grievances and formulating strategy, had profound implications for race and class relations on the docks over the next four decades. The *Daily Picayune* may have exaggerated when it claimed that the "cotton labor question has now assumed the shape of open warfare" several weeks before Hawkins' death, but the statement highlighted vividly that dramatic, unprecedented changes between employers and workers were taking place on the city's docks.[3]

I

Commerce was at the heart of New Orleans' economic life in the nineteenth century. The city shared with other southern ports a dependence, as New South propagandist Edwin DeLeon noted in 1874, on "its great staples, their handling, exchange, and transmission to the Northern and foreign markets."[4] Located in a crescent-shaped bend on the lower Mississippi River some 100 miles north of the Gulf of Mexico, New Orleans served as a transshipment center for the agricultural goods of Louisiana and much of the lower Mississippi Valley. Sugar, molasses, tobacco, lumber and timber products, rice, cottonseed oil and cakes, oil products, and—most important—cotton passed through the port. "New Orleans exists chiefly by reason of the cotton trade it controls," the *New Orleans Times* observed in 1881. "Without this trade the city would actually have no industry to support its population, no reason for such a population to live here. . . . Take away this cotton trade and you take away that which has created and now maintains this city."[5]

No other urban enterprise could compete with transportation and commerce as New Orleans' dominant economic activities. Despite local journalists' and politicians' advocacy of a New South ideology that promoted southern industrialization, the city was home to relatively few manufacturing businesses in 1880. For the most part, non-transportation workers found employment in three categories. The first included numerous small shops of the older crafts, catering to the local market, such as blacksmithing, custom boot and shoe work, carpentry, baking, and printing. Second were larger shops producing for the local or regional markets, such as men's clothing, the new tobacco processing industry, and foundry and machine shops that manufactured and repaired machinery for the cotton presses, gins and steamships. The third and largest category included the labor intensive businesses ancillary to transportation, such as molasses and sugar refining, rice polishing and cleaning, cottonseed oil and cake production, cotton compressing, and the cooperage business.[6] If skilled artisans and craftsmen dominated the workshops and foundries of the first and part of the second category, a far larger number of common laborers and unskilled workers filled the ranks of the transportation and commercial processing sector. As one city guide reported in 1885, "to receive, store, sell, and export" the nearly two million bales of cotton that passed through New Orleans annually "requires an army of men, and furnishes occupation for nearly two-thirds of the population." Indeed, as late as 1911, a British Board of Trade study noted that "[i]ndustrially New Orleans is a town of unskilled or semi-skilled workmen."[7]

The Mississippi River functioned as "the axis on which all the whirl of life in the city of New Orleans revolves," Robert Somers observed in 1870. Few visitors to New Orleans in the years before or after the Civil War failed to marvel at the waterfront, the heart of the city's commerce. Travelers noted with awe the "hundreds of immense floating castles and places" (steamboats), and the "broad forests of masts and steamers' smoke pipes" that filled the river; they wondered at the "acres upon acres" of "pyramids" or "mountains" of cotton bales; from a distance, the visitor might admire the "noble amphitheatric front" composed of miles of rows of warehouses.[8] As impressive to some was the continual motion of the busy season. The loading and unloading of vessels and steamboats and movement of hundreds of drays, one observer remarked in 1845, "strikes the stranger with wonder and admiration." The waterfront offered not just commercial motion but international diversity as well. Another outsider noted the same year that the levee constituted nothing less than a " 'world in miniature'—where one may meet with the products and the people of every country in any way connected with commerce." The city retained its cosmopolitan character: Over four decades later, British traveler George Augustus Sala called New Orleans the "American Amsterdam." It was, another remarked in 1891, a "city of marked contrast . . . [an] historical, ethnical, political gumbo."[9]

Diversity and commercial bustle notwithstanding, a closer inspection revealed the city's many drawbacks. Nineteenth-century wharves often stood in poor condition. The "alleged depots," the *Louisianian*, a black newspaper, complained in 1882, "are nothing but dilapidated shanties in near proximity of the worst dives in this large city." City streets came in for universal condemnation. "Nothing could be more shabby than the streets," one out-of-towner wrote in 1885, "ill-paved, with undulating sidewalks and open gutters green with slime . . . ; little canals in which the cat" became "the companion of the crawfish, and the vegetable in decay sought in vain a current to oblivion." Because New Orleans was situated below sea level, it flooded easily; rainstorms could quickly render the unpaved streets impassable. Protective levees served to hold back the waters of a rising Mississippi River from overflowing into the city. In 1880, a good portion of the city between the cemeteries at the end of Canal Street and Lake Pontchartrain was essentially a swamp, conducive to breeding malaria. The lack of an adequate sewer system until the twentieth century guaranteed the city high illness and mortality rates.[10]

Although New Orleans was the region's largest city and commercial capital, by the middle decades of the nineteenth century it faced several short- and long-term economic challenges. The Civil War disrupted the city's business, as the collapse of agricultural and the Union blockade paralyzed New Orleans' trade. With the cessation of hostilities in 1865, commerce again began a period of renewed growth. But the deterioration of the levees during the war, the collapse of much of the state's sugar production, and a series of floods and epidemics significantly hindered expansion in the immediate postwar years. Nevertheless, despite the collapse of sugar and the weakness of the tobacco trade, cotton provided the basis for a relatively strong recovery. "The wonder is that she makes so good a showing," DeLeon noted in 1874, "and has so rapidly recovered lost ground, commercially as well as agriculturally." From late 1868 to the start of the depression in late 1873, New Orleans commerce could boast of a modest prosperous recovery.[11]

In a "period not long past," the *Democrat* reflected in 1881, the "rivers, bayous and creeks that flow into the great father of Waters . . . poured the wealth of the regions they drained into the warehouses of this city." Now, the paper complained, much of the "domestic commerce has been diverted from the Crescent City." The threat to New Orleans' economic security came from the development of extensive railroad networks and the emergence of new commercial centers. Although New Orleans maintained commercial domination of much of the cotton belt in the decades before the Civil War, by the late 1840s railroads were moving shipments of midwestern flour, pork, and other foodstuffs eastward toward the Atlantic coast. Even control of cotton was not beyond the railroads' reach. By the 1870s, major rail lines included the Southern Pacific, the Queen and Crescent, the Louisville and Nashville, the Nashville, Chattanooga and St. Louis, the Yazoo and Mississippi Valley, the Texas and Pacific, and the Illinois Central.[13]

The establishment of rail lines paralleling the Mississippi River dramatically diminished the old steamboat business in the postbellum era. To observers at mid-century, the city's location on the Mississippi River, that "great inland sea," seemed to have given it the "keys to the Gulf."[14] Steamboats, which first appeared on the lower Mississippi in 1812 and emerged as a major transporter one decade later, carried vast quantities of cotton, sugar, molasses, and tobacco from agricultural producing regions along the Mississippi and its tributaries. Indeed, steamboat traffic provided the basis for New Orleans' golden age that extended from 1815 to 1840, and it continued to constitute the "chief agency" of transportation through the 1870s. But as Mark

Twain observed in 1883, contrasted with its prime vigor, "Mississippi steam-boating may be called dead . . . [killed by railroads] doing in two or three days what the steamboats consumed a week in doing." The change from river to railway transportation, traveler Charles Dudley Warner remarked about his 1885 trip to New Orleans,

> has made her [old steamboat] levees vacant; the shipment of cotton by rail and its direct transfer to ocean carriage have nearly destroyed a large middle-man industry; a large part of the agricultural tribute of the South-west has been diverted . . . and the city waits the rather blind developments of the new era.

The transition was indeed striking, and the city's commercial elite discussed its plight with regularity and alarm in the 1880s. Looking back on the history of American transportation, the Inland Waterway Commission passed similar judgment in its 1908 preliminary report. The 1883 opening of the Louisville, New Orleans and Texas Railroad—later absorbed by the Illinois Central—"went far toward accomplishing the downfall of steamboat traffic on the lower Mississippi."[15]

Although the rise of the railroads had diminished the southern flow of steamboat traffic into New Orleans, the waterfront remained the center of the city's commercial life. Agricultural goods, whether arriving by rail or steam-boat, found their way to distant northern or European markets via ship. Thus workers on the city's docks constituted the vital link in the transportation process. But there was no such thing as a generic, all-purpose dock worker. Dock work involved a wide variety of jobs, each associated with a particular aspect of the transportation business. As John Lovell has noted in his study of waterfront labor in London, the "port worker was a specialist *par excellence,*" who usually handled one class of goods for a small number of firms.[16] In every major port city, a complex and often unique division of labor characterized the work of loading, unloading, and transporting agricultural staples and other products. New Orleans was no exception. Each step in processing and transporting cotton, sugar, and other goods constituted a distinct trade, or separate class of labor. By the 1880s, approximately 13,000 roustabouts, cotton screwmen, longshoremen, teamsters and loaders, and cotton yardmen labored for dozens if not hundreds of steamship agents, contracting steve-dores, railroad managers, boss draymen, cotton yard proprietors, and the other middlemen who produced or repaired the barrels, weighed the goods, and transported the various products between different processing points in the city.

The circuit of trade began with the arrival of cotton in the city by either water or rail. For cotton arriving by water, black roustabouts set the circuit in motion. Roustabouts worked on board the Mississippi River steamboats which transported the bales of cotton, barrels of molasses, or other goods from the plantations and shipping points along the river to New Orleans. Most of their heavy labor ended when the steamboats docked in the city. "The negroes who man the boats running up and down the Mississippi," observed journalist Edward King in 1875, "are not at all concerned in the discharging of cargoes, being relieved from that duty by the regular wharfmen." But the roustabout's job, often filled by former slaves in the decades after the Civil War, was perhaps the least desirable of all transportation-related work. Although wages were sometimes high, roustabouts endured harsh treatment, poor food, and bad living conditions; moreover, as rural, illiterate workers with a reputation for fast living, they were rarely welcomed by the city's established black community.[17]

As soon as a steamboat docked at the landing at the head of Canal Street, cotton screwmen occupying the pinnacle of the waterfront occupational hierarchy took charge. Although their indispensable skills centered on the careful packing of cotton bales into the holds of ships with heavy jackscrews, they also maintained a strict monopoly over the unloading of cotton, a task that required relatively little skill, guaranteeing the elite group of workers additional employment at the same high wages. The largest group of riverfront workers, the longshoremen, received the cotton over the side of the steamboat from the screwmen. They also unloaded all sugar, molasses, and other shipments themselves. A small number of round freight teamsters took charge of all barrelled goods on the docks, while hundreds of teamsters and loaders, sometimes known as draymen, transported bales of cotton from the steamboat landing and railroad yards to "stow row," the elevators and warehouses of the cotton press yards.[18]

By 1873 the city boasted some twenty-six cotton presses concentrated near the river along Tchoupitoulas Street from St. Joseph to Louisiana streets, compressing over 1.2 million bales annually. These "immense brick storehouses" each contained huge presses that received "a bale of cotton as it comes from the plantation and river steamboats, and with one breath of the powerful engines, reduces its size to one-third its original bulk." Employing roughly 1000 men in the busy season, the cotton press yards averaged between 25 and 30 employees each. Numerically dominating the labor force here were the rollers, who worked in gangs of three to move each cotton bale, and those who worked at the press itself. Black scale hands assisted white weighers and reweighers in their work. After the cotton was weighed, classed, purchased, reweighed, and compressed, teamsters transported the cotton back to the wharves, where the longshoremen loaded the cotton onto ocean-bound vessels, and screwmen "screwed" cotton into the holds of smaller sailing ships, or loosely stowed them into the holds of larger steamships. "So well is the handling of cotton organized," remarked Zarcharie's *Guide,* that the details were "expeditiously . . . carried out in a clockwork manner."[19]

A multiplicity of employers participated in these commercial transactions. As one journal put it, since its founding New Orleans had been "a middleman's city." Mississippi River steamboat captains employed crews of black roustabouts to make the journey up and down the river to pick up cotton. Ocean-bound transportation entailed a wholly distinct process. Large steamship lines might hire longshoremen or screwmen directly to load and unload their sea-going vessels; some companies employed contracting middlemen,

called stevedores, to arrange the work for them. Approximately sixteen stevedores hired regular crews of workers by 1881. Numerous boss draymen provided teamsters and loaders with floats drawn by teams of mules or horses.
Cotton factors, the merchants who advanced money to the rural planters and
in turn received their cotton, hired their own cotton weighers, who worked at
the presses selected by the factors. The scalemen, however, actually put the
cotton on and off the scales, where the cotton was checked by the weighers
and reweighers. Cotton classers were a small and relatively highly paid group
who sorted and graded the cotton by its color, cleanliness, and fiber length.
After a reweigher in the employ of a purchaser of cotton determined the
accuracy of the weight, the cotton compresses substantially reduced the bale's
size.[20]

Railroad freight yards constituted a world unto themselves. By the late
nineteenth century, freight handlers unloaded cotton from rail cars at the terminals and depots of the expanding railroad lines near the river—Illinois Central's Stuyvesant Docks at the head of Louisiana Street, and Texas and Pacific's
docks across the river at Westwego. After 1880, few steamship agents whose
vessels loaded at the rail docks bothered to employ contracting stevedores at
all, preferring either to hire their own workers or, increasingly, to rely upon the
docks' own freight handlers. With vastly greater economic power at their disposal, rail employers maintained separate enclaves along the city waterfront.
While hardly free from the threat of unionization, large rail and steamship
companies like the Illinois Central and the Southern Pacific were in a far better
position to resist workers' demands than were the smaller and more economically vulnerable stevedores and agents. Indeed, in New York, Galveston, Mobile, New Orleans, and elsewhere, freight handlers' unions were consistently
much weaker than those elsewhere along the waterfront. Despite union efforts
in every decade from the 1880s to the 1920s, New Orleans railroad dock workers continued to occupy a lower position on the occupational waterfront hierarchy than perhaps any other group save roustabouts.[21]

In the late nineteenth and early twentieth century, New Orleans shared
with most ports several general characteristics identified by W.H. Beveridge
in his 1909 study of the London waterfront. "The cardinal features about the
dock and wharf industry," Beveridge observed, include "first, considerable
irregularity in the arrival and departure of cargoes by ship or by barge; [and]
second, the small extent to which machinery has been able to displace more or
less unskilled manual labor in the work of loading, unloading and otherwise
dealing with goods in transit."[22] Utilizing few mechanized devices until the
early twentieth century, New Orleans employers depended on their workers'
physical strength and familiarity with particular cargos.[23] But from the laborers' perspective, it was less the physical difficulty of the work than its irregularity that proved most problematic. Three elements contributed to the irregularity of employment. First, for all of the careful orchestration of the many stages
involved in transporting and processing agricultural goods, the world of the
waterfront was inescapably vulnerable to the vicissitudes of commerce and
shifting trade routes, producing a chronic economic insecurity for many. Sec-

ond, even during the cotton season, few dock workers were guaranteed daily employment. Strict timetables and the advent of steamships notwithstanding, ocean currents and winds affected the arrival of transatlantic vessels. Hired by a contracting stevedore or agent for a specific job, dock workers did not get paid when they did not work, and work depended on the ship's arrival. Third, waterfront trade was highly seasonal. While the first cotton might arrive in port in August, it was not until late September and early October that the commercial season got under way, lasting roughly until March. Summer months in New Orleans were not only epidemiologically dangerous (throughout the nineteenth century, many members of the commercial elite left the city for the duration of the unhealthy summer season), but times of unemployment as well. While dock wages were often higher than those paid for other unskilled work, the lack of summer employment cut into small savings, sending some workers out of the city in search of temporary employment.[24]

In ports where unions did not exist or were too weak to control the size of the labor force, the irregularity of employment and the manual, non-mechanized nature of the work encouraged employers to utilize a system of casual hiring. The ubiquitous shape-up—the very symbol of casualism—provided employers with as many unskilled workers as they required for a given job; when that job was finished, the labor force was dismissed, only to be reassembled when new ships arrived in port. Yet unlike the ports of London and New York, where the shape-up was deeply entrenched, evidence suggests that New Orleans' employment practices were less competitive and less ritualized. In none of the descriptions of the waterfront and the movement of goods on and off of ships does there appear an account of large gatherings of men on the docks awaiting employment. In the union era after 1880, it is likely that individuals or groups of workers contracted directly with stevedoring offices, eliminating the day-to-day shape-up altogether.[25]

The waterside, Eric Hobsbawm has observed, "is an industry with fluid frontiers and no very exact shape," having "no obvious and pre-destined core for its union." Certainly no two ports were exactly alike, and waterfront workers experienced different employment practices and diverse conditions of labor which depended upon geography, the racial or ethnic character of the labor force, the nature of the cargo, the degree of capital concentration, and the unique political and economic histories of the ports themselves. Even the names given different groups of workers varied considerably from port to port: New Orleans employed its longshoremen, screwmen, draymen, yardmen, and roustabouts; London its shipmen, stevedores, coal whippers, overside corn porters, dockers, quay laborers, hand truckers, and warehousing laborers; Liverpool its stevedores, riggers, quay porters, fruit porters, coalheavers; and South Wales its coal tippers and trimmers, timber porters, grain workers, and iron-ore workers. The variety of sub-divisions could be as numerous as the range of goods loaded, unloaded, and stored. In most ports, the pronounced division was between men who worked on board the ships and those on the docks, quays, and in the warehouses. Whatever the actual

structure of employment in a given port, different types of dock work constituted different rungs on complex occupational ladders.[26]

The history of waterfront labor in the late nineteenth and early twentieth
century shows that the multiplicity of tasks and the diversity of work experiences could generate a number of different patterns of labor relations. In
many ports, the diversity of work groups produced widespread fragmentation
of labor along trade, ethnic, or racial lines. In some cases, unions of the most
skilled dock men at the top of a port's occupational hierarchy emerged.
Among the organized, powerful dock workers were New Orleans and Galveston cotton screwmen (1850s and after), upper Great Lakes iron ore trimmers
(after 1880); coal trimmers in Cardiff and stevedores who unloaded ships in
London (in the late nineteenth and early twentieth century). In Marseille,
France, a minority of dock workers exercised tremendous power in the first
half of the nineteenth century.[27] These skilled men's strength, in part, lay in
their tight control over access to their trade, made possible by their possession
of often indispensable skills. Thus small, sectoral unions of skilled or otherwise valuable men could exist in the midst of large numbers of unskilled
workers who dominated often overcrowded casual labor markets. Skilled
groups were often likely to view those outside their crafts more as a threat to
their economic security than as potential allies.[28]

New Orleans' white cotton screwmen were one such group. Unlike their
less skilled counterparts in the cotton yards or on the docks, nineteenth-
century screwmen were virtually irreplaceable. The work was skilled and
physically arduous. Ships transporting cotton required that the bales be
packed as tightly as possible. Gangs of screwmen, using large metal jackscrews and wood posts that could each weigh up to several hundred pounds,
compressed rows of bales in ships' holds. Since different holds required different methods of screwing cotton, the screwman's skill rested in both his
strength and judgment. Not surprisingly, screwmen commanded tremendous
power on the levee. Their union, the Screwmen's Benevolent Association
(SBA), provided sickness and death benefits to members and their families,
and utilized collective action to maintain high wages and enforce union work
rules.[29] Union relief committees regularly visited sick or injured members and
cared for the immediate needs of the families of deceased union men, while a
labor committee made sure that all members abided by the organization's
rules (which later served as a model for other dock unions). The SBA prohibited stevedores, superintendents, and contractors from discharging any gang
that performed its work faithfully; fined or expelled any union member found
working with a foreman who was not a member of the screwmen's association;
established precisely the working hours and breaks in a nine-hour day, and
strictly regulated the amount of work each member performed. From 1878
until the early 1890s, each gang of four screwmen and one union foreman
stowed a maximum of 75 bales of cotton per day; when it finished that work,
the gang was free to leave for the day.[30]

The screwmen's union protected its members' privileged position as the
"aristocrats of the levee" by carefully regulating the number of men it admit-

ted to the trade, thereby ensuring steady work during the cotton shipping season. According to its 1884 constitution and by-laws, no person under twenty-one or over forty-five years could work as a screwman; the entrance fee for new members was $50; and a special investigative committee judged the qualifications of applicants before the full membership balloted on admission. Moreover, the SBA barred its members from working with non-union screwmen, under penalty of fine or expulsion. Not only did the SBA regulate the size and behavior of its own membership but it exercised control over a smaller number of black screwmen as well. When blacks challenged the white monopoly at some point in the 1870s by securing work as screwmen, the white SBA imposed a strict limit of twenty gangs, or 100 men, on the number of blacks who could work on a given day.[31]

A second and very different pattern of unionization involved alliances across sectoral lines and alliances between a port's more skilled elite and its less skilled workers. The primary goal of these alliances was often the control of the labor supply. Given the generally unskilled character of the work and the largely artificial lines separating the so-called crafts, union success required complete control to counterbalance managerial power. In New Orleans, the Cotton Men's Executive Council, from 1880 through the mid-1890s, and the Dock and Cotton Council, from 1901 through 1923, constituted such alliances. Other ports experienced similar forms of inter-sectoral cooperation. In New York, the Knights of Labor provided an organizational umbrella under which numerous groups of freight handlers, longshoremen, boatmen, grain handlers, and bag sewers (as well as other unskilled and skilled workers) assisted striking coal handlers in New York and New Jersey in 1887; the Knights also played a role in facilitating intergroup and interracial cooperation in the mid-1880s in Galveston, Texas.[32] In Canada, the Halifax Labourers' Union of the 1880s and its successor, the Port Workers Union of Halifax in the late 1890s and early 1900s, bound together permanent and casual dock workers. In 1889, Great Britain's New Unionism manifested itself in a momentary explosion of multi-sector waterfront organizing that gave birth to the Dock, Wharf, Riverside and General Labourers' Union of Great Britain and Ireland, which was quickly crushed by aggressive employer counteroffensives. The next upsurge, between 1910 and 1914, placed a more durable waterfront alliance, the National Transport Workers' Federation, squarely on the map.[33] With the exception of New Orleans, however, most of the waterfront alliances before the twentieth century were short-lived, emerging in periods of labor unrest, subsiding in the aftermath of success or failure. Thus, the pattern of waterfront union development in New Orleans—explored in the pages that follow—shared certain features with other ports, while also departing from them in significant ways.

From the Civil War until 1879–80, however, there was little to distinguish the New Orleans experience from those of many other ports in and out of the American South. During this period, only the screwmen maintained a strong and influential organization. But the union impulse affected other groups as well, as some dock workers in different riverfront trades established benevolent societies and unions during the 1870s. In April 1872, black levee laborers formed the Longshoremen's Protective Union, and white workers established the Longshoremen's Benevolent Association a year later. A small number of black cotton screwmen established the Screwmen's Benevolent Association, No. 2, in 1870, which they incorporated in 1877. Without exception, these unions were far less successful than the white Screwmen's Benevolent Association. During the lean years of the 1870s depression, glutted labor markets seriously reduced their bargaining power. Moreover, the few existing riverfront labor associations extended little help to each other in moments of need.[34] Always a minority on the docks, divided from other groups of workers by trade and race, and adamantly opposed by employers, the municipal government and police, the small and largely black unions of unskilled longshore workers fought a series of uphill, and often losing, battles.

This weakness and isolation changed dramatically at the end of the 1870s, as the waterfront became the center of a reborn labor movement in the city. Black and white dock workers—organized into separate union locals, each representing a distinct "class of labor"—banded together in the early 1880s, forging a powerful organization to coordinate workers' struggles and mediate relations between blacks and whites. In the decade and a half following the Civil War, however, few would have predicted such an alliance. That period, marked by a prolonged economic depression and the demise of Reconstruction (hence most black political power), was hardly conducive to promoting interracial or inter-trade alliances. The economic crisis generated wage cuts, severe unemployment, and job competition; fragmentation, not unity, had characterized workplace interactions. Politically, the Reconstruction experience sharply divided blacks and whites at the voting booth and in the halls of government. The struggle against Republicans and black political power, for many whites, involved the assertion of an ideology of white political and social supremacy. Living and laboring in proximity of, but not with, each other along the riverfront, black and white workers struggled separately and achieved little success during the 1870s.

II

For many New Orleans workers and their families, the national depression that began with the financial panic of late September 1873 produced widespread hardship. "New Orleans has suffered with the general depression of the whole country," the *Picayune* noted in 1877, "but the peculiar difficulties have made the sufferings of her people greater in proportion and more general than in other sections. Here . . . the distress [has been] more extensive, the change from wealth to want more sharp and bitter. Money is scarce, and times are hard with all but the favored few. . . ."[35] Louisiana's economic difficulties, however, began before the collapse of Jay Cooke's financial empire. The previous spring, currency had become scarce and by late summer unemployment had begun its steady upward climb. Although a "depression psychology was developing in New Orleans well ahead of the panic," as historian Joe Gray Taylor put it, little could have prepared the Crescent City for the magnitude of the economic crisis that lasted until late in the decade. New Orleans' banks suspended currency payments in October 1874, and much of the city's black population lost whatever savings it had—$300,000 across the entire state—when the Freedman's Savings and Trust Company failed the following year.[36]

The human dimensions of the depression were most evident in the suffering of the poor and working classes, exacerbated by skyrocketing unemployment and the downward spiral of wages. "Unquestionably there is a great deal of distress among the laboring classes in New Orleans," the *Picayune* observed in 1875. "Perhaps there has never been more in the history of the city." Although there are no reliable statistics for those out of work or underemployed, numerous projections were offered at the time. Administrator of Public Works, Major E. A. Burke estimated the number of New Orleans unemployed at 5000 men in 1875; the *New Orleans Republican* believed that 3000 to 4000 jobless men meant that there might be as many as 15,000 destitute family members. An "absolute destitution—not a thing of words but a hard reality"—according to the *Picayune,* could be observed among between 2000 to 3000 unemployed, and some predicted that the "starvation . . . and the suffering" would only increase with time.[37] Noted one journalist:

> There are thousands who wear old clothes, eat cold and scanty food and tearfully expostulate with landlords as needy as themselves. They have no revenue, no resources, and their future is seen through the reverse lens of that field glass with which hope usually scans the horizon. . . .

One group of white workers concurred, stating publicly that "[we] are now in extreme want, which may bring us to the verge of desperation if not relieved promptly."[38]

If the economic crisis reduced trade union influence to a minimum and, in many cases, destroyed unions altogether, it hardly spelled the end of a labor agenda and the struggle between social classes. In numerous northern and midwestern cities, the onset of the 1870s depression gave rise to coalitions of unemployed workers, trade unionists, and radicals to demand large-scale public works projects. City officials responded by ignoring petitions, breaking up demonstrations, and arresting leaders and marchers, while the urban upper classes denounced the movement as violent, communist-inspired, and a violation of modern political economy.[39] In New Orleans, a movement for public works emerged only after the collapse of efforts elsewhere in early 1874. Like its counterparts in New York, Chicago, Philadelphia, Paterson, Detroit, and Cincinnati, New Orleans' white unemployed demanded that municipal government take responsibility for the economic crisis and allocate jobs in the city's various administrative departments.

Petitioning local government for employment represented a shift in strategy and tactics. In July 1874, several hundred white workingmen, mechanics, and laborers gathered in Lafayette Square to organize a union among the laborers and establish a labor agency to direct the unemployed to jobs. Eight months later, a Mechanics' and Workingmen's Association argued that much of the difficulty which had fallen upon the working people of New Orleans was due to the "inefficiency and fraud of men in office." In a petition to the City Council, it implored local legislators to "relieve our distress and want by giving us our share of the city work."[40] When no relief was forthcoming, unemployed whites adopted an increasingly aggressive posture. In the spring of 1875, the Council of Mechanics, Workingmen, and Laborers' Association bitterly protested the discharge of 300 men on a city public works project, calling instead for the suspension of all city officials "drawing large salaries and doing very little for it"; the retention of only those who were "actually necessary to carry on the affairs of the city"; and a reduction of all salaries to pay for the employment of more laboring men "whose services are of benefit to the whole city." In addition, the Council proposed that those in need replace all property owners "not dependent" at work on city carts. Not trusting city officials to implement such changes, the Council demanded a list of all employees, their occupations, salaries, and addresses for each municipal department. Later that spring, the Council vowed that its members would not deposit their fares in the cash boxes of the street railways, in order to compel the companies under their charter provisions with the city to employ more drivers and collectors. Swift police action ended that strategy.[41]

Neither petitions nor street car protests had much effect on municipal officials. For the duration of the depression, local officials, even those committed to providing patronage jobs to loyal Democratic voters, claimed that the economic crisis necessitated deep financial cut-backs. In early June 1875, Administrator Burke drastically slashed the size of the Department of Public Improvement and reduced its officers' salaries.[42] But persistent unemployment kept protest alive. In March 1876, some Second Ward white workers—"tired of throwing away their votes in favor of those who never labor at all"—broke from the "central dictation" of the Democratic machine to form an Independent Workingmen's Club. That May, the "mechanical interests" denounced aristocratic landlords and renewed their demands for public works. Again, the City Council ignored their protests. (The Workingmen's Association communication, remarked the *Republican,* "will probably rest in a dusty pigeonhole for all time to come.")[43]

In early July 1877, the City Council resorted to a novel budget-slashing technique of delegating the responsibility (as well as the carts and machinery) for street repair, garbage collection, and drainage work to private contractors. With roughly one-third the number of men and half the carts previously used by the Department of Public Improvement, the contract system provided even poorer services than had the widely denounced city department. Not surprisingly, the new system evoked verbal protests from white workers who correctly feared job reductions and wage cuts. They added one further objection: Under the contract system, "colored men might obtain employment on the public works where now the steady Democracy found their bread and butter." Petitions notwithstanding, city officials implemented the contract system with little difficulty. Amid an ailing economy and a glutted labor market, the unemployed appeared wholly unable to influence municipal affairs.[44]

In the end, white workers' organizations failed to force the city government to assume responsibility for the able-bodied unemployed. Instead, what the unemployed received was advice and a small degree of charity. One solution to the unemployment crisis favored by the press and city elite was migration to rural Louisiana. As a transshipment and processing center for agricultural goods, New Orleans remained dependent on the volume of agriculture production. For the duration of the depression, numerous editorials exhorted the poor to become plantation laborers, because "the country is languishing for want of the very thing they possess," namely, labor. The black elite proposed a similar solution. In response to cries by some whites that employers should discharge their black workers, the *Louisianian* suggested that "the surplus of white labor in the city" might better be utilized "in the country—upon whose products the city lives." One organization in particular, the New Orleans branch of the Louisiana Relief Association, concentrated its energies on finding rural jobs for the urban unemployed, providing free transportation out of the city. Not surprisingly, entreaties for the enlargement of the agricultural proletariat found few recruits among unemployed urban whites.[45]

Voluntary relief was an inadequate substitute for a systematic policy, but city officials relied upon the well-to-do to implement that which the city could not afford. Rejecting the Workingmen's Association's demands for city jobs, Administrator of Public Works Burke proposed that leading citizens and merchants fund a vast, and desperately needed, waterfront improvement project that would provide work for up to 1000 men. But the voluntary Relief Association—comprised of the city's "most reliable and reputable citizens"—failed to adopt the project. Instead, it distributed some charity in an effort to dampen what the New Orleans business elite feared was the growing appeal of "idle men bent on agitation and influence by reckless demagogues." At least one organization of white mechanics, however, met these relief efforts by the prosperous with hostility. In May 1876, the Workingmen's Association which was demanding public works jobs also denounced elite relief committees: "[W]e look upon the appointment of any committee for the purpose of collecting any funds to be distributed among the working classes as injurious," a meeting resolved, "as it will be the production and promotion of idleness among the laboring classes."[46]

New Orleans' resilient Democratic political leaders remained firmly in control, withstanding the depression-era challenges as they had the earlier political challenges of Reconstruction. Despite the disfranchisement of some Confederate leaders and the registration of the black male population in 1867, Democrats utilized legal and extra-legal techniques to expand their influence in New Orleans, where whites outnumbered blacks three to one. In the presidential election of 1868, for example, the paramilitary Knights of the White Camellia intimidated black voters and perpetrated ballot box fraud. In local elections that year, Democratic standardbearer John Conway, a wholesale grocer and merchant, narrowly beat his Republican opponent and overwhelmingly crushed a small Workingmen's ticket; in addition, Democrats won eleven out of the fifteen aldermanic seats. But Republican control of the state government constrained New Orleans Democrats, limiting their power for the next several years. Republican legislators transferred effective control over the city's police to the governor by creating the Metropolitan Police Force based in New Orleans; they also stripped local officials of control over the educational system. The Metropolitans, the schools, and federal operations at the Customs House constituted major sources of political patronage for the Republicans, and many blacks fround employment in these large federal and state government institutions. Republican officials made other efforts to secure their tenuously held local power. Through legal maneuvering, the governor installed his supporters on the Board of Aldermen. Then, in 1870, the state legislature approved a new city charter which abolished the aldermanic board, replaced it with a government run by a mayor and seven administrators, and gave the governor the authority to appoint the first council. But by 1872, local Democrats succeeded in electing their candidate, L.A. Wiltz, to the mayoralty and in recapturing much political power in the city. Full "restoration" would depend on defeating the Republicans at the state level.[47]

While intimidation, violence, and fraud heightened the Democrats' challenge to Republican state power, two additional factors contributed to the Republicans' fall. By the early 1870s, internecine conflicts over patronage and political direction shattered Republican unity, splitting the party into warring and weakened factions. Second, the Democrats succeeded in capturing the electoral loyalty of most of the state's white voters. Conservatives articulated a powerful politics of white supremacy in opposition to carpetbagger and "Negro rule," and they enlisted widespread white participation in Democratic ward clubs and paramilitary organizations. ("There are more . . . [political] club presidents and other officers in the city than there are princelings and scions of nobility in all the petty Germany States put together," complained the *Republican* in 1876.) Only the power of the federal government enabled Republican administrations to maintain power. In the state elections of 1872, for example, both Democratic candidate John McEnery and Republican William Pitt Kellogg claimed victory at the polls. Two governments were organized until, two years later, President Grant intervened decisively in favor of Kellogg.[48]

The so-called Battle of Liberty Place, the climax of a political power struggle that occurred on September 14, 1874, was a stark illustration of the Republican government's inability to survive without federal protection. The White League, formed in April of that year, invited all white men, without regard to former party affiliations, to unify under a common anti-Republican and anti-black banner. Where whites rule, the League's program read, "the negro is peaceful and happy . . . [Blacks] have become maddened by the hatred and conceit of race, and it has become our duty to save them and to save ourselves from the fatal consequences of their stupid extravagance and reckless vanity by arraying ourselves in the name of white civilization. . . ." In early July, the Crescent City Democratic Club, an armed group formed six years earlier, joined the fast-growing, not-so-secret society and changed its name to the Crescent City White League; other Democratic ward clubs quickly followed the lead and joined as well. League clubs were nothing less than "politico-military organizations," noted the editor of the *Southwestern Christian Advocate,* "made up of the rebel element . . . organized into companies, regiments and brigades. There is a complete military force, and one armed to the teeth." Relying upon economic coercion and violence to intimidate black Louisianians, the clubs also pressured some urban employers to replace black workers with whites. In August, a handful of steamboat agents temporarily established a labor bureau to furnish white roustabout crews (at the "same wages as are now paid to blacks for like labor"). Early the following month, the *Republican* complained that the White Leaguers "are knocking at all the doors of the colored people and warning them to leave all the public and private employment (except plantation work)." But the New Orleans League is most remembered for its attempt to overthrow violently the Republican state government. In August and September 1874, League members and sympathizers began procuring arms shipments from the North. When, on September 14, 500 Metropolitan policemen, 100 other police, and 3000 black militiamen attempted to prevent the unloading of White League weapons from the steamship *Mississippi,* over 8000 White Leaguers and their sympathizers engaged in armed battle with their Republican opponents near the New Orleans waterfront. After a pitched battle, the white forces found themselves the clear victor. The following day, an embarrassed President Ulysses S. Grant ordered federal troops to New Orleans to restore the deposed Republican Governor Kellogg to power.[49]

"Liberty Place" subsequently became a powerful symbol of white unity and Radical misrule, which the next generation of white political leaders, many of whom had participated in the battle, repeatedly and often successfully invoked. In 1877, Democrat Francis Nichols replaced the Republican Kellogg as governor, marking an official end to Reconstruction in Louisiana. This final transition to Democratic power was facilitated by the national Compromise of 1877, which put Rutherford B. Hayes in the presidency in exchange for the removal of federal troops from the South. Although black Louisianians continued to exercise the franchise—albeit in fewer numbers—until the end of the century, they found their political influence drastically reduced. For rural blacks in particular, the loss of political power was accompanied by an escalation in economic intimidation and personal violence.[50]

III

Political and economic developments shaped the experience of New Orleans blacks in ways that differed sharply from those of whites. Confronted with narrowing political options, dwindling patronage, and increased violence and poverty, the black community found few allies in the municipal government.[51] Throughout these years of economic depression and political violence, the black social network in New Orleans drew upon its own resources to protect its members and advance their social, economic, and political goals. Fraternal organizations, or secret societies, were vital in integrating ex-slaves into a community of free men and women, helping them to brave hard times as well as celebrate good ones. Many local societies were affiliated with particular churches or reflected some common political, social, or occupational basis; others had ties to national organizations (such as the Odd Fellows, Knights of Pythias, the Druids, and Knights of Honor). In general, fraternal associations offered a wide range of services to their members. Many provided sickness or death insurance; social clubs offered a variety of entertainments including picnics, dances, sporting events, literary discussions, and excursions to other cities. Successful and often financially secure, these benevolent societies were, in the opinion of the *Louisianian,* a black newspaper, "harbingers of mercy to the poor and destitute of the community, especially to those of their membership. . . . The manner in which the sick are cared for, and the respectable interments given to their dead, are acts not only worthy of note, but of special pride to our city."[52]

New Orleans blacks had a wide array of organizations from which to choose. Historian John Blassingame has identified more than 226 Reconstruction-era black societies from listings in newspapers and the signature books of the Freedman's Savings and Trust Company. Undoubtedly, the names of many more never found their way into print. The *Southwestern Christian Advocate,* another black newspaper, observed in 1883 that such organizations "for the relief of the distressed and the burial of the dead, are so numerous among our people, that they can scarcely be enumerated." Secret societies would remain a central feature of black life well into the twentieth century. In late 1913 the *Daily Picayune* concluded that it "is doubtful if there is any city in the country which is so great a fraternal center among the colored people as is New Orleans."[53]

Reflecting the heterogeneity of black New Orleans, fraternal organizations reproduced the class and cultural divisions within the black community. Black Creoles, who had constituted the prewar free black elite, may have appealed to former slaves for support in the political arena in the early Reconstruction years, but many maintained a separate realm of social and fraternal life. Their exclusive associations, such as the Dieu Nous Protégé, Jeunes Amis, Société d'Economie, Société des Artisans, and the Louisiana Association for the Relief of Colored Orphans, restricted membership by wealth, education, and culture, reflecting the desire of free blacks to maintain a distinct social and class identity.[54] At the other end of the economic spectrum, many unskilled black wage earners had developed benevolent societies based on their occupations and workplaces by the 1870s. Blassingame has identified at least fifteen workingmen's clubs and unions among longshoremen, steamboatmen, draymen, waiters, letter carriers, screwmen, cigar makers, teamsters, and porters in existence between 1865 and 1880. Along the waterfront, the small number of black screwmen organized in December 1870, established an account with the Freedman's Savings and Trust Company in May 1872, and incorporated their organization in January 1877. In June 1872, black activists established an account for the Workingmen's Cooperative Association of Louisiana, and in September of that year, the Colored Draymen's Benevolent Association, No.1, opened its account at the bank. Some black longshore workers formed the Longshoremen's Protective Union Benevolent Association in April 1872, and another black longshoremen's association may have been established two years later.[55] Like numerous other benevolent societies, the Longshoremen's Protective Union Benevolent Association aimed "to relieve the distress of their members, care for the sick by providing physicians, nurses, and medicine, bury their dead and provide for their widows and orphans." Yet it differed in one important respect as well: as a union, it sought "to regulate the time and fix the price of labor of working upon the shores or levees."[56] Trade unions of the unskilled, then, were part and parcel of the institutional infrastructure of New Orleans' multi-layered black community.

By the end of Reconstruction, the black press had accorded recognition and respect to black unions, acknowledging the important role that autonomous organizations of black workers played. As independent economic institutions, unions provided their members with crucial services and, more important, offered limited job protection for blacks. Yet black workers and the black middle class approached the issue of trade unions from somewhat different perspectives. The black press appreciated the mutual support and benevolence efforts of black workers, but it also viewed workers' actions through the lens of an individualist ideology. The most worthy object of organized labor, the *Louisianian* argued, was the "persistent application of individual effort to individual improvement. Moral, intellectual, and artistic improvement, in some degree, is within the reach of all." Middle-class blacks invoked the old doctrines of racial uplift and morality—persistence, hard work, moral living, and self-improvement—and argued that proper living would go a long way toward raising each workman out of the wage-earning class and making him "a master of his business." Collective action, they conceded, was valuable but limited. The real goal—the "only basis upon which the dignity of labor can be fairly asserted and permanently maintained"—was "individual effort and improvement." New Orleans' black elite shared with most African-American leaders of the era a middle-class vision of progress which, in the words of historian Judith Stein, "made capital accumulation the race's first order of business."[57]

The *Louisianian* also counseled black union members toward caution, patience, and conservative action. Such advice stemmed both from the editors' nineteenth-century belief in the harmony of interest between labor and capital as well as their awareness of the urban black worker's precarious position in the postbellum economy. In their view, white workers as well as employers could threaten the gains of the black working class. For example, in the aftermath of the September 1881 waterfront strike—which the *Louisianian* neither endorsed nor condemned—white dock workers ostracized a number of black longshoremen and screwmen who had broken ranks and returned to work, and deprived them of work for a period of time. The editors sharply denounced the white unions. Advocating racial solidarity over support for the city's labor movement, they adopted a stance prefiguring the economic gospel of Booker T. Washington: They called on the city's merchants, farmers, steamboatmen, capitalists, the Cotton and Merchants Exchange, and the Committee of Safety to "protect and encourage" black labor.[58] The elite's solution, however, held little relevance for most of the port's black laborers. It had missed the fact that black teamsters and loaders and black cotton yardmen stood firm in their alliance with white dock workers, and that the harassment of black longshoremen and screwmen was grounded not only in racial antagonism but in charges of undercharging and strikebreaking. Rather than adopt a program of job acquisition in which employers offered paternalistic protection in exchange for black loyalty (which few employers seemed willing to do), the ostracized longshoremen and screwmen found their interests better advanced by adopting uniform rules and wage rates and joining the city-wide waterfront alliance. Only in the 1890s, when whites totally repudiated that alliance, did most black workers find such arrangements with employers necessary.[59]

Black workers on the city's riverfront disagreed with the black middle class on how best to secure the dignity of labor. The actions and words of union members revealed that they did not share the elite's belief that individual effort was more significant than the reduction of working hours, the advancement of wages, and the regulation of the labor supply through the closed shop. For it was only through control over their working lives that any degree of "self-culture" became possible. If both groups employed a language common to nineteenth-century urban black communities in the North and South, the different emphases placed on collective and individual action revealed contrasting class perspectives. While the black middle class employed individualistic notions of mutual aid, the unions formed by black workers incorporated many of the traditional functions of fraternal orders while expanding the very boundaries and meanings of self-help. Drawing on a shared language, working-class black activists contributed distinct ideas about the relationship between labor and capital that contrasted sharply with the ideology espoused by middle-class black politicians and journalists of the era.[60]

Black longshoremen, who had struck in 1865 and 1867 without a union, proved to be among the most militant of the city's black laborers during the decade of the 1870s. Putting forward their own agenda, based squarely upon a conception of themselves as working men with certain rights not yet accorded them by their employers, these waterfront activists defended their right to strike and to compel non-union men to abide by their actions, depression or no depression. Their struggle to maintain wage rates and improve working conditions before the 1880s revealed a commitment to active organization and a vision of mutuality and solidarity among black wage earners. At the same time, those experiences demonstrated the very real limits of their power. During the 1870s, their numbers remained small, white workers stood aloof, and the city government, under both Republican and Democratic administrations, never hesitated to use police power to impose order and restore stability. Like their counterparts in Charleston and Pensacola, who met with repeated defeats, New

Orleans' black longshoremen proved unable to implement tneir vision on the docks. That failure highlighted both the overwhelming power of waterfront employers during the depression and the need to unify labor's forces across craft and racial lines. Until workers could overcome the sectoralism of the docks—the trade and racial fragmentation—their militancy would bear little fruit.[61]

One year before the depression began, black longshoremen had organized a union for the first time. In 1872, over 200 black workers joined the Longshoremen's Protective Union Benevolent Society for the express purpose of "protecting ourselves and widows, as fellow laborers should, and also to have a regular set of wages for this society to work under, as all chartered societies do."[62] The president of the new union, R.T. Matthews, issued a public statement in late October of that year:

> We, as poor colored laboring men, and many of us unlearned, being as it were but a short time liberated from what we call the ungodly yoke of bondage. We work along the shore many of us, and after some considerable time expired, we, as fellow laborers, some of us at least, after seeing the condition of many of our brethren who follow the river and longshore for a living, took it into consideration to form ourselves into an incorporated body of laboring men. . . . We were scoffed at and rebuked by white men who work along shore, telling us constantly that the negroes broke the wages down, and it caused all to suffer.[63]

Testimony offered by individual black workers at their meetings emphasized abusive treatment and a lack of rights. There was no "harm for any poor man to ask for his right. That was all they were doing now," union member Victor Light noted. "Heretofore they had been cheated. They had been driven at all hours, and at the stevedore's office they had to take what he chose to give them or be driven out like dogs." Another union member, Robert Hopkins, similarly noted that for "eight years past . . . [black longshoremen] had been cheated and defrauded . . . but since they really had been men, they had never gotten their rights. The stevedore had fattened on them and they were as poor as ever—poorer! because money would not get now what it would a few years ago. But they would have their rights now. They would be quiet and orderly, but they would have their rights."[64]

Longshore union goals reflected the thinking of these men. Taking advantage of the renewed riverfront activity at the start of the 1872 commercial season, the black union (also called the Protective Union) and a white association, the United Laborers,[65] issued three key demands. First, stevedores, coal merchants, and contractors should increase wages from $2.50 to $4 a day. In earlier years, read a statement signed by black union president Matthews and white president William Crotty, "rents were low, and taxes low, and provisions and groceries very cheap; but now it is just to the contrary; our rents, groceries and everything are so dear, we poor men can not take so much as a single dollar from our daily wages to aid or support a sick wife or child." Moreover, the seasonality of dock work—a theme that for decades would continue to

dominate discussions of wage scales on the waterfront—meant that longshoremen could obtain employment only for a "few months in the year." Second,
the two unions insisted upon regular hours of work. A printed notice distributed along the levee stated that longshoremen would no longer tolerate a
working day that lasted from 6:00 a.m. to 7:00 p.m.; in its place, a new
working day would begin an hour later and end an hour earlier. Finally, for
the first time, the unions insisted that stevedores and contractors sign one-
year contracts with their men.[66]

Black longshoremen struck on October 17, dispatching large delegations
of union members, including one that numbered 500, to request that stevedores comply with their demands. By ten a.m., roughly 1000 blacks had
assembled on the levee for a public procession. Led by brass bands and
marshals holding the American flag and shouting "Four dollars a day or
nothing" as their battle cry, the marchers, the *Republican* noted, carried
themselves "like men who 'meant business.' " Non-union men at work
"knocked off" and joined the procession as it passed their ships along the
levee. Over the next several days, roving committees of strikers enforced
strict discipline by approaching those still at work on ships and in coal yards,
and in most cases, convincing them to halt work. Hundreds of non-union
workers, by either "fear or enthusiasm," joined the nucleus of black union
strikers. If the rather coercive methods of forging a united front were reminiscent of the strikes of 1865 and 1867, the degree of organization distinguished
the 1872 conflict from its predecessors. Identified as union members by the
blue ribbons they wore, the strikers exhibited a "degree of firmness and
resolution . . . maintained themselves free from intoxication to an unusual
extent, and seemed fully under the disciplined control of their leaders."[67]

The strike quickly spread to other groups of black workers. Coal heavers
knocked off near Jackson Square when a stevedore refused to sign a charter
promising to pay $4 for an indefinite period. On October 21, the Portable
Railroad Company's laborers, who earned $10 for a six-day week for filling up
the Orleans Canal near the City Park bridge, struck to increase wages to $2 a
day, which the company's directors described as "exorbitant" and refused to
pay. When the roustabout crew of the steamship *W.S. Pike* joined the strikers'
ranks, demanding $60 per man per month, the ship left port with only eight or
ten men, instead of its normal crew of sixty. Regular longshoremen continued
to insist that new ships, upon arrival, meet their wage and hour demands.
Finally, organized and spontaneous crowd actions supporting the black workers gave them an initial edge over their adversaries. When police arrested two
or three of the strike leaders, for example, a crowd of 1000 to 2000 rescued
them.[68]

Although largely well-disciplined and orderly, the strike was accompanied
by some violence. The most serious outbreak occurred on the first day, producing the only casualty of the conflict. Some longshoremen had continued to load
salt on a Mississippi Valley Transportation Company barge for $3 a day. The
company's stevedore, Captain Joseph W. Barnes, defended his workers against
the strikers who assembled near the barge. According to some accounts, when

the strikers rushed the barge, Barnes picked up a hatchet, threatened to "brain the first man who came on board," and hit one of the strikers. Immediately he became the target of several gunshots, as well as a "fearful avalanche of stones," from the crowd, and was killed in the attack. The crowd quickly dispersed, but its outbreak appeared at least somewhat successful. By that afternoon, some of the steamship companies and stevedores who could afford no delays surrendered—if only temporarily—to their workers' key demand, agreeing to pay 40 cents an hour for a ten-hour day.[69]

Despite disclaimers from union president Matthews that the black union, as a society, "had no hand in the disturbance," employers, police, and press placed the blame for the violence squarely upon the strikers. Newspapers denounced both the killing of Barnes and the "conspiracy" among black levee laborers whose "avaricious and unreasonable demands" and "base and suicidal exactions" threatened the port's commerce. At a well-attended meeting, steamboat captains, merchants, and other men of commerce vowed to resist the "extortionate" wage demands of the "murderous mob," and called on municipal officials to provide immediate police protection. "We have the most expensive city government and police in the world, and the taxpayers certainly have the right to demand the protection of life, property and industry, in return for their enormous contributions to its maintenance and support," businessmen complained. Republican Mayor Flanders offered immediate protection. Before a gathering of 300 merchants, steamboatmen, and stevedores at the Chamber of Commerce, he admitted that the strikers had taken the police by surprise. But now, he made clear, he would permit no further breach of the public peace, promising the full cooperation of city authorities in protecting strikebreakers and preventing violence. Colonel A.S. Badger, a Republican party leader and chief of the Metropolitan Police (who would be seriously wounded by white paramilitary forces in the "Battle of Liberty Place" two years later), assigned between 250 and 300 policemen, armed with Winchester rifles, to the waterfront. Large reserve forces stood ready at every police station, and squads of police quickly appeared whenever the longshoremen threatened stevedores, shipping agents, or their workers. On one occasion, police charged the strikers with fixed bayonets; they arrested alleged "ringleaders"; sometimes they threatened to fire into the crowds unless the latter dispersed. Simply a rumor of a strike at the Morgan railroad works led officials to dispatch 100 policemen across the river to Algiers to provide protection for that company's non-striking workers. Police, "armed with rifles and revolvers, having more rounds of ammunition than soldiers going into battle," quickly suppressed the strike of laborers at the Orleans Canal. The "display of lawful power," noted the *Republican,* seriously dampened the ardor of the striking longshoremen.[70]

Striking longshoremen found no sympathy from New Orleans' political leaders or the Republican-dominated police force. In a classic expression of the Republicans' conception of the role of trade unions, Mayor Flanders defended the right of laborers "to strike, to organize, make regulations, combinations and terms, and to use all proper means to have them allowed, just as the printers have and the merchants." But all outbreaks of violence would be suppressed speedily. Colonel Badger similarly lay down the law in no uncertain terms to union presidents Crotty and Matthews: While laborers had every right to strike for $4 a day, that right "extended no further than to quit work when they became dissatisfied with their pay." Strikers could not interfere with those who were "willing to work, no matter upon what terms they may engage." To enforce the strikebreakers' right to work, the Republican colonel barred any further union demonstrations or parades on the grounds that they would only provoke more disorder and excite "the worser passions of the uneducated masses." Both labor leaders had no choice but to accept the Colonel's terms, which in effect deprived the unions of any leverage and power. Although the strike lingered on for another week, large numbers of policemen harassed, dispersed, and arrested delegations of blacks congregating on the levee or meeting docking ships. By October 29, the Chief of Police reduced the detail on the waterfront, having concluded that all was quiet on the levee and that the strikers had been defeated. The longshoremen, he reported, were laboring at their old rate without problems; daily wages remained at $2.50, and in some cases, $3.[71]

If police repression contributed substantially to the strike's collapse, the persistence of racial and organizational divisions did little to bolster union ranks. Unfortunately, there is little surviving information on the alliance between the black Longshoremen's Protective Union Benevolent Society and the white United Laborers' Association. The two unions published a joint statement of purpose at the outset of the conflict, and one observer noted that the levee laborers, "colored and white, seemed to be equally interested, and a mutual understanding seemed to be had." But while the initial gathering of strikers "contained all colors," it was clear that blacks overwhelmingly dominated the strikers' ranks, while white laborers appear to have played a passive role in the conflict. On October 17, for example, squads of white longshore workers appeared at various points along the levee, "not engaged in doing anything but looking on." Similarly, few members of the United Laborers appeared in the procession that day. Only 40 whites in a feeder march of 100 joined the parade of 1000 blacks along the levee. "We had set the wages of our society, and the white society had given us letters that they would be with us heart and soul, which they failed to do," black union president R. T. Matthews complained in a public statement. "But there was one society [that] joined us near Thalia street, whether incorporated or not, we do not know." While evidence concerning the nature of racial dynamics is sketchy and inconclusive, it is clear that black workers found the level of white participation disappointing.[72]

One last example ironically underscored these divisions. As the black longshoremen's strike was collapsing under police repression, one group of white riverfront workers immune from harassment celebrated its anniversary in style. The Screwmen's Benevolent Association—which apparently had ignored the longshoremen's efforts altogether—marched proudly through the city in force, followed by four brass bands.[73] Waterfront occupational hierarchies remained firmly intact, and solidarity between races or different groups remained fragile or non-existent. The striking unions' inexperience, the lack of communication between different groups of waterfront workers, and traditional racial divisions contributed to the persistence of fragmentation among riverfront working men.

The depression of the 1870s merely exacerbated the traditional problems of irregular work and unemployment for unskilled black and white waterfront workers. With little bargaining power in an overcrowded labor market, they were subject to the vagaries of commerce and the whims of their employers. "The market value of labor on the levee," the *Picayune* noted in 1877, "on plantations, in shops, in mills, on steamboats, in mines, on railroads or at sea, or wherever else men work for wages, can never be long regulated and controlled by strikes, leagues, society resolutions or any of the methods resorted to by those who only have labor to sell. . . ." The lesson of the depression, it continued, proved the validity of these principles of political economy:

> When there are hundreds, perhaps thousands of willing hands now idle who would be glad to work for small wages . . . it cannot be insisted on as strictly right to exclude such men from selling their labor for what they can get. When there are ten thousand men competing for labor it is much better to give such wages as will give employment to the greatest number than to fix an arbitrary high rate which will give remunerative employment to a few and leave the majority unemployed.[74]

If black workers proved unable to control the supply of labor available to shipping agents and stevedores before 1873, the depression made a difficult task even harder. As the full brunt of the depression began to be felt, the four-dollar day became a demand of the past, as workers struggled to preserve what they had or restore wage cuts by employers. For the rest of the decade, black waterfront activists followed a strategy of collective action based on persuasion and coercion. These short-lived, rolling strikes continued periodically for the duration of the depression, but at best they produced little more than fleeting success.

The first of the depression-era strikes occurred in October 1873, when the black Protective Longshoremen's Union again struck for the four-dollar day and the exclusive hiring of union workers. In the middle of the month, a group of union men singled out the ship *Eliza Stevens,* which was being loaded by non-union men. Recalling the fate of Captain Barnes the year before, the contractor immediately sent word to the captain of the Harbor Police for assistance. But by the time the police arrived, the contractor had surrendered to the strikers and hired the Association men. A committee of strikers petitioned now-General A. S. Badger a week and a half later. In their statement they insisted that black longshoremen had a right to ask whatever wages they pleased, to compel men accepting less than the Association rate to quit, and to parade and enforce discipline without police harassment. (They concluded that "the police were paid to fight against them by the captains and steve-dores.") Badger, having no more sympathy for the longshoremen's positions now than he had the previous year, acknowledged that he had no right to interfere with the workers' right to set their own rate of pay. But he also promised to protect any strikebreaker who continued to work. Permitting union men to parade peacefully along the levee, Badger refused to tolerate any interference with "peaceable workmen."[75]

The warning had no impact on the longshoremen's behavior. On October 29, a crowd of more than 100 black longshoremen gathered at the head of Jackson Street and surrounded a large number of men employed by W.G. Coyle & Company to load coal on the steamship *Caldonia*. The workers knocked off when the union men warned them against working for less than $4 a day. With the element of surprise still on their side, the crowd moved down the levee to Race Street and convinced another gang employed by Coyle & Company to quit work; the "same scene was transacted" at the yard of the Gas Company on Robin Street, and at Tyler's yard on Freret Street. But General Badger backed up his threat, guaranteeing protection to all men working for the coal dealers, "even if he [had] to use bullets to disperse the mob." The strike quickly collapsed. Badger and the police hastily suppressed another small strike the following year. In early September 1874, a group of black dock workers "set upon the crew" of the steamship *Paragon*, demanded the employment of union men, and briefly stopped the loading of its salt cargo. Under police protection, however, the ship's captain quickly resumed the loading process.[76]

A more serious, if brief, outbreak of waterfront labor conflict came in January 1875, motivated by standard wage complaints as well as by a fluid political situation. Months earlier, President Grant had sent federal troops to New Orleans in the aftermath of the "Battle of Liberty Place." Black dock workers likely drew a distinction between the Republican party's local and national branches. Unlike local Republican officials, who regularly repressed riverfront strikes, federal troops, representing a national Republican interven-tion in local affairs, might afford strikers a protective umbrella of sorts under which they could exercise their collective muscle. In late January 1875, spokes-men at a meeting of the Longshoremen's Protective Association informed their audience that the federal troops would protect them if they again struck to restore their wage scale.

Union president Joseph Harris attributed the conflict to the contractors' failure to pay longshoremen loading barges the agreed-upon wages of $2 a day; laborers instead received between 75 cents and $1. In a scene that fol-lowed a familiar pattern, some 350 black men—possibly led by two men with badges who could have been Metropolitan Police officers—marched along the levee, joined by new workers they had notified or warned to quit loading and discharging steamboats. But not all workers appreciated the union's efforts or joined the strike. The depression had made finding steady work along the shore, which was always difficult, even harder. The regular hands of the steamship *Natchez* resisted the strikers' entreaties and found themselves the target of the crowd's anger. At one point, the strikers rushed the crew and forced it to take refuge on the steamship. Union men similarly resisted the efforts of the Harbor Police to disperse them, attacking the commanding officer with stones and clubs. The police drew their revolvers, shot and wounded several blacks, and finally dispersed the crowd. The *Natchez* continued loading under a heavy guard of police.[77]

Accused of rioting, the black union sought to "place the matter in its true light before the people" and arrive at some understanding with the steamboatmen and bargemen by other means. In early February, a meeting attended by 350 longshoremen unanimously adopted a series of resolutions on the matter. Concurring that the violent outbreak on the steamboat and barge landing on January 30 was "injurious alike to capital and labor," they declared that the welfare of steamboat and barge owners, contractors and laborers alike required that "fair dealing, honest work and honest pay, according to contract, should be strictly adhered to." They agreed to work for $2.50 a day, or 25 cents an hour for a 10-hour day, and to ask the captains, owners, and contractors to "see that we are fairly and honestly dealt with." In a new approach, they resolved that

> in order to see on our part that our contracts are faithfully and honestly carried out, two persons shall be selected to witness all contracts and settlements, such persons to receive five cents from each laborer for witnessing and seeing that said contract is faithfully carried out.

The union appointed a committee of five to present its resolutions to the various riverfront employers. Neither the earlier show of collective strength in the streets nor such peaceful appeals, however, achieved the union's goal of restoring wage rates.[78]

Subsequent efforts by black longshoremen met with similar results. In mid-April 1875, laborers discharging the steamship *Great Republic* briefly walked off the job, demanding a return to the pre-depression $2.50 a day rate. But the contractor stood firm at $2 a day, and the men, after much arguing but no fighting, returned to work. Two years later, in November and December 1877, a Union Laboring Society—composed "mostly of colored laborers" and numbering about 500 members—complained bitterly about their wages. Citing five contractors and two mates who paid between 10 and 15 cents an hour, the black union resolved to work for not less than 25 cents an hour. But aware of a record of almost unbroken failures, union president Joe Harris, who had been a leader of the 1872 and 1875 strikes and currently worked as a timekeeper on the docks, cautioned his members at several meetings against striking.[79] Relatively few in number and isolated from their white counterparts during the depression years, black union men remained unsuccessful in imposing some degree of order on their chaotic working conditions.

At the end of the 1870s, transforming waterfront workers' expectations into reality awaited two related developments. First, given a racially divided labor market, some mechanism for ensuring interracial cooperation, or at least neutrality, was absolutely indispensable. Second, the strict sectoral identifications associated with a multitude of employers and employment situations had to be broken down, and bridges between different groups of largely unskilled laborers constructed. In other words, riverfront men had to develop what Hobsbawm has called the "habit of industrial solidarity" in order to become an " 'effective' labour movement."[80] Precisely such a movement emerged in the last year of the old the and first years of the new decade, fundamentally transforming both race and class relations in the region's largest city.

IV

It "strikes me that the New Orleans of February, 1880, is an extremely prosperous city," remarked British traveler George Augustus Sala, "and that somebody must be making an immense deal of money." The return of commercial prosperity following the 1870s depression promised higher profits for businessmen; it also created new expectations among the thousands who labored on the riverfront and in the city's workshops, warehouses, and factories. Across the United States, the end of the depression witnessed an explosion of organizational energies on the part of industrial and craft workers. The expansion of market relations and the rise of industrial capitalism by the beginning of the Gilded Age had called forth new forms of labor struggle that often involved unprecedented, if temporary, solidarity between immigrant and native-born workers, men and women, and blacks and whites. Although the depression had wiped out or reduced to impotence many unions and city-wide labor assemblies, in scores of cities and counties after 1878–79 new trade unions quickly formed and new federations of trade unions tenuously united skilled and unskilled workers, culminating with the remarkable growth of the Knights of Labor by the mid-1880s.[81]

In the summer of 1881, the president of the New Orleans Typographical Union, the city's oldest association of skilled workers, initiated a "grand fraternization of the labor organizations," the Central Trades and Labor Assembly, with the assistance of the newly formed Cotton Men's Executive Council (discussed below). Representatives of the city's old and new trade unions met in August "in order to amalgamate all men who earn their living by labor into a solid compact combination for the purpose of mutual aid and support in all controversies between labor and capital." Declaring its opposition to allying with any political party, the Assembly pledged to "use its influence with the law-making power" to secure legislation regulating child labor and the hours of work, the reformation of prison labor, and the enforcement of all laws beneficial to labor. Although it favored the arbitration of differences between workers and employers, it committed itself to the "amelioration of the condition of the laboring classes generally" and pledged its members to "assist each other in securing fair wages . . . by all honorable means," including the use of boycotts against any unfair employer. The constituent unions invested the Assembly with general legislative and executive powers. "Any one of the constituted bodies could strike, if it saw fit," observed labor reformer George McNeill in 1887, but "it could receive the support of the entire body only after the complaints had been considered by the central body and approved as sufficient grounds for a strike." By 1882, some twenty trade unions, representing 15,000 workers, had joined the Assembly, making it a veritable "Army of Artisans," in the words of the press. The Assembly included not only the new riverfront cotton unions but associations of coopers, iron molders, bricklayers, boilermakers, foundry workers, and typographers as well.[82]

Given the role of commerce as the city's lifeblood, it is not surprising that many of the new workers' organizations, and indeed the very heart of the new labor movement, came from the New Orleans docks. The upsurge began in 1879, as workers along the shore and in the pressing yards organized on an impressive scale for the first time. Leading the new union drive was the white Cotton Yardmen's Benevolent Association formed in December 1879 under the direction of Democratic party ward boss and Administrator of Police, Patrick Mealey; two years later it boasted a membership of 986 and a bank account of $13,000. Black cotton rollers, with the assistance of the leaders of the white yardmen, formed the Cotton Yardmen's Benevolent Association, No. 2, in January 1880, and both organizations agreed to work "in full harmony" with each other. In April 1880, the black teamsters and loaders established their benevolent association, as did black coal wheelers the following month. White coopers organized in August 1880, black coopers following suit in July 1882. In September 1880, cotton weighers and reweighers formed a mutual aid association that aimed to secure a uniform system for weighing and re-weighing cotton and to regulate wages and arbitrate disputes. Black and white freight handlers on the city's expanding railroads organized racially separate unions in February 1883.[83]

Race and location defined the contours of various longshore organizations. Some 500 members of several white longshoremen's associations in the First, Second, Third, and Fourth wards joined hands in October 1879, forming a "permanent body representing their craft throughout the city." The older black Longshoremen's Protective Union Benevolent Association (organized in 1872) was joined by a branch across the river in Gretna. Black workers laboring on the basin established the Magnolia Longshoremen's Association in January 1880. Jurisdictional and organizational competition kept relations tense between some four small black branch associations (which had formed a Grand United Union of Longshoremen of the city of New Orleans in 1883) and the larger and more powerful black Longshoremen's Protective Union Benevolent Association through the decade.[84]

Widespread unionization itself did not guarantee success. In fact, the decade began with a number of unsuccessful efforts by laborers to raise waterfront wages. Black longshoremen and roustabouts—who had been little more than a "source of considerable annoyance" to steamboat captains and police for years—responded to the economic upturn with a new militancy. Under the control of "unscrupulous leaders," the *Picayune* complained, these blacks were "most insolent in demands for higher wages." In

early 1880, some 1000 mostly black longshoremen resurrected an old demand and struck again for the four-dollar day (which would increase their
wages from the depression rate of 25 cents to 40 cents an hour). Some of
their tactics resembled those of the previous decade: a group called the
Union Laboring Men's Association, for example, paraded down the levee
from Tchoupitoulas and Galentine streets, driving off the men loading the
steamship *Golden Rule*. One steamer under a strict contract deadline temporarily acceded to the laborers' demands. Then, in a new move, strikers
marched to the State House, where a committee of workers, including at
least one member of the city's small socialist party, presented a memorial to
the General Assembly, "praying for legislative assistance." According to
another account, a delegation of strikers met with Mayor Patton, complaining of police interference with their peaceable efforts to secure wage increases. Later that day, the laborers, led by a field band, marched to Lafayette Square for a public meeting at which union leaders and local socialists
reported that their petition had been "courteously received" by a legislative
committee and praised the determination of the crowd.[85]

Neither the petition for legislative relief nor the isolated strike had the
desired impact. Steamboat captains and owners demanded that Democratic
party Mayor Patton protect those riverfront laborers working for less than 40
cents an hour. Although Patton was initially reluctant to assume the role of
strikebreaker, he ordered the police to arrest anyone found disturbing the
peace and dispatched a new detachment of police to the riverfront "to impress
upon the minds of the strikers that they would not be permitted to interfere"
with strikebreakers. The police performed their task faithfully, arresting those
allegedly interfering with non-striking waterfront workers. Steamship operators were able to continue loading and unloading cargo by using crews and
"outsiders." There were "plenty of men willing to work" at the old rates, the
press announced in March, "and there are many who do." Although the
strikes dragged on for months, they proved to be little more than an annoyance to the targeted employers.[86] Despite the upturn in the trade cycle, isolated sectoral black unions were unable to overcome the difficulties posed by
a still glutted labor market, staunch employer opposition, and municipal government repression.

Events in the fall of 1880 proved decisive in transforming the nature of
waterfront labor struggles. When conferences in July between the newly
unionized Cotton Yardmen and the owners of the cotton presses produced no
agreement on wage increases, both the black and white unions prepared for a
joint strike. On September 1, some 500 white and 300 black yardmen closed
every cotton press in the city. What was surprising, the owners noted, was that
"in a strike of cotton rollers, all the men connected with compressing of cotton
have joined in, although some of them are not affected, one way or another,
by the rate paid the rollers." The yardmen's associations embraced almost the
entire workforce of the cotton presses, including the cotton rollers (the largest
number of yard workers), press and yard hands, and truck men; only clerical
workers and superintendents remained outside their ranks. Unprepared for

this unprecedented interracial and multi-trade strike, and unable to secure sufficient replacement labor quickly, press owner after press owner capitulated to their workers' demands within a few days.[87]

Members of the newly organized black Teamsters' and Loaders' Union followed the yardmen's lead, striking on September 6. They demanded that their employers, the boss draymen, increase their wages from roughly $12 to $18 a week and hire them not by the day but by the week. Most employers expressed a willingness to meet the wage increase for hours actually worked, but resisted hiring men by the week, reluctant to provide steady employment in a field characterized by a highly fluctuating demand for labor. But like the cotton press owners, boss draymen could not afford a delay in business. Most were small employers with little capital, contracting for railroads, shipping lines, and press yards. And by early September, they already had signed contracts with these larger concerns. Efforts to secure strikebreakers proved futile, as strikers' sharp warnings reduced the number of new recruits significantly. Within days, most teamsters and loaders had won their demands. The walkouts of the yardmen and draymen were followed by an "epidemic of strikes" that affected roustabouts and iron foundrymen.[88]

Impressive as the first victories by unskilled riverfront workers, these strikes also inaugurated a new era in the history of New Orleans labor. In mid-December 1880, a wide range of dock unions established the Cotton Men's Executive Council. Drawing together non-industrial, common laborers engaged in the cotton trade, the Council sought to minimize longshore workers' economic insecurity and vulnerability. This "solid organization of the labor element embracing every class employed in handling the staple from the time of its reception until it is stored in the ship's hold" counted in its initial membership the white and black unions of screwmen, longshoremen, yardmen, white weighers and classers, and black teamsters, representing approximately 13,000 men. As a coordinating body that respected the institutional integrity of its constituent unions, the Council ended the fragmentary nature of waterfront unionism, making solidarity, or at least collaboration, between the different dock trades possible for the first time.[89]

What explains the formation of such an alliance in 1880, given the previous pattern of hierarchical, sectoral unionization, sharp occupational fragmentation, and political violence against blacks? The structure of the industry, as well as the patterns of unionization in other ports, suggests a certain logic. The overwhelming number of dock workers possessed no critical skills with which to bargain successfully. Overcrowded labor markets during and after the depression had reduced further their power vis-à-vis employers. Unskilled, easily replaced, and divided along sectoral and racial lines, the port's workers—the screwmen aside—possessed little capacity to impose their demands. The experiences of the small, black longshore unions in the 1870s showed that divisions among workers led to failure; in contrast, the example of the cotton screwmen demonstrated that success depended upon control over the labor supply. Shortly after the Council's formation, an alliance of white screwmen, black and white longshoremen, and black teamsters imposed a single standard for screwmen's work on the port. In late January 1881 they struck a number of ships using sailors or black screwmen, who were paid a dollar a day less than their white counterparts, to load cotton. Whites refused the offer of two ship captains to discharge their black workers and employ whites exclusively, insisting instead that employers pay the same wage to all workers performing the same tasks. That is, they demanded not the dismissal of blacks but the raising of blacks' wages to the level of union whites. Within days, they had achieved their goal.[90]

Given the industry's structure and the existence of a racially divided labor force, alliance in the form of the Cotton Men's Executive Council made excellent sense. Logic, however, is hardly a sufficient explanation to account for the formation of such an unprecedented organization. Yet there are no union records, first or second-hand accounts, or testimony from union leaders, ordinary workers, or the press to provide precise insight into the calculations and bargaining that contributed to the Council's formation. Historians must be careful not to place words in subjects' mouths; historical ventriloquism will not substitute for absent documents. Hence, our understanding of the alliance's formation must necessarily remain incomplete and speculative. But if its founders' exact motives remain shrouded in some mystery, the Council's structure, goals, and subsequent actions are plainly visible.

The new organization possessed considerable authority and power. While individual unions formulated their own wage rates and work rules, the Council's stance was often indispensable to the success or failure of a given struggle. On the one hand, by refusing to sanction changes in an existing tariff, Council delegates could exercise limited veto power over member unions by offering them no support in their moment of need. On the other hand, the Council's endorsement would commit all waterfront unions to support actively a union in need. German economist August Sartorius Freiherr von Waltershausen captured the essence of the Council's role in his 1886 book:

> The closely-knit connection among the workers makes it difficult for the employers to break a strike. If, for example, the longshoremen have declared a strike with the approval of the central authority, their rather simple work could be performed by workers of similar occupations; but as the latter usually belong to the "ring," the employer must renounce his intention. He can only count on non-members of the union. But the latter are in the minority and do not dare oppose the union. But even if substitutions are made, the merchants are not rid of the difficulty. For even if the cotton is loaded from the river boats by other longshoremen, there would be a lack of draymen whose union would strike at the command of the central authority. And thus it would continue successively, and the merchant would have eight such strikes on hand, from which he would emerge victorious only with difficulty.[91]

Relying upon this network of institutionalized solidarity in the decade after its formation, the Cotton Men's Executive Council imposed a new order on riverfront employment and work practices. It helped to codify and enforce work rules and wage rates, regulate the size of the labor force, and oversee interunion relations.[92] The Council never succeeded in wholly eliminating the economic insecurity of waterfront work, nor did it entirely erase divisions between black and white workers and among the different waterfront crafts. But in providing the framework necessary for the formulation of strategy, the resolution of differences, and the promotion of a limited solidarity or collaboration across trade and racial lines, it brought some degree of stability and predictability to the docks and substantial improvements in workers' lives.

The opening of the 1881 cotton shipping season presented New Orleans waterfront workers with their first opportunity to test their new-found commitment to interunion and interracial solidarity. The Cotton Men's Executive Council coordinated the efforts of the various branches of waterfront labor, administered the strike fund, and approved a joint strategy for the fall season. "The laborers in every branch along the Levee appear to be ripe for a strike for higher wages," the *Picayune* reported in August, "and this spirit is fostered by the success attending the efforts of others in the same direction last year." Although most dock unions made new demands that August, the key fight centered, as it had the year before, on the cotton yards.[93] Unlike earlier conflicts, however, the fundamental issue revolved not around wages but control over the labor process. "There is no great difference in the tariff offered by the employers and that demanded by the employes," noted the *New Orleans Times*.

> What the employers demand is that they shall control their labor in the cotton press yards and everywhere else. Under the demands of the employes the employers are deprived, in a measure, of the direction and control of their business. The cotton press owners haven't much to say about the management of their presses. They have very little authority in their press yards. Employers cannot engage, discharge or direct their employes. The labor associations do all that kind of business.[94]

But if workers believed that a strike in 1881 would result again in a speedy victory for the unions, two developments rendered such an outcome unlikely. First were the recent changes in municipal politics. In the fall of 1880, a newly formed People's Democratic Association, a reform movement representing the business interests of the city, challenged the city's regular Democratic party, known as the Ring. Ironworks owner Joseph Shakspeare, the reform candidate running on a platform that sharply condemned "machine rule" of the city, narrowly won the mayor's race in the November 1880 municipal elections. Although the reformers succeeded in electing only their mayoral candidate and failed to oust the machine-dominated administrative officers, the *Mascot,* a journal of political satire, concluded with only some exaggeration in 1882 that "[t]he Cotton Exchange of this city has been running its politics." The businessmen's mayor posed a limited challenge to the patronage power of the Ring and indeed proved sympathetic to the plight of his fellow employers confronting a growing challenge from their workers.[95]

The growing organizational power of the city's business elite presented the new union movement with its second key problem. Merchants and middle-men, who vowed that 1881 would not see a recurrence of the previous year's experience, had strong resources and allies upon which to draw. Well before the labor question assumed "the shape of open warfare," New Orleans cotton merchants had developed institutions for regulating their affairs and rationaliz-ing their business. Back in February 1871, in an attempt to adapt to new patterns and methods of trade in the postwar period, roughly 100 local traders had established the New Orleans Cotton Exchange, for the purpose of "mu-tual protection and co-operation in preserving and building up the cotton trade of New Orleans." The Exchange produced needed statistical data on annual cotton crops, arbitrated conflicts between buyers and sellers, elabo-rated strict and systematic rules governing cotton sampling, handling, and sales, regularly advocated physical improvements of the city's transportation facilities, and lobbied for federal aid to deepen the Mississippi River. The impulse toward self-regulation soon spread to other merchants and business-men in the Crescent City. In 1880, for example, exporters and importers formed the New Orleans Maritime Association to protect their interests in navigation.[96]

Businessmen's organizations assumed an explicitly anti-union stance in re-sponse to the formation of the Cotton Men's Executive Council and heightened labor activism. On July 1, 1881, the city's associations of cotton factors, bro-kers, buyers, and ship agents created an Executive Council, composed of the chairmen of each association, "for the protection of its members," and empow-ered to formulate and recommend collective strategy. By late August the chair-men of the associations of cotton pressmen, boss draymen, and ship brokers joined the Employers' Council, thus uniting almost "all the storage and com-pressing capacity in the city." "[W]e do most strenuously urge," the new organi-zation declared, "that no arrangements be made with members of organiza-tions which seek not only to dictate the wages of the laborer, but also the terms on which he may be employed. It is easy to see that such a dictatorship must be in the last degree fatal to the commercial interests of this port." On August 27, this Executive Council recommended that the port's employers discharge all workers refusing to accept their tariff. Workers must understand that in making contracts "business must be free and unrestricted, subject to no rules or regula-tions" of any outside organization. With these reasons in mind, the business-men's Council's first statement of purpose declared: "We shall resist to the last the attempt to create a tyranny that will surely destroy the energy and self-respect of our commercial community."[97]

The employers' Executive Council rejected all calls for mediation made by the press, the city council, the mayor, and even the Produce Exchange and Chamber of Commerce. Thomas L. Airey, president of the Cotton Exchange, spoke for his fellow businessmen when he wrote to Mayor Shak-speare that compromise and mediation were unacceptable. He recognized that any delay in the handling of cotton would "militate against the commer-cial interests" of the city, but employers believed a "far more serious blow would be struck" by acceding to the workers' demands. In asserting the absolute right of capital to control work relations in the transportation busi-ness, Airey contended that the

> differences [between the employers and their employees] are of such a na-ture that they do not admit of arbitration. The issue to be fairly and squarely met is not so much the price of labor, which in this controversy has become a matter of secondary importance, but whether the laborers' union shall or shall not control and direct the cotton trade of New Orleans. The question must be settled definitely and beyond possibility of misunderstanding, and any arbitration or compromise instead of effecting a settlement will only delay the evil until a further period.

Moreover, he concluded, employers

> can not and will not surrender the control of their capital and the produce entrusted to them by their constituents to irresponsible parties. Nor are they willing to treat or confer with such parties or their representatives so long as they set up such claims. It is indispensable that every merchant should have complete control of the employment and discharging of his clerks and assistants.[98]

Not everyone in New Orleans saw the matter in these terms. Although the port's commercial elite possessed considerable economic power, it did not command automatic respect and subservience from other social and political groups. Indeed, many concerned observers did not find a war be-tween labor and capital at all warranted. The *New Orleans Times,* which was sympathetic to the employers, decried the deadlock as a "public misfortune" and concluded "the merchants and laborers, the employers and the employ-ees cannot afford to have the cotton business of this city further impaired by engaging in a conflict which must bring commerce to a dead stand or drive the cotton to other ports where it can find expeditious handling . . ." The *Picayune* editorialized that labor "asks increased wages, capital asserts its right to control. This is all. Is there then no basis upon which to agree?" During the strike in early September, the City Council passed a resolution appointing a five-member committee, with the mayor as chair, to bring the workers' and merchants' associations together. Seeking some compromise that might spare the city a prolonged and bitter strike, reform Mayor Joseph Shakspeare approached the Cotton Exchange with an offer to arbitrate. Similarly, the Produce Exchange and the Chamber of Commerce both of-fered their services toward resolving the conflict. While recognizing that the waterfront employers might be justified in resisting labor's "attempt to dic-tate," many important political and business leaders put the port's peace and prosperity before principle, and counseled the employers' associations to find some ground for settlement.[99]

Merchants, remaining adamant in their refusal to "surrender the control of their capital," rejected all offers of mediation or arbitration, which, they argued, would merely "delay the evil until a further period." Only the "uncon-ditional withdrawal by the laborers from the positions they now assume" could serve the interests of the port. The Employers' Council took unprecedented steps to meet the waterfront crisis. In late August, its Labor Committee met with steamboat owners and captains and railroad managers at the Cotton Exchange to coordinate their efforts; they unanimously determined to import labor in preparation for the impending strike. Captain F.O. Minor assumed responsibility for securing large numbers of strikebreakers and quickly dispatched recruiting agents to Mobile and Savannah and advertised for replacement labor throughout the rest of the South. In New Orleans, Minor himself registered hundreds of applicants for "all classes of laborers." The employers' plan called for dispatching Minor's men, as the strikebreakers were called, to two protected cotton presses that the boss draymen had arranged to keep operating.[100]

The crisis broke on September 1. As all union members ceased work, between 100 and 200 Mobile strikebreakers arrived at the Canal Street depot, assembled at the Union Press, and were soon joined by 300 unemployed New Orleans men. "It was a conglomerate crowd," the *Picayune* observed. "Whites, negroes, old men, boys, shabby-genteel looking men and the woolen shirted laborer were all huddled together. . . ." After feeding the Mobile recruits, Captain Minor dispatched most black scabs to the Liverpool Press, most white scabs to the Orleans Press, and assigned the New Orleans strikebreakers to the presses closest to their homes. Protected by police and armed guards, the strikebreakers toiled beyond the range of the union men's threats. Although many companies confronted substantial numbers of desertions, the supply of strikebreakers was sufficient.[101]

Despite careful planning, the employers' experience with scab labor proved to be mixed. For the duration of the two-week strike, the black and white strikebreakers, who kept some yards open, operated the cotton presses at less than full strength. Although employers boasted satisfaction with the new men, the unions calculated that the cost of maintaining the replacement labor—referred to as "stripling boys and tramp-looking negroes"—were high: functioning cotton presses were turning out scarcely 200 bales a day instead of the normal 800 to 1000, and the factors had to pay the imported workers even while they were not working. Moreover, the strike did not affect only the presses. The absence of the longshoremen and screwmen left thousands of bales of cotton piling up on the levee. Employers believed that all "that is wanting is screwmen," the *Picayune* reported, "then victory will be theirs." But the screwmen remained on strike "through a feeling of 'honor' " and refused to abide by their contracts. Ship captains frequently employed their own crews, but lacking the tools and the numbers of men needed, they only made a dent in the growing mountain of bales on the docks.[102]

Waterfront workers did not maintain a completely solid front during the strike. Two small groups of black union men rejected the interracial alliance and broke rank with the Cotton Men's Executive Council. On September 2, rumors spread that black screwmen belonging to a longshoremen's union had bolted and opened independent negotiations with the factors and other employers. On September 5, several gangs of black screwmen attempted to store cotton on board a Spanish steamship, but strikers "hooted . . . and jeered . . .

until they were compelled to desist." Later, accompanied by labor bureau head Captain Minor and three policemen, the black laborers resumed stowing cotton. Their efficiency was limited, however, by a lack of necessary jackscrews, which had been removed earlier by the white screwmen who owned them. Another ship captain temporarily rejected the use of black screwmen on the grounds that without the necessary screws only half his cargo could be stowed. Employers secured proper tools by the end of the strike's first week, however, and ten gangs of black screwmen found work loading cotton on the British steamship *Cella*. [103]

Resentment over racial restrictions motivated this minority of black union members to violate the strike. The well-paid, largely Irish cotton screwmen had no intention of sharing more than a small number of jobs with their black counterparts. While cotton yardmen and longshoremen had developed work-sharing plans during the 1880s whereby blacks and whites would divide the existing jobs equally, the white screwmen placed a daily ceiling of twenty gangs (100 men) on the number of black screwmen. Not until the first years of the next century would the white screwmen's union agree to a half-and-half division of work. Some black Third District longshoremen apparently shared the black screwmen's anger, protesting their lack of equitable representation in the Cotton Men's Executive Council. On September 8, both unions, which had joined the Cotton Council in February 1881, ended the "unholy alliance," withdrew from the Council, and "compromised with the merchants with whom [they were] desirous of establishing the most harmonious relations." The establishment of "harmonious relations" with employers, however, meant strikebreaking. "No sooner had we done this," union presidents P. S. Jackson and M. Sparks later wrote, than "personal violence [by the whites] was resorted to." After the strike had ended, the *Louisianian* protested in November that the "colored longshoremen and cotton screwmen are [still] deprived of work because they stood by the merchant, the white associations refusing to give them work or to work with them because they stood firm by the business community during the recent strike." [104] No other group of black workers, however, had broken with the alliance. Black cotton yardmen, teamsters, and other branches of the longshoremen's union remained committed to the new Cotton Council and stood firmly behind the strike.

The union strikers received much credit for maintaining order during the first week of the strike. "To them, no less than the police, belongs the credit," one reporter noted, "for had they been disposed to become rioters considerable trouble, if not bloodshed, would have resulted, as has been the case in other large cities of the United States where labor was arrayed against capital." Strikers often assisted local police in dispersing threatening crowds, and on one occasion, union leaders promised to arrest any man found creating a disturbance. Union representatives met twice with Mayor Shakspeare to complain of armed strikebreakers and foremen. They also financially assisted strikebreakers who deserted their new employers and wanted to leave the city. (Some 500 blacks gathered at the Louisville and Nashville rail depot to wave good-bye to 40 black strikebreakers who had abandoned their work.) Press operators, shipping agents, and captains continued to call on city officials and police to enforce order and disperse "mobs," though some observers grew weary of the constant alarms raised by the port's employers. Every day the police were "sent scurrying to and fro from the Levee to the presses and back again . . ." but in few instances was there any problem. "The cry of Wolf! Wolf! when there is no wolf," the *Picayune* complained, "has been raised so frequently of late on the part of those in the employ of the Factors' and Employers' Council, that should an emergency really arise, but little attention may be paid to their call for assistance."[105]

Serious violence eventually erupted. On the afternoon of September 11, Sergeant Thomas Reynolds spotted James Hawkins, a black teamster whom he believed had escaped from custody that morning following an arrest for disorderly conduct. As Reynolds tried to arrest Hawkins on the corner of Thalia and Tchoupitoulas streets, a "general melee ensued" and Reynolds' "head became the target for frying pans and other utensils which were thrown at him by the colored women." When Hawkins, claiming innocence, resisted arrest, Reynolds drew his revolver, fired twice at Hawkins, and killed him. Hawkins was a "law abiding, peaceful man," the *Louisianian* later reported, "with honest convictions and a Christian character, and was shot down for no other cause than that a negro has no rights which a police officer is bound to respect." Immediately following the shooting, "the bloody corpse for an instant held the spectators spellbound," the *Times* reported. "Many colored and white women" taunted the officers, and "the men formed a solid mass and marched" on them. According to the *Picayune,*

> intense excitement was created by this unfortunate affair and great indignation was manifested by the negroes, who swarmed into the yard. The negro women in particular were exceedingly insolent and demonstrative, and what few policemen were in the vicinity were soon surrounded by a howling mob of infuriated negro men and women.[106]

Hawkins' death raised the stakes in the strike, as well as the specter of greater violence. As news of the killing spread like wildfire through the black community, the strikebreakers and police became the target of popular indignation. Black residents and neighbors on Tchoupitoulas Street—in particular, boys between the age of ten and fourteen, black women and a few men— stopped every dray and float loaded with cotton in the vicinity and sent the scab drivers running. A policeman who had cornered a boy accused of pelting a driver with rocks found himself showered with stones by local residents; the boy escaped. One crowd chased another officer who had fired into a crowd that had "hooted and yelled at him"; they "then fell upon the unfortunate policeman, and wrenching his pistol from him, began beating him with rocks until he lay bleeding and insensible on the sidewalk." Despite police protection along the wharves, strikebreakers on steamships suspended work when they heard the rumors of riots and disturbances.[107]

This popular resistance to strikebreakers and law enforcement officers led

Mayor Joseph Shakspeare to take stronger measures. The mayor closed all the barrel houses near the river (where strikers and their supporters congregated), and he ordered police to clear the wharves of all idlers. But the businessmen's mayor feared that the city police would be unable to handle a rioting mob; in a real emergency, he believed that only military force could clear the streets. Declaring that the city was on the verge of "anarchy" and that his police force had been "overpowered," Shakspeare called on Governor Louis Wiltz to intervene. Wiltz complied, ordering General William J. Behan—a local businessman with close ties to both organized labor and the Democratic party's political machine—to mobilize his forces. Assembled at their armories, some 2000 state militiamen remained on alert.[108]

The death of James Hawkins infused the strikers' struggle with new energy, led to a reaffirmation of interracial solidarity, and broadened the conflict into a community affair. Hawkins' funeral drew together black trade unions and benevolent associations and white dock unions, including white cotton yardmen, longshoremen, white weighers and reweighers, and the white cotton screwmen. These laborers, the *Picayune* observed, "are more determined than ever that they will not be whipped into submission without raising an arm in defense of their cause." As the employers' Labor Bureau finalized plans to put newly imported strikebreaking screwmen to work, strikers prepared to stop any loading of ships, clear the levee of all scabs, and tie up the movement of cotton by preventing the boss draymen from sending out their teams.[109]

Over the next few days, union men and supporters halted most of the port's cotton traffic. Groups of strikers and sympathetic crowds of boys and women attacked some lone teamsters, ordered others to return to the yards with their drays, and demanded that yards cease operations and steamships stop loading and unloading operations. Ignoring the presence of a small police squad, a large crowd gutted the Atlantic press, attacked and barricaded the laborers at the Natchez press, and "ejected the press hands and others then at work, cut the traces of teams and did considerable damage" at the Levee press. Only the intervention of John Delaney, president of both the white Screwmen's Benevolent Association and the Cotton Men's Executive Council, prevented further violence. But the crowd actions proved effective. By the morning of September 12, no press below Canal Street remained open and virtually no cotton moved through city streets. Only where police protection was available did strikebreakers continue to load cotton, but such protection could not cover the entire riverfront. Even where it existed, it could not guarantee the strikebreakers' safety. In several cases, union sympathizers ignored police guards, boarded ships, and ordered strikebreakers to quit work. For two days, the *Galveston Daily News* observed, the strikers and their allies were "virtually in possession of the levee."[110]

The deepening crisis brought all parties to the bargaining table and produced a settlement. The Cotton Men's Executive Council had already accepted the Chamber of Commerce's mediation offer. Hawkins' death and the crowd actions that ensued convinced employers to reverse their opposition and submit the outstanding issues to arbitration. The final settlement granted to waterfront workers most of their wage demands, while sidestepping the question of the employers' "right to control"—that is, to hire and discharge whom they pleased.[111] Although the unions failed to eliminate all non-union labor from the waterfront, conceding to work with the small number of "other employees" currently employed, they did succeed in imposing an unprecedented degree of uniformity on the docks and in the cotton yards. At the ratification meeting, labor leader John Delaney stressed that the agreement was "just as binding on the employers as on the laborers." It established one wage standard, which employers could not change unilaterally, for each class of labor; barred employers from paying non-union men at lower rates; and mandated that in the future foremen could hire only association members. Although many strikers protested against working with non-union laborers, Cotton Council delegates overwhelmingly approved the new tariff. Bitter feelings toward strikebreakers intensified, however, when some union members found themselves replaced by the new men. For a short time, they resumed their attacks against scabs, though on a smaller scale. Union leaders imposed order on their ranks, and work resumed in full.[112]

The strike raised an immediate, powerful lesson for the port's commercial elite. Mayor Joseph Shakspeare summarized it well when he observed that the strike had illustrated the "weakness of the power of the state when a few labor organizations could do as they pleased, and set at defiance law and order, and stop the business of such a large city as New Orleans." Indeed, neither the employers' Executive Council nor the businessmen's reform mayor possessed sufficient political or police power to crush the new interracial waterfront alliance. In the post-depression era, the close connections between the Democratic party Ring, through such ward politicians as Patrick Mealey and John Fitzpatrick, and the ranks of organized labor, constituted a formidable obstacle to the employers' attempts to mold a workforce unfettered by unions or interracial alliances. Businessmen continually sought to build the political power required to impose their agenda by forming, financing, and fielding reform tickets in every election through the turn of the century. But the Democratic Ring and its working-class constituency proved to be durable and resilient opponents. At times, the commercial elite's failure to capture full municipal power and to control the police apparatus enabled labor—and especially waterfront labor—to consolidate its influence, bolster its strength, and maintain and expand some control at the workplace.[113]

The 1881 dock workers' strike raised the labor question to a prominent place in the calculations of the city's commercial elite and made the relationship between labor and capital a subject for careful and regular scrutiny. While employers on the docks did not "surrender the control of their capital" to their workers, neither did they fully direct the labor process. The compromise of 1881 and the elite's failure to impose an anti-union settlement merely postponed of the day of reckoning; or, from Cotton Exchange president Thomas Airey's perspective, delayed "the evil until a further period." The strength of the Cotton Men's Executive Council and many of its constituent unions continued to grow over the decade of the 1880s. In subsequent struggles, waterfront employers would have to contend with powerful unions exercising considerable control over the labor process, with one exception, for the next four decades.

3

Testing the Limits:
Politics, Race and the Labor Movement
1882–1892

Many workers on the New Orleans waterfront resolved the labor question in their own favor during the 1880s. Over the course of the decade, the key levee unions in the Cotton Men's Executive Council wrested control of the labor supply from their employers, implemented complex conference rules defining the conditions of their labor, and received what were probably the highest longshore wages in the country. Interracial and inter-trade solidarity constituted the foundation of their remarkable power. Writing at mid-decade, James T. Newman, a black doctor, noted in the *Southwestern Christian Advocate* that the "great tidal wave" of workingmen's associations that reached New Orleans in 1880 blended

> all classes and colors . . . into one great whole. The white and black . . . stood shoulder to shoulder. Capital succumbed to the inevitable law of justice, and hence we have in the city of New Orleans a splendid organization of men, which is second to none in point of a well regulated system of labor.[1]

Employers hardly shared Newman's celebratory stance, but they had to acknowledge the strength of the waterfront labor movement that confronted them. Many steamship agents, contracting stevedores, press owners, and boss draymen, who had bitterly opposed the unionization of their workers and had complained of the labor unrest in the 1870s and early 1880s, reluctantly accepted the biracial unions as an unpleasant fact of life.

Two conditions set the stage for the dock workers' success between 1882 and 1892. First, biracial unionism benefited from a political culture receptive to the white working-class electorate's efforts to organize unions, even across racial lines. Generally adopting a hands-off approach to labor conflicts, the Democratic party machine permitted an interracial union movement to confront employers in an unfettered way. Second, autonomous black trade unions, deeply embedded in the city's black social network, offered black workers considerable protection against white employers and employees alike; any white strategy to displace black workers would have encountered sharp, concerted resistance from the black workplace organizations. Unionized black workers, however, shared with whites an interest in regulating the supply of labor, reducing racial competition, maintaining wage rates, and extending job control. Entering into formal alliance with each other proved the only way for blacks and whites to achieve their common goals.

But it would be misleading to focus only on the biracial structure, union conference rules, and the favorable political climate for labor, as important as these were. The experiences of New Orleans dock workers in this era were shaped by a wide variety of factors. The location in the port's complex occupational hierarchy, the degree of employer opposition, the character of labor alliances—all influenced labor's capacity to organize and resist sucessfully. Freight handlers on the city's rail docks and Mississippi River roustabouts were two important groups that fell outside the powerful circuit of trade and solidarity that linked the Cotton Men's Executive Council's members. Despite the adoption of a biracial structure, freight handlers' unions made little headway against determined and powerful railroad companies. Roustabouts at the bottom of the port's hierarchy had no allies and few resources to challenge their brutal treatment. Extensive as it was, Gilded Age solidarity on the waterfront had clear limits. Even within the Cotton Men's Executive Council, the commitment to inter-trade solidarity and biracial unionism could not eliminate craft divisions; their reemergence led to strange alliances and a fatal redefinition of the organization's purpose. Throughout the decade, the diversity of workers' experiences along the riverfront continued to defy any single formula.

I

To the outside observer, New Orleans' politics in the Gilded Age could be hard to comprehend. "One might make various studies of New Orleans," wrote Charles Dudley Warner in his *Studies in the South and West,* including one of "the city politics, which nobody can explain, and no other city need covet." Yet New Orleans residents and subsequent historians had little trouble identifying the social actors behind the arcane political factionalism. Chief among them was the Democratic-Conservative party's political machine, the Old Regulars. Also known as the Ring, the machine was a tightly knit, well-organized hierarchy with power firmly anchored at the ward level. With several exceptions, it dominated municipal politics from the mid-1870s through the early 1920s.[2]

With the collapse of Reconstruction, the machine had emerged as New Orleans' dominant political force. The Republican-inspired city charter of 1870 concentrated legislative power in the hands of a mayor and seven administrators, each of whom directed one of the city's seven departments (finance, commerce, improvements, assessments, police, public accounts, and waterworks and public buildings). When the Democratic-Conservative party recaptured municipal power from the Republicans in 1872, it spent the next decade developing a powerful political organization. The mayor served as titular head of government, but, as historian Joy Jackson has noted, he was "never the 'boss' of his party, but only a compromise candidate chosen by equally power-ful ward leaders." In 1882, a new city charter replaced the small council form of government with a larger, aldermanic system. But the new system did little to disturb the control that professional politicians and ward bosses wielded over municipal affairs.[3]

"It is a travesty of statesmanship to say these men make laws for a peo-ple," the *Mascot,* a local journal of political satire, argued in 1886. "New Orleans is the most and worst governed city in the country, if not the world." Opposition politicians, newspaper and magazine editors, and religious leaders regularly charged the Ring with widespread corruption and election fraud, and blamed machine politicians for the poor condition of city streets and schools, inadequate policing, and rampant hoodlumism.[4] That corruption and violence characterized elections and politics-as-usual did not make New Or-leans politics unique in the South; neither did the fact that social services deteriorated dramatically after Reconstruction's downfall. Indeed, much of southern politics during the period could be best described as corrupt and violent, and the Bourbon reaction produced slashed budgets for social ser-vices in most places.

What distinguished New Orleans from many of its southern counterparts was the social base of its municipal political divisions. The city's commercial elite, which regularly dominated the ranks of municipal reformers, often did not command political power commensurate with its economic position. The Ring drew its electoral support from native-born and immigrant white working-class voters and from the myriad small contractors who received city franchises. The Louisiana state constitution of 1879 facilitated the alliance of politicians and immigrants by granting men of foreign birth the right to vote. Elite reformers and the press did not always appreciate this fact of political life. "[N]o class in America have proved themselves so plastic in the hands of corrupt politicians as Irishmen," the *Mascot* complained. "They seem to think that God made the Democratic party as immaculate and infallible as the Blessed Virgin Mary. They would stand up and vote the Democratic Party ticket straight if the devil and all his angels were at the hell-m [sic]." In addition to immigrants and native white workers, black voters were some-times cultivated by Ring politicians. Shortly before the municipal election of 1884, for instance, administrator John Fitzpatrick appealed to that electorate by adding a number of black workers to the Department of Public Works' payroll as cart drivers for the city.[5]

The professional politicians who ran the machine were often self-made men who boasted of working-class backgrounds. The son of Irish immigrants, John Fitzpatrick was born in 1844. Orphaned when he was only a few years old, Fitzpatrick was raised at the St. Mary's Orphan Boys' Asylum. He received a basic education in the city's public schools, worked as a newspaper boy, and eventually apprenticed as a carpenter in the Third Ward. With the collapse of the Reconstruction government in New Orleans in 1872, politics offered him a ladder out of poverty. He was elected clerk of the First District Court, appointed clerk of the newly created Superior Criminal Court in 1874, and elected state legislator in 1876, criminal sheriff in 1878, administrator of improvements in 1880, commissioner of public works in 1882, and mayor in 1892. Fitzpatrick the politician "seems never to have forgotten his working class origin," radical Covington Hall later wrote in his unpublished manuscript on New Orleans labor.[6]

Ward boss Patrick Mealey's biography reveals intimate ties to the city's labor movement. Born in County Galway, Ireland, in 1845, Mealey came to New Orleans at the age of eleven and worked at the cotton presses as a boy. After serving as a volunteer in the Confederate army, he worked briefly on the docks before Mayor John Monroe named him to the city police force. That patronage post ended following the Mechanics' Institute riot of 1866, when General Sheridan ousted Mayor Monroe. Mealey labored as a cotton roller at a press yard and later opened a small grocery. Like Fitzpatrick, Mealey found politics the avenue to success. Active in his ward organization, he was appointed captain of the city workhouse in 1874 and was elected Commissioner of Police and Public Buildings in 1878. The following year he helped establish the white Cotton Yardmen's Benevolent Association No. 1, and served as the organization's first president. While a city administrator, Mealey used his position to promote his union's affairs. He regularly marched at the head of the yardmen in annual Labor Day parades, and in 1881 and 1887 he was instrumental in negotiations between dock workers and their employers.[7]

Like political machines in many northern cities, the New Orleans Ring catered to the needs of its lower- and working-class constituents. Its electoral strength was rooted in its control of patronage, its support of labor (or neutrality on issues important to the city's unions), and the hostility of a working-class electorate to the pretensions and programs of the "silk-stocking" elite. Using their patronage powers liberally, machine administrators turned out workers on election day. In particular, as Administrator of Public Works during most of the 1880s, Third Ward leader John Fitzpatrick used the municipal work rolls to build Ring support. "The thoughtful liberality of Capt. John Fitzpatrick," the *Mascot* complained in July 1882,

> is evidenced by the happiness and contentment of the city employees in the tenth ward, where . . . the lotus land of laborers is found. In this latter-day Utopia, the oppressive heat of summer is unflinchingly endured by the hardy sons of toil connected with the bureau of improvements, since wherever work for one man can be found, the services of ten are engaged, and this apportionment lessens the severity of the task. Baseball is among the healthful means of recreation practised by the street cleaners during their arduous daily labors, to keep their muscles up to the requisite pitch of firmness desirable on election day.[8]

Fitzpatrick and other machine leaders came under regular fire from reformers and businessmen for squandering the city's financial resources to sustain their own organization. "It's astonishing what a sudden frenzy John Fitzpatrick has taken for cleaning up," the *Mascot* noted before the 1882 municipal election.

It looks as if he would like to embrace the whole world of tramps and hoodlums in his army of diggers and scrapers; and the nearer the time draws toward the election, the greater becomes the enthusiasm of Johnny for arming the exile of Erin with the shovel and pickax . . . Bayonets may have triumphed in the days gone by, but Fitzpatrick swears spades shall be trumps at the next election.[9]

The city's labor movement benefited enormously from the Ring's political power. Dependent upon labor's vote, the machine generally did little to alienate it, often adopting a supportive or at least a hands-off policy toward strikes and other labor conflicts. In such a political environment, white and black workers in the longshore trades had ample space in which to develop a powerful interracial movement to implement complicated conference rules and work-sharing agreements. While machine politicians did maintain "law and order," dispatch police forces to scenes of violence, and offer protection to employers utilizing strikebreakers, the lack of enthusiasm—and promptness—with which they performed these tasks proved an almost never-ending source of annoyance and cost to city businessmen. The commercial elite's relative lack of influence over the municipal government during labor conflicts was most visible during the streetcar and general strikes of 1892 and the riverfront race riots of 1894 and 1895. In these cases, machine mayor John Fitzpatrick adopted either pro-labor or pro-white labor stances to the dismay of employers. In several cases, it required the intervention of the governor to end the labor insurgencies.[10]

The existence of a powerful political machine, of course, did not go uncontested. In the late 1870s and early 1880s, a small, interracial Greenback-Labor party denounced the Democratic-Conservatives and the Republicans as nothing but "the servants of the monopolies," and argued that "any poor man who voted for them would vote for his own enslavement." Yet despite its claim to be the "real champion of the laboring man," this party never succeeded in getting more than a small fraction of the vote. And while workingmen's political parties rose under the auspices of the Knights of Labor and other radical movements in hundreds of towns and cities during the 1880s and early 1890s, New Orleans' unionized white workers offered no serious, independent electoral challenge to the regular Democratic-Conservative party. Inflated figures in corrupt elections aside, the Ring succeeded in capturing the majority of working-class votes in almost every election during the Gilded Age. To a great extent, the machine represented the institutionalization of a distinctly non-radical working-class politics.[11]

On occasion, local labor leaders opposed the machine in municipal campaigns, making common cause with elite reformers to oppose corruption and inefficiency or to promote their own interests. In 1882, the president of the Cotton Weighers and Reweighers Association, Fendal Horn, ran against machine Administrator of Police Patrick Mealey, who was also president of the white Cotton Yardmen's Benevolent Association. Motivated perhaps by tensions between his union and Mealey's in the aftermath of the 1881 strike, Horn ran on the citizen's ticket. "The industrial associations that politically made Pat Mealey what he is, and without difficulty secured for him two successive terms as Administrator of Police, [now] are arrayed against him," noted the *Mascot*. "Fendal Horn is a very popular man." Not popular enough, evidently, for Horn saw his ticket crushed on election day. He eventually made his peace with the machine, and in 1886 he accepted a patronage post as constable of the first city court.[12] In 1884, the leader of the white screwmen and the Cotton Men's Executive Council, John Delaney, made two unsuccessful forays into the electoral arena on reform tickets, as a candidate for criminal sheriff in the April municipal elections and for Congress from Louisiana's first district later that fall.[13]

The only union leader to challenge the Ring successfully in this era was Thomas Agnew, president of the white screwmen's union. A reform candidate for Administrator of Public Buildings and Police, he was elected to a four-year term when the anti-Ring Young Men's Democratic Association swept into power in 1888. Unlike Horn or Delaney, Agnew was a man with whom employers could feel comfortable. Born the eldest son of a sea captain in Belfast in 1846, Agnew came to the United States as a young man, commanded a sailing vessel between New York and Savannah, and around 1870 settled in New Orleans, where he joined his maternal uncle, Captain Sam Donald, a prominent stevedore. Agnew became a cotton screwman and worked as a foreman for his uncle, supervising the loading of his ships for several years. Following his four years as Commissioner of Police and Public Buildings, he briefly returned to the riverfront. In 1896 he was appointed to the new dock commission overseeing city wharves and landings, serving as one of four harbormasters until his death in 1901.[14]

How was it possible for labor leaders like Horn, Agnew, and Delaney to enter into temporary political alliances with the business elite, or conversely, for the business elite to accept former opponents into their camp? First, elite opponents of the Ring clearly recognized the importance of working-class voters and sought to broaden their electoral appeal by placing important union men on their ticket. Second, the Ring's patronage powers did not necessarily ensure the labor movement's support. Immigrants on the municipal payroll in the Department of Public Works, for example, were often unskilled workers with no direct connection to the union movement. Third, despite the machine's general support for the city's trade unions, organized labor sometimes maintained a critical distance from the municipal government. Issues of law enforcement, corruption, poor schools, and unpaved streets had a non-partisan quality to them; union workers, as well as businessmen, at times demanded widespread improvements. Moreover, the diverse membership in trade unions led labor organizations in New Orleans and across the nation to ban official discussion of political issues as unnecessary distractions from unions' true task—organizing the workplace.[15]

While the Republican party posed little threat on the municipal level, a succession of short-lived independent citizens' associations challenged the Democrats' hegemony in every election from the 1870s through the 1890s. Dominated by the city's commercial and professional elite—cotton merchants, sugar planters, and financial brokers—these groups held the Ring responsible for electoral fraud, widespread graft and corruption, high taxes, and the retention of large numbers of workers on the municipal payroll.[16] In an 1884 campaign statement, reformers argued that the election of the machine candidates would fasten upon the city

> an administration which from its very weakness and incapacity will more than neutralize the most strenuous efforts of merchants and people for the advancement of New Orleans. The democratic ticket . . . shows so clearly the dominance of bossism, such a disregard for merit and integrity as to the nominees, and such sacrifice of every consideration of the public weal to the advancement of the political fortunes of a few leaders, that no reform, no good results, can be expected from it, if placed in power. . . . The simple fact is, that under ring rule our city government has been a failure. . . . These are not political questions. The matter of reform is neither one of democracy nor of republicanism. It involves exclusively business questions—economical and faithful administration of our financial affairs, as in any business corporation, and vigorous enforcement of the laws for the protection of life and property without fear or favor.[17]

Historian Samuel Hays' assessment of Progressive era reformers holds equally true for New Orleans elite reformers in the Gilded Age through the 1920s. The "business, professional, and upper class groups who dominated municipal reform movements were all involved in the rationalization and systematization of modern life; they wished a reform of government which would be more consistent with the objectives inherent in those developments."[18]

The Ring's constant opposition to professional business methods and standards led reformers to explore ways of reducing the machine's power. In the late 1870s and early 1880s, the commercial elite identified as a key political evil the very structure of a city government that concentrated vast powers in the hands of a small number of professional politicians uncontrolled by the mayor. For example, during his mayoralty from 1880 to 1882, the businessmen's reform mayor, iron-works proprietor Joseph Shakspeare, often found that powerful, uncooperative machine administrators set clear limits on his power. By early 1882, elite reformers and Ring politicians alike focused on devising a new city charter called for by the state constitution of 1879. Each group offered a proposal that reflected its needs and aspirations. A "citizens charter," proposed by a committee representing the Cotton Exchange, the Chamber of Commerce, the Produce Exchange, and the Mechanics', Dealers' and Lumbermens' Exchange, advocated strengthening the mayor's executive power, investing in that office broad appointive powers to name administrative municipal officers. Recognizing that the Ring maintained strong electoral support in the city's working-class districts, elite reformers sought a governmental structure that they could dominate. The city's conservative commercial exchanges, Jackson argues, "felt that their best chance of securing a city government sympathetic to their point of view was to have only one elective executive, the mayor."[19]

A variant of the Ring's proposed legislative charter, which provided for the popular election of all executive officers, was adopted by the state legislature in June 1882. The new charter replaced the older administrative system with a new council of thirty representatives—most elected from districts and some at large—vested with full legislative authority, and created an executive department run by an elected mayor, treasurer, comptroller, commissioner of public works, and commissioner of police and public buildings. The new city charter broadened the base of legislative power but did nothing to address the reformers' concerns; the Ring dominated both the executive and legislative branches from 1882 to 1888 and from 1892 to 1896. Almost immediately reformers raised their voices against this system in favor of a commission form of government. From the 1880s to the 1920s, historian Matthew Schott has concluded, New Orleans businessmen hoped "to create more efficient administration with centralization of city government and to eliminate the ward, with its concentration of lower class elements, as the basis of department administration and tax assessment."[20]

If the machine proved an easy target for rhetorical attacks, it remained a formidable foe at the polls. With several exceptions, the city's elite fared poorly in the electoral arena through the turn of the century. William J. Behan, a merchant and a major-general in the state militia who had supported the waterfront unions during the cotton strikes of 1881, was elected as the Ring candidate in 1882, easily defeating reform Mayor Shakspeare, candidate of the Independent Citizens' Association. Yet the Ring did not always control its own candidates. Once in office, the independent-minded Behan clashed with party regulars over the distribution of patronage and certain municipal reforms. During his two-year term, he experienced problems similar to those of his predecessor. "No city could have been in worse or more unpromising condition in time of peace," he remarked in his message to the Ring-dominated City Council. He ran for reelection in 1884 as the reform candidate of the elite's newly formed Citizens' Democratic Parochial and Municipal party, waging a vigorous anti-machine campaign.[21] Neither the Ring nor its largely working-class electorate was ready to surrender political power to the commercial elite. With between 800 and 1000 pro-Ring special deputy sheriffs guarding the polls, Democratic regular ward boss J.V. Guillotte overwhelmed Behan in an election marked by widespread fraud and violence. A commander of a White League company that fought in the battle of September 1874, Guillotte had climbed the political ladder quickly from court clerk in the 1870s to administrator of police and public buildings under Shakspeare to city comptroller under Behan. During their terms in office from 1884 to 1888, Guillotte and Fitzpatrick exercised their patronage powers by considerably expanding the city's floating debt, as the elite looked on in distress.[22]

But elite reformers did not remain idle during Guillotte's mayoralty. A Citizens' Committee of One Hundred, composed of prominent men, formed in May 1885 to denounce government corruption and to advocate thorough reform, clean elections, and merit appointments. The Committee pored over voter registration lists and purged hundreds of dead or unaccounted-for names from the rolls, invisible men who presumably cast their posthumous votes for Ring candidates. Yet the Committee, and its successor, the Law and Order League which formed the following year, accomplished little else.[23]

The Young Men's Democratic Association (YMDA), having inherited the mantle of the Committee of One Hundred and the Law and Order League in 1887, focused on the 1888 municipal elections. Technically a non-partisan movement, the YMDA received the enthusiastic backing of New Orleans' commercial elite and the more pragmatic support of black Republican voters. Its platform, a condemning litany of the machine's sins, promised to

> strike from the payrolls political deadheads, and to give employment to those who can and are willing to work . . . to have the police force purged and remodeled and so fairly paid that proper men may be induced to serve . . . Vice and corruption should be suppressed and all legitimate enterprises fostered and encouraged.[24]

Learning from previous elections, the reformers copied the methods of their opponents. Well-drilled and armed representatives of the YMDA stationed themselves at polling places around the city to ensure a fair vote, accompanied the ballot boxes to the Criminal Court, and remained present until the ballots had been tallied. The 1888 election saw an almost complete rout of the city machine. Most of the YMDA's municipal candidates, including former Mayor Shakspeare for that same office and the white screwmen's union president Thomas Agnew for commissioner of police and public buildings, were elected. The greatest celebrations for the YMDA's electoral success were held at the commercial exchanges.[25]

The Young Men's Democratic Association succeeded for a time in holding the reins of New Orleans' government, but it ultimately failed to break the Ring's power or radically transform the city's politics. During their four years in office, reformers implemented some of their campaign promises: they initiated investigations into the former administration's executive departments; cut municipal expenditures for schools, public improvements, and the police; ended the politically oriented, volunteer pro-Ring fire department and replaced it with a paid, more professional one; reduced the tax rate; and established the Orleans Levee Board and the Police Board as important regulatory and oversight bodies. But the YMDA proved unable to establish itself as an enduring reform organization, and its elected officials quickly adopted the style of the men they had defeated. By the end of Shakspeare's second administration, Jackson concluded, "he found himself in the same position he had occupied in the early 1880s—facing a group of legislators who consistently tried to push through scandalous special privilege bills and franchises over his vetoes." In the 1892 municipal elections, the Ring again emerged victorious, and John Fitzpatrick, campaigning as a workingmen's candidate, soundly defeated the incumbent reformer.[26] Events during Fitzpatrick's term would give the commercial elite even more cause for regret.

II

Despite the collapse of Reconstruction, New Orleans' black citizens had reason to feel optimistic at the outset of the 1880s. For many riverfront workers, membership in the Cotton Men's Executive Council enabled them to advance wages, protect jobs, and resolve racial and trade differences in an institutionalized manner. At the same time, independent black unions protected their members and participated fully in the social and political life of the black community. Even city race relations seemed to suggest the possibility of peaceful co-existence with whites. "In New Orleans, as perhaps in no other American city," historian John Blassingame argued about the Reconstruction era, "there were many cracks in the color line. . . . Jim Crow did not erect a monolithic barrier between the races; instead, race relations in the city represented a very complex and varied pattern of complete, partial, or occasional integration and intimacy in several areas."[27]

Contemporaries would have agreed that the such conditions held true for the 1880s. At the turn of the decade, the *Southwestern Christian Advocate* suggested optimistically that in

> New Orleans the commingling of the races, the associations in business relations of its exceptionally cosmopolitan population from all the States and nations, and its commanding political supremacy, all combine to make it a center where of all others in the South, the antagonisms of race and sections will be first fully solved, and pass away.[28]

There were numerous examples of racial tolerance to justify this prediction. During the 1885 World Cotton Exposition, Charles Dudley Warner noted that in New Orleans "the street cars are free to all colors; at the Exposition white and colored people mingled freely, talking and looking at what was of common interest."[29] Until the mid-1880s, black baseball teams regularly competed with their white counterparts, and visitors to Lake Pontchartrain's beaches shared interracial bathhouses and picnic grounds. Most of the city's black Catholics worshipped in the same churches as their white brethren, although they were restricted to separate seating.[30] The city's residential patterns also promoted the "commingling of the races," as the *Advocate* put it. While the white commercial elite owned homes along affluent St. Charles Street and in the Garden District, most unskilled workers and their families lived within a short walking distance from the docks and riverfront warehouses. While some distinct ethnic neighborhoods, like the Irish Channel, had emerged as early as the 1840s, for the most part immigrants and blacks were widely dispersed throughout the city's wards. In the early twentieth century, investigators for the British Board of Trade found that "along the belt between Magazine Street and the river bank, white and coloured people live in close proximity."[31]

Flexible racial codes, the absence of overtly discriminatory legislation, and racial proximity in riverfront neighborhoods produced little in the way of substantive integration. If black and white men occasionally frequented the same barrel houses or gambling dens in certain riverfront neighborhoods, there is little evidence that blacks and whites led anything but segregated associational lives. Often participating in racially separate churches, fraternal and benevolent associations, unions, and political clubs, New Orleans whites and blacks, like those in other southern urban centers, inhabited separate worlds.[32] Urban blacks "have nominal equal civil and political rights," one paper observed in 1887, "and they are in sufficient number to be independent of the whites in matters of social enjoyment." Charles Dudley Warner concurred. New Orleans blacks, he observed in his 1889 book, "are no more desirous to mingle socially with the whites than the whites are with the negroes."[33] Scores of black churches and fraternal organizations constituted a strong infrastructure that bound the black community together by responding to its material and recreational needs as well as by creating forums for protest and political action.

By the 1880s, organizations of black workers had emerged as a central part of the city's black social network. In addition to maintaining wage rates, regulating labor competition, upholding work rules, and protecting black workers against both employers and white workers, the black unions provided essential social functions. Each summer, for example, the black Teamsters' and Loaders' Association sponsored a popular excursion from New Orleans to Mobile, Alabama. Teamsters, longshoremen, yardmen, screwmen, and black Knights of Labor all organized numerous parties, picnics, and balls, especially during the summer months when work was scarce. The Ladies Aid Union Longshoremen's Association and the Screwmen's Aid Circle, both composed of the wives of unionists, sponsored minstrel performances, concerts, and balls to benefit union relief committees.[34] As financially stable and secure organizations, the black cotton unions frequently contributed to other organizations in monetary need. One of New Orleans' oldest black churches, the St. James A.M.E. Church, sought to erase a $3,500 debt in February 1887 by sponsoring a series of rallies, each aimed at a different waterfront union. Preachers from around the city offered all-day sermons to the audiences attending "Cotton Yard Men's Day," "Teamsters' and Loaders' Day," and "Longshoremen's Day," all designed, in addition to spiritual uplift, to raise the requisite sum. The Longshoremen's Protective Union, along with the Equal Justice Marine Benevolent Association, the Justice Protective Educational Club, and Amis Sincères, contributed heavily to the renovations of the city's black Central Church. In April 1887, the longshoremen's union and the Ladies Aid Union Longshoremen's Association, headed by the Onward Brass Band, marched to a grand rally at the New Hope Church across the river at Algiers, where they raised $116 on the church's behalf. Praising the union's "high character, integrity and zeal," and hoping it would "ever continue in the service of the Lord," N.H. Selico, who served on the church's finance committee and was the longshoremen's chaplain, presented the union with a cane as a token of the church's appreciation.[35]

The longshoremen's union in particular involved itself in the educational affairs of the black community. When the black Southern University's board of directors dismissed its white president, the Reverend George W. Bothwell, in 1887, the longshore union protested sharply and offered support to Bothwell's new educational project, the establishment of Columbia University for black students. Two years later, longshore leaders E. S. Swann and William Penn joined members of the city's black elite in the Indigent Orphans' Institute and the Justice, Protective, and Educational Society, to demand that the city provide sufficient public schools for black children in the city's Third Ward. At the end of the century, Penn and I. G. Wynn, then president of the black cotton yardmen, served on the board of directors of a new industrial school for black children.[36]

No clear distinction can be drawn between working-class union activity and the larger world of black social life. The union, the fraternal association, the local church, and the ward political club all drew upon the loyalties of black workers and their families; membership in numerous organizations was common. Black labor leaders held leadership positions in other organizations comprising the social and political networks of the black community. The biographies of three union officers—James Porter, E. S. Swann, and William Penn—reveal how inseparable were union activities and the larger world of black social and political life.

James Edward Porter, the perennial secretary of the Longshoremen's Protective Union Benevolent Association, was born in Mississippi in 1856. At the age of ten, he enrolled in the New Orleans public school system which he attended until his high school graduation in 1872. Unable to attend more than one semester of college at New Orleans University because of his father's death, he served as a messenger in the Republican-controlled state House of Representatives in 1874 before apprenticing to a baker the following year. His long career in the labor movement began in 1876 when he was elected secretary of the longshoremen's union. In addition to his union activities, Porter received a political appointment as clerk in the post office's registry division, a job he held for two years before becoming a full-time labor leader. In the mid-1880s, Porter also served as secretary of the joint conference committee of the black and white longshoremen's unions and as secretary of the Cotton Men's Executive Council. In the early 1890s, Porter became an official organizer for the American Federation of Labor. Later, working as a clerk in the U.S. Customs House, he continued to serve as secretary of the longshoremen's union and remained active in forming new black labor organizations in the city until the eve of World War I.[37]

Porter was deeply involved in other aspects of the black community as well. A founder of and a first lieutenant in the city's Larendon Rifles, Porter drilled regularly with the city's Attucks Guards and the Summer Rifles of Mobile, Alabama. Religious affairs also occupied a great deal of his time. A devout member of the Felicity Street Baptist Church, Porter was a delegate to the Baptist State Convention in Opelousas, Louisiana, and the Baptist National Convention in Mobile, Alabama, in the summer of 1887; he served as recording secretary of a Baptist Sunday School Convention, and performed as a vocal soloist at the First Street M.E. Church's Emancipation Day celebrations in 1894. According to a fellow labor leader, Porter belonged to nine associations by the late 1880s.[38]

Longshoremen's Protective Union Benevolent Association president Everett Samuel Swann was born in 1846 to slave parents in Virginia. From 1863 to 1865, he served in the U.S. Army, after which he apprenticed as a blacksmith to his father in Washington, D.C. Unhappy with the trade, Swann moved to New Orleans, where he received an appointment to the Metropolitan Police, an arm of the state Republican party, from 1871 to 1877. Following the collapse of the Reconstruction government in Louisiana, Swann found work as a longshoreman, served as chair of the black union's sick committee and as a member of its board of trustees until 1884, when he became the union's president. Like Porter, he participated in community affairs outside the labor movement. In 1887, he belonged to at least seven secret and benevolent organizations.[39]

Active in union politics and functions during the 1880s, William H. Penn was elected president of the black longshoremen's union following the 1894–95 violence that destroyed the interracial labor alliance forged in the 1880s. At the age of forty-six, he suffered a fatal stroke while mediating a strike of Illinois Central freight handlers in December 1902. The *Picayune* reported that as a labor leader, he was a "power for good in the colored labor world . . . always conservative, fair minded, level headed." Perhaps reflecting lessons of caution learned the previous decade, Penn possessed the "utmost confidence in the businessmen with whom his unions had dealings. Steamship agents were always ready to do business with William Penn." His motto, the paper reported, held that "contracts with labor unions are sacred; they must not be violated at any cost." Once made, "nothing could induce him to break them."[40]

Like Swann and Porter, Penn was a community leader who, in the *Pelican's* words, gave "proof of what the world may expect of the rising generation of colored men." During the 1880s, he worked as a business manager for the Excelsior Brass Band and was co-owner of an undertaking establishment. Joining the Odd Fellows in 1875, he served as treasurer of the organization's joint committee during the following decade. He held membership in numerous fraternal societies as well—the *Pelican* estimated twenty-nine in 1887—including the Pure Friendship Association, the Harmonial Association, the Young Men's Mutual Benevolent Association, the Ladies and Gents Perseverance Association, Pilgrim Tabernacle Number 4, the Young Men's Hope Benevolent Association, the Ladies' Friendship Association, the Daughters' Friendship Association, the Junior Benevolent Association, the Equal Justice Benevolent Association, the Delachaize Benevolent Association, the Jeunes Amis Benevolent Association, and the De Gruy Lodge Number 7. His funeral in 1902 was "one of the largest among the colored population of New Orleans in recent times": hundreds of mourners gathered at the First Street Methodist Episcopal Church and followed the procession to a tomb of the Friendship Benevolent Association at the segregated Lafayette Cemetery, Number 2.[41]

Not only fraternal affairs but politics and protest activities engaged the energies of black union leaders and members in the 1880s and 1890s. Union representatives, black and white, frequently visited the chambers of the City Council, where they expressed strong stands on issues affecting their livelihood and the city's commerce. In August 1887, various labor organizations lobbied against the Council's plans to grant new wharf privileges to the Illinois Central Railroad. The following year the Cotton Yard Men's Association No. 2 and the Sanctuary District Assembly No. 102 of the Knights of Labor petitioned the City Council against a switch ordinance that permitted laying spur tracks to the cotton yards—a development that would have eliminated jobs.[42] More often black leaders concerned themselves with issues of party politics and racial injustice. Longshore attorney J. Madison Vance served on the state Republican party executive committee in the late 1880s, while longshore leaders A. J. Kemp and Chaplain N. H. Selico were active in Republican party affairs, as president and secretary, respectively, of the black Lincoln Republican League in Gretna. Longshore president E. S. Swann took to the road on behalf of the national Republican party during the 1896 presidential election, promoting William McKinley's candidacy before black voters in Illinois, Indiana, Wisconsin, and Michigan. On the local level, black workers constituted the urban electoral base of the party of Lincoln. Louisiana Republicans regularly presented themselves as the party of the working man. (The *Weekly Pelican,* a staunchly Republican black newspaper, described itself as "the laboring man's friend.")[43]

Racial demographics informed much of New Orleans politics. With blacks constituting only a quarter of the city's population, the Republican party generally did not field candidates for local elections; instead, it backed Republicans for state and national office who usually lost to more corrupt and violent Democrats. In New Orleans, black voters were forced to choose between the white candidates of the Ring and the reformers, and in close elections, white politicians actively courted their support. Thus, despite their numerical disadvantage, black urban Republicans did exert some limited influence on municipal elections during the Gilded Age.[44]

If black and white riverfront workers forged a successful alliance in the economic realm during the 1880s, they continued to follow separate paths to the polling place. White workers in the 1880s appreciated the Democratic Guillotte administration's relatively benign approach to labor conflicts; the city's black Republicans, in contrast, could find little positive to say about the Ring. Blacks criticized its corruption, inefficiency, and attitude toward black residents. The City Council, they widely believed, was composed of "patriots serving the city for the privilege of voting away her franchises for exceedingly low sums, but on the outside whopping up large amounts that come in the shape of rebates." Blacks similarly denounced the government's moral character. "Unfortunately for New Orleans," the *Pelican* noted in 1887, "her Mayor is devoid of conscience. Instead of helping to build up the moral tone of the city, he, upon every conceivable occasion, makes himself a party to all measures directly opposing her moral and political reformation."[45] Black Republicans also applauded the elite Committee of One Hundred's removal of

the names of dead and unqualified voters from the official voting lists—names
the Ring utilized to stay in power—and they resisted harassment by white
Democratic voter registrars when local legislation made re-registration necessary.[46] But the most scathing denunciations dealt with the government's treatment of the city's African-American population. The *Pelican* summarized the
situation:

> Now, if there is one class more than another, who have suffered at the hands
> of the hoodlumistic element conducting the city government, it is the Ne-
> groes. Upon every turn they have been made to feel the effect of placing in
> office the ignorant and vicious, the thug and the ward bum.

The redemption of the city from such "antagonistic elements," black Republican leaders felt, was a duty black voters owed "to their children, their wives,
their own manhood." In the municipal election of 1888, they threw their full
support behind the anti-Ring Young Men's Democratic Association.[47]

Yet white reformers proved no more receptive to black aspirations than
the machine. Shortly after the election, the *Mascot* noted,

> the "colored man and brother" that was implicated in causing the boost into
> office of the heterogeneous political mass known as the Y.M.D.A. has now
> an opportunity to ruminate on the treachery and "onsanity" of white folks in
> general and reformers in particular. But in all earnestness, the contimptible
> [sic] manner in which the reformers in the Council attempted to shirk their
> ante election pledges to the poor negro dupes who were their allies.[48]

The black press agreed. Although 8000 to 10,000 of the votes for YMDA
candidates came from mostly black Republicans, the *Pelican* complained, the
reformers' movement "immediately turned its back upon its stalwart deliver-
ers." Neither the mayor nor the City Council gave the Republicans "the
slightest recognition nor acknowledged the debt" owed to them. This was
especially evident in law enforcement. "The rough and brutal treatment visited upon petty colored offenders in their arrest by policemen in this city, is
heartless and murderous," the *Southwestern Christian Advocate* complained
in 1890. "We had hoped that with the reorganization of the force, such treatment would become a thing of the past."[49]

By the late 1880s, black Republicans and labor leaders had grown concerned at developments foreshadowing the anti-black, anti-labor turn that
Louisiana politics would take in the 1890s. Reflecting the national Republican
party's waning interest in black civil rights and its desire to build a white
electoral base in the South, some white Lousiana Republicans attempted to
seize control of the party apparatus and create a lily-white organization. Any
future success for their party at the polls, they believed, required minimizing
racial questions and emphasizing economic issues like the tariff, instead.
Black Republicans fiercely criticized these efforts, and successfully blocked
their disfranchisement within their party, postponing for another decade white
Republican efforts to oust them from party councils.[50]

New Orleans blacks complained of mounting segregation, racial tension, and "color-phobia" even before Louisiana state legislators voted the first Jim Crow statute on the books in 1890. Civil rights legislation passed in the 1870s did little to end segregation in reputable hotels, restaurants, and theaters; in contrast to streetcars, some trains and steamboats segregated black passengers.[51] Religious affairs also demonstrated a similar tendency as some white priests and parishioners had "drawn the color line" in Catholic churches by the early 1890s. Although he did not bar black Catholics from attending other churches in the city, New Orleans Archbishop Francis Janssens turned over Saint Katherine's Church in 1895 for exclusive use by black Catholics—an act, the *Southwestern Christian Advocate* noted critically, that "only was in obedience to a pressure upon the bishop and priests of the city by that class of white people who cannot endure the idea of worshipping in the same church and at the same time with Negroes."[52]

The tendency toward segregation found formal expression in the law and public policy in the last decade of the nineteenth century. In 1890 the Louisiana state legislature enacted its first Jim Crow legislation, which required railroads to provide separate and equal accommodations for black and white passengers; four years later it barred interracial marriages. In 1898 delegates to the state constitutional convention mandated segregated schools. Much earlier, however, Redeemers had repealed Reconstruction laws sustaining New Orleans' integrated public school system. "The colored children of New Orleans are treated in the matter of public school facilities as they are in no other city in the South," the *Southwestern Christian Advocate* complained in 1890. "Our children are crowded into worthless and dilapidated shanties, and are given incompetent white teachers having no interest in them whatever." At the turn of the century, the New Orleans school board ordered that no city public school for blacks would offer courses beyond the fifth grade.[53]

If urban blacks found reason to complain about their deteriorating condition, rural blacks confronted significantly greater problems. Rural Louisiana Democrats, always anti-black in their rhetoric, perfected the art of political deception and adopted ever more aggressively racist practices in the late 1880s. Undisguised electoral fraud, especially in the heavily black cotton parishes of northern and central Louisiana, involved packing party conventions, doctoring voter registration lists, outright vote theft through ballot box stuffing, and widespread violence and intimidation. Moreover, white conservatives responded to black labor unrest in the sugar parishes in 1887, subsequent smaller strikes, and black campaign activity in the 1888 state elections by forming an armed "Regulating Movement," in which Democratic "bulldozers" attacked, whipped, beat, and sometimes killed black and white Republicans in rural parishes.[54]

Perhaps nothing illustrated better the particular dangers facing rural black workers than the response of Louisiana planters and the state government to the strike of sugar plantation laborers in 1887. Sporadic strikes had plagued sugar planters since Reconstruction, but it was not until the Knights of Labor united 9000 black and 1000 white sugar workers into an unprecedented interracial movement that the elite felt its political and economic hegemony dangerously threatened. The strike late that year by Knights' District Assembly 194 met with the combined opposition of the planters, state militia, and armed white vigilante groups, who attacked the strikers and drove them from their homes. One Knights of Labor supporter observed the crisis:

> there has sprung into the existence a mushroom crop of bull-dozers all over the troubled section, who are exercising unauthorized vengeance upon un-armed negroes—such a sight sickens sympathy and destroys all regard for law. . . . The object of these intimidators seems to be two fold: first to break up the lodges of the Knights of Labor and scatter its membership, and secondly to make use of intimidation for political effect.

The militia essentially ended the strike with the Thibodaux Massacre of No-vember 23, 1887, which left at least fifty blacks dead. In historian Jeffrey Gould's words, the massacre "initiated a reign of terror in the Louisiana sugar region" enabling the planters to "solidify mechanisms designed to maintain a submissive and stable work force."[55]

Workers in New Orleans immediately protested the rural repression. The interracial Sancturary District Assembly 102 of the Knights of Labor de-nounced Governor Samuel McEnery for ordering out the troops and overriding

> the rights of her citizens to peaceably assemble and ask for a fair compensa-tion for their honest labor . . . the lawless destruction of human life by the armed force employed was assassination, and we pronounce against it and demand that the legal constituted authorities, by the process of law, do thoroughly investigate the killing of the laborers at Pattersonville, and bring the slayers before the bar of justice for such punishment as they may be found deserving.[56]

In early December, 300 black workers not affiliated with the Knights held their own mass meeting to denounce the killings in Thibodaux and Patter-sonville. Those attending elected A. J. Kemp and Dr. S. P. Brown, both of the longshoremen's association, as chair and secretary, respectively, of the meet-ing. Representatives of the longshoremen, cotton yardmen, and screwmen made addresses "denouncing the killing as murder, condemning the action of the administration in not using proper civil authority before calling out the military, and calling upon fellow colored people, by all lawful means, to avenge the killing of the fellow blacks by demanding the trial of those who did the killing." The meeting's committee on resolutions reported that

> as the unjust killing of the negroes at Patterson and Thibodaux is deplorable in the extreme, and that this condition of affairs is not only robbing men of the fruits of their labor, but tends to disrupt the labor system and the agricul-tural interests of Louisiana, therefore the constitutional authorities and the congress of the United States be petitioned to speedily investigate the trou-ble and bring the guilty to punishment. . . .[57]

What lessons might urban workers draw from the suppression of the Knights' strike? The *Pelican* linked the state and vigilante violence and en-forced servility of plantation workers to the denial of their political rights:

> Do the workingmen of the country understand the significance of this move-
> ment? The negroes of the [sugar parishes] are practically disfranchised.
> Their votes are of no value and for that reason they can be forced to work at
> starvation wages in the richest spot of land under the American flag.[58]

Closer to home, the *Mascot* argued that the central meaning of the massacre
lay in the dangerous precedent it established. "If Governor McEnery is justified in sending armed janissaries against the colored Knights of Labor in St.
Mary and other interior parishes," its editor argued,

> would he not be justified in sending them in against white Knights of Labor
> in Orleans parish? There is no more reason why a strike of cane-cutters in
> the country is to be put down by gatling guns than is a strike of shoemakers,
> bricklayers or cotton handlers in the city.[59]

The *Mascot* overstated the point. There were many reasons why an urban
strike would not be suppressed in such a violent way. New Orleans was not
rural Louisiana, and the balance of political forces prevented local employers
from adopting a solution similar to that of their sugar parish counterparts.
New Orleans workers were relatively free of government intervention or harassment to devise their own class strategies and build structures of interracial
accommodation. During the 1880s and early 1890s, their unprecedented experiment in interracial cooperation provided a sharp contrast to rural repression, intensifying urban segregation, and racial exclusion.

III

"Two possible but contradictory policies" toward black workers were available to southern white labor, noted C. Vann Woodward in his *Origins of the
New South:* "eliminate the Negro as a competitor by excluding him from the
skilled trades either as an apprentice or a worker, or take him in as an
organized worker committed to the defense of a common standard of wages.
Southern labor wavered between these antithetical policies from the seventies
into the nineties, sometimes adopting one, sometimes the other."[60] From the
early 1880s through the early 1890s, the cracks in New Orleans' color line
extended well into the labor movement, as some white unions forged successful working relationships with black unions. The white labor movement was
by no means immune to the increased racial hostility of the 1880s and early
1890s, but it was slow to adopt the crude racism and exclusionary practices
developing in other sectors of southern society. Sandwiched between a decade
of Reconstruction and one of economic depression, labor and agrarian insurgency, and the rise of segregation and virulent racism, the 1880s constituted,
in the words of Sterling Spero and Abram Harris, a temporary and "unprecedented era of good feeling" between black and white workingmen.[61]

Throughout much of the South in the mid- and late-1880s, the Knights of
Labor served as the vehicle uniting, however tentatively, black and white
workers. Maintaining an egalitarian and cooperative ethos, the organization

promoted an unprecedented if short-lived interracial solidarity.[62] Large numbers of unions of southern black workers had affiliated with the national organization by the mid-1880s, and tens of thousands more urban and rural blacks joined the order as well. Although blacks generally participated in Knights' affairs in separate, all-black local assemblies, joining with whites only at the district and general assembly stage, Spero and Harris concluded in their study *The Black Worker* that "the democratic spirit which the Knights sought to create among the laboring classes was a powerful influence toward creating solidarity between Negro and white workingmen and bringing the former into the labor movement." William I. O'Donnell, editor of *Southern Industry,* a newspaper of the New Orleans Knights, echoed the Knights' official policy when he explained to the local press that the "Knights of Labor is composed of men and women of all political and religious opinions. Our order knows no religion, politics, sex, race or color, and I want it to be understood that there are no politicians engaged in running or controlling the order of the Knights of Labor."[63]

Widespread unionization of black and white workers in New Orleans had preceded the dramatic growth of the Knights in the mid-1880s, and most city trade unions, black and white, remained independent of the national organization. Indeed, by the time the Knights began their dramatic national rise after their triumphant strike against Jay Gould's rail system in 1885, New Orleans already had a reputation as having the "most thoroughly organized and the largest and most influential body of workingmen in the entire South." Still, District Assembly 102 of New Orleans did attract many members: locals of printers, journalists, bricklayers, freight handlers, carpenters, cotton mill and cotton seed oil workers, clerks, and small businessmen had formed and affiliated with the order by the end of the decade. Praised by the press for its conservative and "responsible" orientation, the order avoided strikes, stressed arbitration of differences between workers and employers, and sponsored education forums and classes, banquets, picnics, dances, and other social outings. The Knights in New Orleans were in decline by the end of the decade. In 1894, the Grand Master Workman of District Assembly 102 noted that most of the city's Knights were professionals, clerks, government employees, gardeners, and shop owners.[64]

In New Orleans, the Cotton Men's Executive Council and the Central Trades and Labor Assembly, not the Knights, dominated organized labor in the 1880s. Like that of the Knights, the approach of these federations to race relations reflected a new, somewhat egalitarian ethos. The results were truly impressive. Despite the occasional outbreak of racial conflict on New Orleans' docks, noted German economist August Sartorius Freiherr von Waltershausen, the waterfront unions

> have had some success in reconciling the opposing sides, and by recognizing black labor they have undertaken (albeit driven by self-interest) a civilizing task whose definitive solution will probably not be witnessed by them. The organizing efforts of the dock-workers there . . . have been to incorporate

both races into the union [federation] and to insist that wages for all members of the union be equally high.

At the fifteenth anniversary celebration of the Longshoremen's Protective Union Benevolent Association in 1887, black lawyer and Republican party official J. Madison Vance attributed the "wonderful improvement" in the longshore workers' condition to the biracial alliance. Before the formation of the union, he noted,

> there were no councils and board of arbitrators, and no honest and equal division of labor, whether it was among cotton yardmen, screwmen or longshoremen.
>
> During that period of years you had to bear up under the severest strains and the most unjust discriminations . . . borne down by oppression and confined to a narrow sphere by an intolerant spirit of prejudice and racial difference.

Because of "educational improvement and common contact" between the races, he told his audience, "the white laborer is no longer necessarily your foe, and the color line is no longer a dead line between the races."[65]

To be sure, a biracial union structure did not eliminate racial tensions, but it did provide a framework for their peaceful resolution. And managing racial conflict, as waterfront labor leaders realized in the 1880s, was essential if they were to make progress toward unionizing the unorganized and extending their power and influence. The reason was simple. Although different aspects of the transportation business required some familiarity with work processes and equipment, employers could draw upon a potentially unlimited market for common laborers; or, alternatively, in a racially divided labor market, employers could favor one group over another to mold a more pliant workforce. Without crucial, irreplaceable skills to withhold, the only course of action open to transportation workers was to regulate the character and size of the labor force. And that meant devising some mechanism for mediating conflict and fostering cooperation across racial lines.

The biracial, cooperative spirit extended well beyond the ranks of the longshore unions and the Cotton Men's Executive Council. The 1881 formation of the Central Trades and Labor Assembly, the federation of most of the city's trade unions, opened up new avenues of interracial collaboration that would have been unimaginable during the previous decade. The Assembly, veteran labor reformer George McNeill noted in 1887,

> is confessed to have done more to break the color line in New Orleans than any other thing . . . since emancipation of the slaves; and to-day the white and colored laborers of that city are as fraternal in their relations as they are in any part of the country,—the negroes, especially, taking great pride in their loyalty to their organizations.

The black editor of the *Southwestern Christian Advocate*, at times a harsh critic of union tactics and racial policies, similarly painted the city's labor federation in glowing terms in 1882. Commenting on the Assembly's first anniversary parade, he noted that the

> display has seldom, if ever, been equalled in this city. It was grand in many ways; but in no particular did it appear to better advantage, and reflect so much grandeur as in the recognition of colored as well as white societies as equals, and in the manifest rigid exclusion of anything bordering upon caste, or color prejudice. The celebration was model in this respect. To gaze upon these representatives of every trade, and of every shade of colors, following each other, each division sandwhiched [sic] between the other, in its proper place, and marshalled by white and colored officers, with positions according to official rank, without discrimination on account of race or previous condition, was very gratifying . . . God bless the Central Trade or [sic] Labor Assembly of New Orleans. May it live a long and happy life of usefulness.[66]

This relative "equality" between black and white workers' societies and the absence of color prejudice found concrete expression in the policies and practices of New Orleans unions and labor federations. Both the Central Trades and Labor Assembly and the Cotton Men's Executive Council maintained integrated leadership structures. While whites always dominated the presidencies of both organizations, blacks served as vice presidents and recording secretaries throughout the decade. Moreover, from 1882 to 1888, black and white union members marched together in every annual anniversary parade sponsored by the Central Trades and Labor Assembly.[67]

In somewhat different ways, local unions manifested a degree of flexibility and openness on racial issues. In trades where both blacks and whites labored, it was not uncommon for biracial union structures to govern relations and reduce competition between the races. Union locals were rarely if ever integrated. Usually, separate locals of blacks and whites formed within a given trade. Biracial arrangements ensured equal wages for both black and white workers, and despite segregation at the local level, collaboration across racial lines was not uncommon. Not only cotton yardmen, longshoremen, and screwmen, but black and white freight handlers, shoemakers and fitters, coopers, and carpenters frequently met in the same hall to discuss common problems and formulate united strategy; black and white coopers and shoemakers operated under the same set of union work rules.[68] The House Mechanics Grand Council, modeled on the Cotton Men's Executive Council in early 1884, joined black and white building trades workers—white gas fitters, plumbers, plasterers, tinners and slaters, black bricklayers, and white and black carpenters. The mechanics' association was far less successful than its waterfront counterpart in sustaining unity across trade and racial lines in the 1880s. Organized into two racially separate locals, 1500 house carpenters, among the worst paid skilled laborers, struck to increase daily wages from $2 to $3 in April 1884. While interracial committees visited building sites to pressure non-union workers to quit work, the strike collapsed when white carpenters, who had won a wage increase, returned to work, leaving blacks to resume work at the old rate. Nonetheless, by 1886, wage rates for black and white workers in the same trades had been equalized, and that pattern endured at least into the second decade of the twentieth century.[69]

The experiences of longshore workers in the 1880s most dramatically illustrates how the expansion of union power depended on maintaining biracial collaboration. The goal of longshoremen, like that of most union workers, was the regulation of the labor supply, which promised steadier work, higher wages, and improved working conditions. But regulating the labor supply required, as we have seen, that white and black workers in certain trades share some mutual understanding about organization, aims, and strategy. In the 1870s, black longshoremen had faced a glutted labor supply that enabled employers to ignore their demands and replace striking workers. In the 1880s, union workers understood well that employers could manipulate real racial divisions to their own advantage, by using black workers to frustrate white efforts at advancement, and vice versa. For organized labor to control the labor supply in a racially divided labor market necessitated a coming to grips with the question of racial division. During the 1880s, longshoremen addressed these issues successfully. Their efforts in that decade did not eliminate racial tensions but made their mediation possible. The biracial structure they imposed was deliberate and carefully designed and, while it lasted, ensured that longshoremen exercised considerable power on the docks.

The biracial union structure in the longshore trade rested upon three foundations—work-sharing agreements between the black and white unions; the maintenance of a united front in contract negotiations; and adherence to a single, port-wide set of work rules and wage rates. Unions of cotton yardmen and longshoremen arrived at work-sharing agreements that guaranteed each union half of all available work, effectively regulating racial competition and forming the basis of interracial cooperation. A joint executive committee of black and white union leaders debated goals and devised negotiating proposals, subject to their members' ratification. Once agreed upon and implemented, all union members, black and white, were bound by the conference rules. These rules operated at several levels. Some addressed issues of workers' behavior, while others dealt with questions of managerial authority and the conditions under which workers performed their jobs. But even after the 1881 cotton strike, the biracial alliance did not cover all New Orleans longshoremen, and divisions between different groups of longshore workers remained pronounced. Before union workers could gain control over the workplace, they needed to extend control over the union and non-union labor supply.[70] Unity, they learned, was difficult to achieve.

Consolidation on the waterfront came only gradually. As early as October 1879, white unions of the First, Second, Third, and Fourth wards consolidated into a permanent, single, city-wide body. But until the mid-1880s, some five different longshore unions—each "to a certain extent an independent body" with a separate set of officers—represented black workers in New Orleans and Gretna. In the aftermath of the 1881 cotton strike, four small associations (representing some 800 to 900 black longshoremen in the Laborers' United Benevolent Associations and the Protective Unions of the Third District,

Algiers, Gretna, and Middletown) withdrew from the Cotton Men's Executive Council. The following year, they jointly reapplied, without success, for
membership in the Cotton Council. The Council recognized only the oldest
and largest black union, the Longshoremen's Protective Union Benevolent
Association (LPUBA), which had formed in 1872 and received a state charter
in 1874; the LPUBA refused to deal with the smaller unions, demanding that
they disband and join the larger association. In 1883, the four small unions
loosely joined together in the Grand United Union of the Longshoremen of
the City of New Orleans, each body represented by five elected delegates;
those men elected the black longshoremen's representatives to the the Central
Trades and Labor Assembly. Tensions between the smaller branches and the
parent body persisted, however, and the Cotton Council continued to recognize only the LPUBA.[71]

In December 1885, the white longshoremen's union (the Stevedores' and
Longshoremen's Protective Association), and the black LPUBA agreed to
abide by a common set of "fair and equitable" work rules. While the new rules
covered the majority of the port's longshoremen, the four small black
longshore unions remained outside of the alliance's jurisdiction. The Cotton
Men's Executive Council objected to those branches' independence and autonomy. Delegates argued that members suspended by one union could join
another; "one branch did not know the disqualified members of the other, and
could not guard against working with the delinquents." Perhaps more important was the difficulty in maintaining a single, port-wide union wage standard
without a single union agreement. In late February 1886, for example, a few
stevedores began employing non-union men at union wages, while several
others employed non-union black longshoremen at wages below the Association tariff rate. The Cotton Council successfully retaliated, declaring a boycott
of those particular stevedoring firms. Union dock workers correctly interpreted any deviation from the official tariff as an opening wedge which threatened to undermine the entire system of union labor and wage regulation in the
port. An overriding concern with permitting only union labor to load or
discharge vessels was shared by both the vigilant white union and the black
LPUBA.[72]

In January 1886, the Cotton Men's Executive Council moved to bring all
longshoremen under its jurisdiction by ordering the black longshore unions to
consolidate into a single body under the leadership of the LPUBA. The four
non-recognized unions protested that the consolidation plan would deprive
them of their institutional identity; force them to pay new initiation fees; bar
their members from voting until a two-year membership probation period
elapsed; and render their leaders ineligible to hold office for five years. The
Cotton Council found these arguments unpersuasive. The following month it
prohibited any Council member from working with non-LUPBA black longshoremen. While this move had the immediate effect of swelling the LPUBA's
ranks, officials of the four other associations refused to consolidate. Days later,
the Council successfully threatened contracting stevedores loading three cotton
vessels in Gretna into discharging the men in the Gretna longshoremen's union

and employ in their place officially recognized union men from the New Orleans side of the river. The Gretna men resisted, the *Times-Democrat* noted, "and a free-to-all fistfight was the result." Gretna's black women participated, observed the *Picayune,* "siding with their husbands, brothers and townsmen. The result was that the New Orleans men were defeated, and were compelled to seek safety in flight." Soundly and physically beaten on the docks of Gretna, the Cotton Men's Executive Council declared a boycott of stevedores employing the Gretna men. Negotiations between the black branches continued over the next five months. In April, the four unions consolidated and in July the Gretna branch surrendered to the Council at last.[73]

The port's white and now-merged black longshore unions struck for six days in September 1887 in an effort to extend their control over the labor process and supply. At the center of the conflict was a new set of conference rules stipulating that union members would not work with outsiders; that any foreman found blacklisting a union man would be fined $15 for each offense; that any member found "ill-speaking or influencing any foreman to discharge a member" would be fined $5 for the first offense and expelled from his association for the second; that no person would be recognized as a foreman unless he was a member of one of the two unions; that in all disputes regarding overwork and overloading, the foreman and laborers would each choose three members who would together choose a seventh member, and their committee's decision would be final; and that all violations of conference rules would be tried by a conference committee, and all fines levied for infractions would be divided between the two associations. Other rules addressed particular work concerns: the tariff of any member ordered to leave the city to go to the quarantine or any landing along the Mississippi River to load or unload any seagoing vessel would begin from the time that man left the city and would continue until he returned; no man could leave a ship on which he was working to go to another ship to work unless the work had been stopped for the space of five hours; and in order to give the men a "breathing spell," stevedores would increase the size of work gangs.[74]

At the heart of the new conference rules was a demand with potentially far-reaching implications for both race relations and union power on the city's docks. The two unions insisted that contracting stevedores officially abide by the unions' work-sharing agreements and employ only half-and-half crews, composed equally of blacks and whites. Fear of renewed racial competition lay behind the demand. As one white union official explained, the "real cause of the trouble" was the stevedores' efforts

> to array the white against the black longshoremen, to compete for the work which they have been doing previous to the adoption of the new conference rules, which were approved by both associations and the Cotton Men's Executive Council in February, 1886, but never invoked until the recent re-adoption.
>
> The stevedores have made advances to the colored association direct, that if they would break up their compact with the white association they,

the stevedores, would give all of the work to them, knowing that a complete reduction of wages would follow.[75]

In response to such moves, longshore unionists sought a structural solution to the problem of racial competition, proposing contractural recognition of the half-and-half division of work as the means to resolve the enduring tensions between the black and white unions. Toward this end, the unions demanded control over hiring: stevedores would employ only union foremen, who would themselves hire members of the black and white associations and "distribute them equally inside and out" of the ship. Any foreman failing to comply with the rule would be fined $25 for the first offense and would be expelled from his union for the second offense.

Some of the new conference rules encountered stiff resistance from employers. Stevedores offered no objection to hiring only union members at union wages, but they condemned the "obnoxious and unjust laws" passed by the associations, in particular the rules concerning the half-and-half crews and the foremen's powers. It was essential, they declared, "that the control of our work must be in our hands and to have the right to hire and discharge men as necessity demands." One stevedore explained:

> The reason we object . . . is that they place in the hands of the foremen what is our power, and we desire to put to work any man that we may see fit, either white or black, if he is a union man.[76]

With the steamship agents and cotton brokers firmly behind them, stevedores made overtures to black and white screwmen, offering them the longshoremen's work at screwmen's wages. Although both screwmen's unions opposed the longshoremen's strike as unreasonable and remained at work, they rejected the stevedores' offer and instead arranged a series of conferences of the striking longshoremen, their employers, and the ship agents.[77]

The settlement gave the longshoremen a great deal of what they demanded, yet the controversy over work rules and job control persisted. Stevedores agreed to all but two rules. They rejected any increase in the size of work crews and insisted on sharing the powers extended to the union foremen. On the issue of overwork or overloading, the final adjustment statement gave the stevedore *or* the foreman power to appoint three representatives to the temporary arbitration committee whose judgment was final. Thus, while the settlement greatly increased the foremen's, and union's, powers, the stevedore also retained a great deal of authority. But most important was the precedent set by the strike. Black and white workers implemented their work-sharing agreement, despite strong opposition from employers. Stevedores accepted the longshoremen's conference rules, explicitly acknowledging the extensive control union members exercised over the performance of work.[78] The two issues, of course, were closely related. For unskilled workers in a racially divided labor market, union power—codified in work rules—and racial cooperation—made possible by the biracial union structure and the new half-and-half rules—were simply two sides of the same coin.

IV

Racial competiton was an important but by no means the only obstacle to union power. If the case of the longshore workers above illustrates how union strength depended on biracial collaboration, an examination of several other groups of riverfront workers reveals that a variety of factors influenced the outcome of particular labor struggles and shaped the contours of workers' lives on the job. Of particular importance were the extent of employers' strength and resources, the nature of the labor market, workers' location in the port's occupational hierarchy, and the degree of support from allies in the labor movement and larger working-class community. Indeed, many of these factors proved more decisive than race in the struggles waged by two groups at the bottom of New Orleans' occupational hierarchy. Newly formed freight handlers' unions discovered in 1883 that, despite interracial cooperation and popular support, common laborers were no match for powerful railroad companies. In very different ways, black Mississippi River roustabouts found that while certain tactics might win them temporary wage increases, their lack of ties to organized labor and the black social network, strong employer hostility, and police repression prevented them from making substantial changes. For these waterfront workers, then, racial competition was only one variable in a very complex equation.

The emergence of new railroad terminals in the Cresent City in the 1870s and 1880s provided new sources of employment for thousands of black and white workers. But the men who labored in the new freight yards did so under conditions that differed substantially from those of the other waterfront businesses. Through the 1920s and after, dock workers employed in New Orleans railroad freight yards occupied a domain defined by geographical separation from other dock workers, low wages, and harsh working conditions. The wage gap between regular Cotton Men's Executive Council riverfront workers and their counterparts in the freight yards was vast. During the commercial season, a cotton screwman received wages of $5 a day, while longshoremen, paid 40 cents an hour, brought home $4 after ten hours (the wage would increase to 50 cents by 1887). In contrast, Illinois Central freight handlers in the early 1880s worked as day laborers earning 15 cents an hour, roughly $1.50 a day, the lowest wage of any group of transportation workers; by mid-decade, Southern Pacific freight handlers earned 25 cents an hour. Promised no definite amount of work and subjected to discharge at a moment's notice, the railroad workers suffered with the seasonal nature of their employment.

The nature of freight work and the character of railroad employers in part account for the disparity in conditions between freight handlers and most other riverfront workers. "The class of labor employed in handling the cotton and sundry freight received and forwarded by our railroads," the *Times-Democrat* observed, is "to a certain extent skilled labor." In the handling of cotton, "familiarity with the cotton-hook, knowledge of the flags and marks facilitate the discharge and delivery of cotton, and like skill is acquired by men engaged in handling sundry freight." But familiarity with the intricacies of the transportation process hardly constituted an indispensible skill, and the rail-road companies easily expanded and contracted the size of their labor force at will to accommodate their widely fluctuating demand for labor. Railroad companies relied on a large reserve army of out-of-town tramp laborers who journeyed to New Orleans during the winter months in search of work. The often urgent and sudden demand for labor led rail superintendents to dispatch agents to local barrel houses and similar resorts to secure whatever help they required.[79]

More important in determining the poor working conditions of freight handlers was their employers' size and strength. The new railroads arrived on the scene as large corporate concerns in their own right, wielding power far disproportionate to their numbers. With vast financial and manpower re-sources unavailable to the smaller steamship operators, railroad corporations were in a strong position both to establish freight rates injurious to the port's commercial elite and to resist attempts by employees to secure the benefits won by other cotton laborers. In contrast, regular dock workers' membership in the Cotton Men's Executive Council strengthened their bargaining position considerably, for individual groups of employers confronted not individual unions but, at least in theory, most classes of dock labor. As a result, these men often were able to implement work rules and command wage rates that were the envy of the city's freight handlers.

Railroads initially constituted an alien presence in the city, and their prac-tices came under sharp attack from many quarters. The unprecedented eco-nomic power possessed by these corporate giants disturbed workers, business-men, and the press alike. Addressing a crowd of strikers in 1883, Edward McCall, a white freight handler, suggested that it was only the illegitimate appropriation of their workers' labor "that has enabled these corporations to accumulate their riches and gain their vast influence in the politics and over the resources of the country." The previous year the *Mascot* complained about that "Southern Juggernaut," the Morgan's Louisiana and Texas Railroad and Steamship Company, asking who "amongst us can successfully fight a corrupt corporation, backed by money wrung from this suffering community?" In 1888 the satirical journal extended its criticism to the Illinois Central Rail-road, a "grasping soulless corporation [which] stands without rival. It wants everything in and out of sight, and if need be, would put up a fence around the world." The *States* concurred. Denouncing the efforts of "railroad kings and princes and their ministers" to "bulldoze or overawe the Freight Handlers into submission" during the 1883 strike, it predicted that the inevitable conflict "between labor and organized and grasping capital may be expected to begin" in New Orleans.[80] Yet press critics of the railroad were often mindful that the city was increasingly dependent on the very giants that were also diverting a portion of the city's potential freight to other markets. While the railroad lines threatened to eclipse New Orleans' commercial position, they also repre-sented its economic future.

That fact did not stop freight handlers from turning to collective action to challenge low wages and harsh working conditions. With the biracial union structure pioneered by longshoremen and cotton yardmen as their model, freight handlers established two benevolent associations, one black and one white, in October 1883. (In their task they were aided by M. E. Brower, Alexander Paul, and John James of the black Teamsters' and Loaders' Union.) While each branch would maintain its own membership, elect its own officers, and hold its own meetings, both agreed to abide by the decisions of a joint executive committee; in the ensuing months, interracial meetings of Associations No. 1 and 2 were not uncommon. In order to win the fight against the railroads, black union leader Sumpter Watts exhorted the black and white members to be "united and harmonious," assuring whites that the black union would abide strictly by the orders of the executive committee. By late November, the white local numbered between 800 and 1000 and the black local 600.[81]

Although their initial purposes were mutual benefit and relief, the railroads' labor practices were the main object of the Associations' attack. Following the example set by other waterfront cotton workers, the new unions of freight handlers sought similar control over the character and supply of labor. Their initial demands centered not on wage increases but on union recognition and steady employment. As family men who were "bona fide residents" of New Orleans, they fiercely objected to the "invasion of tramp labor" from the North and West that arrived each winter in search of work. Yearly "there is always an influx of strangers from other parts," the joint executive committee explained in a letter to the superintendents of the various rail lines,

> who, having no one dependent on them, nor any expenses attendant, fill our places, remaining only during the rush of the season, in consequence of which the members of our association are thrown aside or are forced to accept the terms under which they labor. The injustice is done them by permitting this transient labor the right to enter into competition with them, and in order to protect themselves against these encroachments they have organized into an association numbering over eleven hundred men.[82]

The union men demanded the right to handle the entire rail freight of the city, promising in return to provide sufficient labor at all times to the companies. For their part, railroad superintendents rejected the demands, asserting their right "to employ whomsoever they desired." At an impasse, the handlers' associations struck all freight yards and platforms in the city—the Louisville and Nashville, Morgan's Louisiana and Texas, the Texas and Pacific, the Jackson, and the Northeastern railroads, on November 26.[83]

The railroads immediately imported large numbers of non-union laborers of all "classes, colors and nationalities, the riff-raff of other cities" to handle the accumulating freight. In response, union delegations of "ubiquitous committeemen" greeted every incoming trainload of hundreds of strikebreakers to convince or coerce them to abandon their new-found employment. Railroad companies did their best to prevent desertions by shielding recruits from union delegations; the Louisville and Nashville Railroad, for example, was able to maintain a force of strikebreakers in its brick depot. But a union committee managed to induce strikebreakers working in outside sheds to quit.

Hundreds of new men joined the union, and hundreds more found at least some material support from the local labor movement. Despite mass defections from the strikebreakers' ranks, the rail lines resumed almost normal operations within a few days.[84]

The large number of unemployed out-of-towners considerably strained the unions' resources, and they appealed to the public to assist in feeding and caring for "over 600 deluded and destitute white and black laborers who have been brought here by the railroad officials, under false representations, from their various homes." At the same time, the influx of idle men—"invaders, as dangerous as a hostile army"—generated considerable public alarm. "It requires only a glance at our streets," the *Times-Democrat* noted, "to show that New Orleans boasts a larger tramp population just now than for many years. They have been pouring in for the past week from all portions of the south, invited here by the large amount of work now under way." The *Picayune* issued a similar warning: "these men have not the means to return to their homes and will be compelled to work, steal or starve in the streets." The railroad companies' threats to import 10,000 more men if necessary did little to calm the community's nerves.[85]

Trade unionists and businessmen supported the freight handlers' cause, though they insisted on the cautious and conservative path of arbitration. The Cotton Men's Executive Council and the Central Trades and Labor Assembly appointed a joint committee of five to arbitrate the conflict. Local businessmen, hurt by delivery delays and unexpected price increases, shared labor's interest in mediation. The Chamber of Commerce responded affirmatively to a petition from a large number of bank and insurance company presidents to act as mediator. As the strike entered its second week, committees from the Cotton Exchange, the Chamber of Commerce, the two labor assemblies, the railroads, and the freight handlers' unions sat down at the bargaining table.[86]

"It is an unequal contest," the *Times-Democrat* editorialized on December 1, 1883, "in which corporate power representing five hundred millions of money is opposed by a handful of men." Indeed, arbitration failed to secure what the freight handlers could not win in the streets. The railroad companies agreed not to discriminate against the freight handlers' associations and to give preference to city residents in employment. But they rejected the union shop, and in any given labor shortage, rail superintendents left themselves free to employ any men they could find. Despite considerable opposition to the compromise from the handlers' ranks, the unions voted to end their strike. "The unfortunate freight handlers have just experienced another defeat at the hands of our shoddy aristocracy," the *Mascot* concluded. The compromise was "simply a shallow attempt to cover up the humiliating defeat of a downtrodden class of men by a number of pompous humbugs."[87]

New Orleans freight handlers were in no position to exercise the kind of power available to riverfront cotton laborers involved in the steamship trade. Despite the formation of a biracial union alliance, a disciplined and dedicated rank and file, and the backing of city businessmen and union federations, the freight handlers could not match the sheer power of the railroads. Perhaps a full-scale alliance with the port's cotton workers in the competitive contracting sector might have tipped the scales. A general strike of all freight handlers and riverfront workers affiliated with the Cotton Men's Executive Council that completely paralyzed the city's commercial traffic might have enabled the handlers to wrest significant concessions from the railroads. But no one in labor's ranks seriously proposed such a broad alliance, much less implemented one.[88] While offering moral and tactical support, the Cotton Men's Executive Council was unwilling to take any action that might jeopardize its own members' standing. Although many freight handlers were fellow dock workers, the railroad yards along the Mississippi River constituted enclaves separated from the world of longshoremen, screwmen, and other cotton workers both by geography and the power of their employers. Freight handlers learned that interracial cooperation and discipline could not alter significantly their position near the bottom of the port's occupational ladder.[89]

As difficult as the freight handlers' position was, no group worked under harsher conditions, suffered greater abuses, or provoked more racist commentary than black Mississippi River roustabouts. "It is said that the men who work as roustabouts on the boats are from the lowest strata of the Negro race—that is the reason why they should not be treated as humans," one sympathetic correspondent to a black newspaper remarked at the turn of the century.[90] White observers both appreciated the difficulty of the roustabouts' labor and employed stock racial stereotypes to describe their behavior. To Edward King in 1873, the "army" of roustabouts constituted "an ebony-breasted, tough-fisted, bulletheaded, toiling, awkward mass." To a white marine hospital surgeon in 1874, the

> average "roustabout," or, as he is termed in slang, "rooster," is a strong
> black fellow, who has probably been a slave and leaves the plantation for
> that supposed freedom and rollicking life which this class take enjoyment in
> while their trip's wages last. He soon becomes corrupted by his associates,
> and after making a trip scarcely ever makes another effort at labor until
> necessity compels him to procure food. . . . The property of the average
> roustabout consists generally in what he stands and sleeps in, comprising an
> old flannel shirt, a pair of coarse pantaloons, a pair of tattered brogans, run
> down at heel, and an old ragged hat. He may also have a cotton hook stuck
> in the waistband of his breeches, but more probably he has traded it off for a
> drink before seeking a berth for a new voyage. He wears neither socks,
> drawers nor undershirt, and has no bedding or blanket to protect him from
> the cold when asleep. The usual way their tasks are performed is to have "all
> hands" at work at once, consequently their rest is very broken and irregular;
> frequently they are obliged to work thirty-six hours or longer without rest
> except for meals.[91]

The passage of time barely changed the roustabouts' work or whites' descriptions of these black workers. Echoing earlier depictions, for instance, in 1893 Julian Ralph was struck by their strength, idiosyncrasies, and difficulty of labor on his steamboat trip down the Mississippi River:

The work is hard, and they are kept at it, urged constantly by the mates on shore and aboard, as the Southern folks say that negroes and mules always need to be. But the roustabouts' faults are excessively human, after all, and the consequence of a sturdy belief that they need sharper treatment than the rest of us do leads to their being urged to do more work than a white man. There were nights on the *Providence* when the landings ran close together, and the poor wretches got little or no sleep. They "tote" all the freight aboard and back to land again on their heads or shoulders, and it is crushing work. . . .[92]

Considered drifters and sometimes criminals by more respectable citizens, Mississippi River roustabouts earned an enduring reputation for toughness and rough living. "They earn a dollar a day," Ralph observed, "but have not learned to save it. They are very dissipated and are given to carrying knives . . . . The scars on many of their bodies show to what use these knives are too often put."[93] The rousters generally lived outside the boundaries of New Orleans' established black community, often residing in gambling and prostitution dens, barrel houses along the riverfront, and the "low dives and back dens" on Franklin Street near the docks. Other black (and white) unskilled male workers were not infrequent visitors to these spots, as laborers from nearby railroad depots and sugar refineries would meet to eat, drink, gamble, dance, and engage in other such after-work activities.[94] Yet rousters alone were singled out for such obloquial treatment. As a general rule, the *Mascot* remarked in 1889, the black roustabouts were "a low thieving gang of scoundrelly vagabonds, who squander their money in gambling and on equally vile females and who would not hesitate for a moment to perpetrate any crime, if the opportunity offered." Without deep roots in the black community and with a lifestyle that many stable working-class families found sinful, the riverboat hands essentially were left on their own. Even though some rousters had formed a short-lived Inland Seamen's Union in 1892, the city's black and white unions offered them little support.[95] Riverboat men were the one group of black workers with little connection to the city's labor movement. In the late nineteenth and early twentieth century, the roustabouts had few vocal defenders in the press, the black or white communities, or the labor movement.

Unable to eliminate them through mechanization or replace them with white workers, steamship captains remained dependent on the roustabouts. To ensure that crews exerted the necessary energy when required, captains employed white steamship mates, who used abusive language and violence to keep their crews hard at work. The mates treated roustabouts "worse than the owners of animals are allowed to treat them," utilizing hickory clubs and guns to enforce discipline, one writer found in 1900. Boat officers "appear to regard them as beasts," the *Mascot* noted in 1889, and "most men could or would not stand" their cruelty. Ten years later, one traveler concluded that as the lowest class of labor in New Orleans, roustabouts were "driven like beasts by their overseers—degradation causing brutality and brutality causing degradation."[96] For decades, roustabouts commonly complained about brutal treatment.

But at the same time, steamship captains' dependence on them provided black riverboat hands with an occasional source of strength. The *Picayune* noted in 1880 that the roustabouts

> have a very novel and very often effective method of inaugurating their strikes. Usually they wait until the boat is ready to start out, and then they jump ashore and refuse to come on board until their demands are complied with. As the boat cannot afford to lose any time by hunting for substitutes for the strikers, the latter generally succeed for the time being.[97]

The rousters' behavior plagued steamboat captains for decades. "As long as they have a dollar left," one steamboatman complained in 1901, "there is no Mr. Roustabout at the plank ready to go." By withholding their labor at the last moment, the rousters often succeeded, if only temporarily, in winning wage increases and curbing the abuses of ship's mates.[98]

Steamboat captains developed two strategies for coping with the riverboat hands. First, they sought white workers to replace black strikers. During one strike in 1901, a steamboatman secured a "motley crew" of cheap labor from a nearby circus which was accustomed to the strenuous loading and unloading of circus cars; other employers contracted with a labor agent for 75 Italian workers. Three years later, 200 Swedish workers from Chicago were hired to replace troublesome black roustabouts. "Should the white men replace the blacks," the *Picayune* noted, it would mean the "passing away of a unique character in steamboat history, and the sentimental character of river travel will have been lost." Yet nostalgic whites need not have worried. The experiment in Swedish labor ended in utter failure when the white "green hands" jumped ship, complaining of bad treatment, poor quality food, the severity of the work, and the mates' rough language. A second approach to the roustabout labor problem proved somewhat more effective. Blaming the owners of the barrel houses and riverfront dives for encouraging the rousters to "bring all their earnings and squander them away in drink and debauchery of all sorts," and for providing them with "shelter and subsistence while holding out against going to work," the police raided the riverfront saloons where hundreds regularly congregated, arresting over 175 strikers and their sympathizers in 1904. Faced with prison sentences as an alternative, most rousters returned to their ships. But this approach, too, provided little more than temporary relief to the ship captains, who reluctantly accepted the rousters' volatility as a fact of business.[99]

Neither freight handlers nor roustabouts fared particularly well during this period. When the wages, work rules, and conditions of screwmen, longshoremen, and yardmen (and to a lesser extent, teamsters) in the competitive contracting sector are compared, the situation of both groups looked grim indeed. The biracial cooperation which enabled freight handlers to strike in 1883 (and in 1886 and 1892 as well) could hardly counter the vast power of the city's railroad "juggernauts." Constantly subject to racist scorn and physical abuse, roustabouts periodically managed to exact temporary concessions from their employers, only to find themselves ignored by black and white unions, attacked by the press, and repressed by police. Freight handlers' and roustabouts' location at the bottom of the port's occupational hierarchy and the Cotton Men's Executive Council's refusal to protect them were the two most important factors shaping their experiences. Highly vulnerable to their employers' determination to dominate them completely, freight handlers and roustabouts struggled outside of the circuit of solidarity established by other waterfront cotton handlers with little success.

V

From the strike of 1881 until 1886, the Cotton Men's Executive Council successfully protected its members' jobs and wages, institutionalized the relations among different riverfront unions and between black and white workers, and exercised a stabilizing influence over dock workers' employers. Black and white screwmen, longshoremen and yardmen, black teamsters, white weighers and reweighers, and white cotton classers operated under the Cotton Council's protective umbrella, linked in a network of interracial and intertrade solidarity impressive by any standard. At a time when longshore workers in both New York and San Francisco had suffered major defeats in bloody strikes, and the Knights of Labor was beginning its national decline, the Cotton Council's power was at its height. "No such a monopoly ever existed before," the Chicago *Inter-Ocean* observed in 1887. "Not a pound of cotton could be packed or loaded by any but a member of the Cotton Council."[100] Assessing the Council's power in mid-decade, the German economist Sartorius Freiherr von Waltershausen concluded that through

> firm solidarity and relentlessly concerted action, the "ring" of these unions had become greatly strengthened in recent years, and has become a heavy burden to the merchants engaged in the exporting of cotton. As a rule workers complain about the demands of the capitalists; in New Orleans, the latter complain about the tyranny of the former.[101]

Contemporaries might applaud or condemn the Council's accomplishments, but no one doubted its strength.

But despite its unrivaled achievements, the Cotton Men's Executive Council failed to sustain the inter-trade and interracial collaboration that lay at its core. Between late 1886 and mid-1887, waterfront unions divided along lines of trade and, to a lesser extent, race. Splitting into two competing bodies, riverfront workers battled each other, employing strikes and boycotts in the struggle for organizational and jurisdictional supremacy. When the smoke cleared, the old Cotton Council—and the spirit it had embodied—emerged severely weakened. Its decline as a major force was gradual and incomplete, and its constituent unions continued to exercise substantial power. But by the summer of 1887, the broader notions of solidarity that had inspired the original federation had given way to purely pragmatic alliances among different groups of waterfront unions. What began as a triumph of solidarity over craft ended with the triumph of craft over solidarity.

The roots of the Cotton Council crisis lay in the organization's very success. By 1886 it had become more than a federation of distinct unions, regulating both labor relations between workers and employers and business relations among different sets of employers. In that year, two employers' associations, representing the contracting stevedores (who employed longshoremen and screwmen) and the boss draymen (who employed the teamsters) successfully petitioned the Cotton Men's Executive Council for admission. Their entrance seriously altered the Council's character; by the start of the commercial season, the press could accurately describe the Council as "representing the organizations of labor and capital engaged in handling cotton at this port."[102] Now the Council not only controlled the price and supply of labor but effectively regulated the costs of contracting as well.

Stevedores and boss draymen, unable to defy the Council, now found advantages in allying themselves with their former opponents. As members of the Council, they could utilize its rules to regulate competition in their own ranks. Just as all yardmen, screwmen, or longshoremen were required to abide by the same work rules, all stevedores and draymen would be bound by the same rules as well. This had the virtue of eliminating the possibility that one employer would undercut other employers (by charging lower fees to those companies which contracted for his services). As long as Council employers paid union wages to their workers and abided by the rules, they could depend on the cotton unions to back them up; Council rules would require unions to boycott any employer found undercutting other Council employers. These advantages were not lost on other waterfront employers. Several additional business associations sought entry into the Council's ranks. During the summer of 1886, the Spot Cotton Brokers' Association applied for membership in order to "place the brokers in the same position as workers in other branches of the business," establish uniform brokerage commissions, and "secure harmonious relations with other cotton handlers," as its president explained. (One non-Association broker critical of the move believed that it could "take away from buyers any option as to the employment of brokers," would increase the cost of cotton by the amount paid for brokerage, and would force all brokers outside the Association into its ranks since, presumably, cotton workers would not work for non-union brokers.) Unimpressed with the arguments of the Brokers' Association, the Council rejected its application.[103]

At the same time, the Cotton Men's Executive Council did admit the Cotton Press Association, whose members owned and operated the cotton press yards and employed hundreds of cotton yardmen. That decision would ultimately prove disastrous for the waterfront unions. Representing all but three city presses, the Cotton Press Association hoped an alliance with workers would force recalcitrant, independent businesses into its organization. One critic of the move offered his interpretation:

> When, a few years ago, associations were inaugurated by the laborers engaged in the handling of cotton, these associations were formed for the protection of labor. Did the Cotton Press Association need protection when

it was admitted into the union? or did the Cotton Press Association, doubt-
ing the strength of its combined capital, ally itself to these various labor
organizations simply to augment its strength, strength which will be used as a
coercive power to force all independent presses, whether they will or not, to
pool their earnings?[104]

Events bore out this prediction. When the independent Commercial Press
began its seasonal operations on September 1, 1886, the Press Association
requested the Council to implement Article 15 of its constitution, which
barred any member from working with or for any non-union person. This
caused considerable consternation in the Council, whose labor members had
no problem of their own with the independent presses. The Commercial
Press, for example, while refusing to join the employers' union, hired only
union workers at union tariff rates. Individual longshoremen and screwmen
initially refused to accept any of the Commercial Press cotton; but within
days, most unions had voted to handle cotton from non-union presses that
abided by the union tariff. On September 10, the Cotton Men's Executive
Council formally removed any uncertainty, declaring that members could
work for any employer who paid union wages.[105]

The new alliance of employers and employees was widely condemned in
the press. "The admission of this band of bloated capitalists to the council
board of employees," noted the *Mascot*, is "the entering wedge which we fear
is destined to split the present powerful and hitherto harmonious consolidation" of cotton workers' unions.

> It was a most grievous error in the employees' associations admitting their
> most inveterate enemy into their council and confidence, and that they will
> most grievously repent it, can be seen by the signs already exhibited. When
> an association which was formed for protection against the encroachments
> of the employer, admits that employer into full membership, the association
> is virtually disbanded, its object is defeated and its influence destroyed. So it
> is with a consolidation of council associations. When the enemy is taken into
> the fold, anger lurks in every corner, and when least expected, the crash of
> failure or defeat comes with double force.[106]

Many businessmen offered an additional criticism: membership in the workers' Council enabled the commercial middlemen to maintain artificially high
monopoly prices and to compel all independent middlemen to follow suit. The
independent cotton presses, for example, charged prices below the Association's rates. Had the Press Association succeeded in its attack, the indepen-
dents would have been forced into that organization and required to raise
prices. Until its dissolution in November 1886, the Stevedores' Association
also regulated competition in its ranks by requiring its members to adhere to
uniform prices for contracts for loading and unloading vessels in port.[107]

Complaints over high port charges were nothing new in New Orleans. For
years the local press carried stories of what it considered outrageously high
rates and linked them to the relative decline of New Orleans' cotton trade.
The 1886 cotton press issue raised the question of port charges to a highly

visible level, generating, in the words of the *Mascot,* a "fuss and stir" or, as one pressman put it, "one of the periodical spasms of reform to which New Orleans is given." In November and early December, the city's men of commerce sponsored a series of conferences to pursue a "war on high cotton charges," demanding reductions in costs from business and labor alike.[108]

Workers divided sharply over the transformation of their federation into an association representing both labor and capital. In bringing to the surface considerable dissatisfaction in some unions' ranks, the cotton press issue shattered the Cotton Council, raised craft divisions to a new prominence, and initiated a five-month crisis in the waterfront labor movement. Led by the white screwmen, several white unions (yardmen, classers, clerks, weighers and reweighers), as well as the black screwmen, withdrew from the Cotton Council in November 1886. They recognized, the *Times-Democrat* editorialized approvingly,

> the fact that the organization was for the protection of labor, and that when it was thrown open to capitalists it lost its true aim and purpose, and could be made the instrument of capital in driving out competition, to the injury of the whole community, and especially of the workingmen. . . . [This faction feels] that competition is to the interest of all classes, and refuses to crush it out.[109]

Immediately, the bolters founded a new organization (which, contributing to linguistic confusion, they named the Cotton Men's Executive Council) based on a "strictly labor" basis—that is, with no stevedores, boss draymen, or cotton press owners as members. Its purpose, the preamble to its constitution made clear, was the "amelioration of the condition of the laboring classes connected with the cotton business, in order to maintain proper tariff rates, and to secure thereby wages in their different callings, which can only be done through combination."[110] The secretary of the black screwmen's association put it succinctly: "Let capitalists take care of themselves. If we continue as we have started, the Cotton Men's Executive Council will be what Mr. John Delaney intended it to be when he organized it—an association for the protection of labor."[111]

The reorganization process did not proceed smoothly, as craft rivalries and tensions that had long been kept in check now broke out in the open. Members of the new reorganized Council adopted two measures that would have, if accepted by all unions, subtly altered the balance of power within waterfront labor's ranks. First, they insisted that its delegates had to be "actually employed in the calling of the association they represent," empowering the Council to expel unqualified delegates. This had the effect of punishing several black unions led by men employed in other professions, unions which some whites held responsible for the earlier admission of employers into the original Council. Second, new Council delegates directly challenged the longshoremen, who had struck successfully for a major rule change in mid-November. Under the new rule, longshoremen received payment for all time "consumed in loading a vessel on which they might be engaged," whether or not they actually performed the labor. Longshoremen frequently had to cease working—to stop sending additional bales of cotton into a ship's hold—while screwmen completed stowing the bales on hand. During this waiting period, they had traditionally received no pay; under the new rule, they received payment for "all the time the screwmen are working, although they themselves are idle." In the midst of the reorganization, cotton screwmen came out against this important rule change. Although the press praised them for taking "excellent" action to keep down shipping costs, it was less a sudden burst of cost-consciousness than concern with their own privileged position that led the screwmen to side with employers on this matter: they undoubtedly feared that the longshoremen's rule would lead shipping agents to demand the elimination of the screwmen's 75-bale limit. At a time when public opinion appeared firmly set against any increase in port charges, the screwmen protected themselves at the longshoremen's expense.[112]

These two stances—the delegate qualification requirement and opposition to the longshoremen's rule—constituted an unbridgable barrier to a united body of workingmen. By early December 1886, two cotton councils represented New Orleans dock workers. The new reorganized Council included the white and black screwmen, and white cotton yardmen, cotton classers, and weighers and reweighers. Remaining in the original Council were the black and white longshoremen, black cotton yardmen, black teamsters and loaders, and one group of employers—the boss draymen. The crisis not only drove a wedge between former allies in different unions but, in one case, widened cracks within the black longshoremen's union as well. In early 1886, the Council had ordered the merger of four small black unions into the larger Longshoremen's Protective Union Benevolent Association (LPUBA), suppressing but not eliminating organizational tensions. Taking advantage of the realignment on the waterfront, the remnants of the four smaller branches—representing roughly 100 men—broke from their parent body and allied with the new Council. The old Council longshoremen quickly retaliated, declaring the bolters to be non-union men in early January 1887; that move meant that old Council men would refuse to work with any new Council longshoremen. The new Council responded in kind, declaring all black longshoremen in the old body to be non-union and barring its members from working with them. On January 14, a skirmish on the levee at the head of Seventh Street broke out between the black and white old Council longshoremen and black new Council longshoremen. The fight, which injured five people, was "one of those peculiar engagements where considerable powder is burned, numerous weapons of all kinds and classes used, and very little damage done." Although occasional fights broke out between different longshore factions, the two councils maintained a tense peace for the next two months.[113]

The issue of organizational supremacy erupted in March 1887, creating a morass of jurisdictional entanglements that further divided the riverfront cotton workers. The new Council precipitated the new crisis by declaring *all* old Council workers to be non-union; accordingly, it forbade its own members from working with the now non-union old Council men. Through this action, it hoped to force the dissolution of the old Council and to absorb its members. The matter first came to a head in the cotton yards and presses. New Council weighers and classers refused to handle cotton brought to them by black workers in the old Council's Cotton Yardmen's Association No. 2. Three presses and a warehouse still handling a substantial amount of cotton bowed to the pressure, discharged their black yardmen, and replaced them with white rollers. When one press retained its old Council black yardmen, the screwmen refused to handle its compressed cotton.[114]

Baffled, frustrated, and angered by union threats and boycotts, employers denounced the waterfront unions for the enduring confusion and prepared to do battle. The *Times-Democrat,* speaking for the commercial elite, concluded that

> the frequent interruptions of business, the injurious effect these have had on our trade . . . has rendered it necessary for some action to be taken that will put an end to the present system. . . . The only hindrance in the way of a decided [commercial] improvement is the labor trouble that has hung over the trade for years, like a sword of Damocles.

Members of Factors' and Buyers' Associations, who employed the weighers and classers at the cotton presses, threw down the gauntlet and asserted a position of employer supremacy. Complaining that "they have practically no control over their employees [and] that their business is liable to interruption at any time," the associations insisted "upon the supervision and control of their own weighers and classers." Toward that end, they announced that they would no longer employ any man belonging to an organization which showed a disposition to obstruct commercial operations.[115]

If the *Times-Democrat* applauded the move as essential to reestablishing labor control and profitability, the *Mascot* condemned it as the cynical manipulation of the factional dispute for the employers' advantage. "We think . . . we can see the cloven foot of the Mephistophelian cotton press proprietor in this contest," the journal commented in early April.

> It is to his interest to destroy the labor associations and he will never rest easy until he has succeeded in bringing about that destruction. He bears an intense hatred to the cotton yardmen particularly, and is determined to pay the score he owes them for his expulsion from the Cotton Council last year. It is the cotton pressmen who are stirring the fire and keeping the pot boiling, and it is they that will sup the broth.[116]

Whatever the factors' and buyers' motive, their discharge of the weighers and reweighers accomplished one of their chief goals—"breaking the solid line" maintained by the new Council. Embattled weighers and reweighers, soon followed by the classers, withdrew from the new Council. When the new Council members expanded their boycott and refused to accept any cotton weighed, reweighed, or classed by the bolted associations, now declared non-union, the cotton presses dismissed the new Council men and reemployed black workers of the old Council. The screwmen steadfastly refused to handle old Council "trouble" cotton, and bales piled up on the docks. Despite the animosity generated by the factional dispute, old Council longshoremen rejected the opportunity to bypass the screwmen and refused to deliver cotton to ship crews willing to load the bales themselves.[117]

The war of the Cotton Councils raised not only questions of jurisdiction, power, and control but issues of race as well. "It is an open secret," the *States* reported in March 1887, that the conflict "has its origin in race feeling"; many employers also interpreted the conflict as "simply a fight on the color line." Workers' testimony provides some evidence to support that point. Some whites blamed black union delegates for admitting the employers' associations into the original Council. It was the "colored men who had created all the trouble in the council," one white sampler argued in November 1886, and "the sooner we shut them out the better." For their part, black workers viewed the new Council's actions as a "boycott against the colored people who are in the majority in the Old Council," and one black delegation complained that whites "have for some time tried to shut them out, or give them only such work as they see fit."[118]

Heightened racial tensions, the two councils' reciprocal boycotts, and the presses' selective lockout of various Council members created an atmosphere for potential racial violence. The struggle between the Cotton Councils, the *Mascot* argued,

> threatens to bring about a conflict between the white and black races for supremacy, which must necessarily eventuate in a war of extermination, so to speak, in the handling not only of cotton, but of all kinds of produce received at, or shipped from this port. Such a conflict, besides being suicidal in its results to the participants, will prove disastrous to the commercial interests of the city. . . .[119]

Despite these predictions, no riots erupted. The only incident of racial violence occurred in April after the cotton presses discharged their white, new Council yardmen and substituted the black, old Council men. Patrick Gilchrist, a leader of the white yardmen, threatened Alexander Paul, a leader of the black yardmen and former vice president of the Central Trades and Labor Assembly, in front of the Alabama Cotton Press. Paul shot and slightly wounded Gilchrist in self-defense. It "was feared that a war of races would be inaugurated," the *Pelican* noted, "but luckily this was averted."[120]

But racial division was just one axis upon which the conflict turned; the other was occupation. If craft conflicts and jealousies reinforced racial resentment in some cases, in others they proved the basis for interracial alliance. For example, the new Council's rules affected not just black unions but all longshore workers. As a result, the large white longshoremen's union allied itself firmly with its black counterpart, as well as with black teamsters and loaders and yardmen. Although the new Council was overwhelmingly white, black screwmen and a smaller number of dissatisfied black longshoremen allied with it, instead of with the largely black federation. Two other comments on the role of race are necessary. First, racial tensions may have reflected in part programmatic or strategic differences. While most white unions came to oppose the admission of employers to the Council, the sketchy evidence suggests that some black workers, at least initially, welcomed their presence. It is possible that these blacks hoped that, in a period in which racial tensions in the state and city were beginning to rise, alliances with employers as well as other workers might reinforce stability or promise greater security. Second, the new Council never proposed the elimination of black workers from the docks. In fact, it invited all unions—black and white—to join its ranks, provided they accept certain rules, which were not primarily racial in character. Of course, such a move had obvious racial and sectoral implications: it would have reduced the influence of both black workers and the unions proposing a more cooperative relation with employers.

The war of the Cotton Councils ended without fanfare at the close of the cotton season in early April. Delegates of the two federations attended numerous conferences to discuss reunification, but they failed to agree on the proposals. Instead, it was the sudden, undramatic and unexplained withdrawal of the white cotton yardmen from the new Council that brought the matter to a close. Left essentially alone in the new Council, the white and black screwmen, with no real grievances of their own, ended their boycott against old Council members and returned to work in mid-April.[121] The new Council quietly disappeared. But tensions among different classes of waterfront labor persisted. While the original Cotton Men's Executive Council survived the factional struggles intact, though weakened, the era of stable alliances between waterfront sectors was over. In August 1887, the remnants of the old Council reorganized their association, elected new officers and unanimously renewed their current tariffs for the following year. Delegates from unions of white and black longshoremen and cotton yardmen and the black teamsters elected a white yardman as president, a black teamster as vice president, a white longshoreman as treasurer, and black longshore leader James Porter as secretary. Once again an association of laborers, the Council excluded from membership representatives of the stevedores and draymen. Equally important was the absence of the white and black screwmen's unions. The white screwmen, New Orleans' oldest, wealthiest, and most powerful union, and its smaller black counterpart temporarily adopted a go-it-alone approach.[122]

The biracial union structure continued to function—blacks and whites in unions of yardmen, longshoremen, and screwmen operated under the same work rules and wage rates, and yardmen and longshoremen divided available work equally. But waterfront workers no longer acted in a unified manner. Over the next seven years, the Cotton Men's Executive Council consisted of various combinations of riverfront unions, as different associations affiliated or dropped out. By 1891, the two screwmen's unions had rejoined the Council, while all other unions except the black teamsters and loaders and their employers, some sixteen white boss draymen, had quit. The following year, contracting stevedores had gained admission.[123] After mid-1894, the remnants of the Cotton Council finally collapsed amidst racial recriminations. It was not until the formation of the Dock and Cotton Council in the first years of the twentieth century that waterfront workers reconstituted an interracial labor movement. But before dock workers regained some degree of control over

their labor, both race relations and riverfront working conditions would deteriorate dramatically. Having repudiated the broader vision of interracial and
inter-trade solidarity of the early and mid-1880s, riverfront unions were unprepared to deal with heightened racial competition and their employers' assault
in the depression years. Given their earlier accomplishments, their failure was
all the more tragic.

VI

While riverfront laborers retreated from that broader vision, other workers
experimented with biracial collaboration in an attempt to win union recognition, higher wages, and the closed shop. Following almost a year of organizing
and agitation, the November 1892 general strike marked the climax of the era
of the Gilded Age solidarity. Waterfront unions played little role in what a
labor journal called "one of the most colossal struggles in this country between labor and capital," and Samuel Gompers referred to as "one of the
largest and most general strikes that ever occurred in this country."[124] But the
New Orleans general strike represented a decisive turning point in the history
of biracial unionism in the city. The principle of interracial labor solidarity
provided the foundation for the new union alliance of skilled and unskilled
workers, joined under the banner of the American Federation of Labor.
Solidarity notwithstanding, most of the workers' associations failed to win
their key goals—union recognition and the closed shop—in the November
strike. Crucial support offered to the strikers by local political officials proved
inadequate when powerful employers' associations received military and legal
assistance from the state government and the courts. Unable to defeat a
united capitalist class in a period of relative economic prosperity, New Orleans
workers fared considerably worse several years later during the economic
depression. The New Orleans trade union movement received a drastic setback during the 1890s, and the prospect of biracial unionism in the city's
trades fell victim to the economic crisis and the rise of a virulent racism.

New Orleans experienced a wave of union organizing in 1892 surpassing
even that of the 1879–81 period. Following a May streetcar strike in which
workers won a union shop, the city-wide organizing campaign intensified.
"Labor organization is being pushed to its limits in New Orleans," the *Picayune* observed in July, "and the number of unions is increasing wonderfully
every day." In August, a Workingmen's Amalgamated Council assumed the
mantle of the now-defunct Central Trades and Labor Assembly, joining together some 61 locals of older craft unions, newly organized journeymen and
unskilled workers, and Knights of Labor lodges. A flurry of strikes in the
summer and fall produced a series of small union victories.[125] Toward the end
of a year that had witnessed large, headline dominating and unsuccessful
strikes by steel workers in Homestead, train switchmen in Buffalo, coal miners in East Tennessee, and silver miners at Coeur d'Alene, the New Orleans
labor movement prepared to take a major stand for union recognition.

The New Orleans general strike originated in a walkout that began on November 24 by members of the newly formed Triple Alliance—composed of the approximately 2000 to 3000 black and white recently organized workers in the Round Freight Teamsters' and Loaders' Union, the Warehousemen and Packers' Protective Union, and the Scalemen's Union. These laborers either handled sugar and molasses products for processing in the city or delivered goods consumed by city residents. Supported in their efforts by most local unions, they succeeded in partially paralyzing New Orleans' non-cotton freight traffic for nineteen days.[126]

While higher wages and shorter hours were important issues for the strikers, the key demand was the union shop. The "sentiment prevails among the men that the union demand is the most important one urged, and the last one to be relinquished," a statement of the three unions' executive board read. "They take the position that the principle of unionism is a basis on which the government is founded and administered, and they are only struggling to secure for themselves that recognition to which they are properly entitled as law-abiding people and respectable citizens." It was precisely this notion that the commercial associations refused to entertain, and the Board of Trade, assuming command on behalf of the wholesale grocers, boss draymen, and the city's employers in general, adopted an intransigent line. Declaring that its members were willing to discuss all grievances with their employees, Board chairman F.J. Odendahl sponsored a series of widely publicized meetings with workers opposed to the strike, barring from attendance all union representatives. The Board refused to recognize the authority of the laborers' organizations. This was the "principle issue," Odendahl explained. Merchants "would not be dictated to in regard to the selection of the men whom they employed. All the other issues were side issues . . . and could readily be adjusted."[127]

Confronted by a highly organized and determined association of local businessmen, the laborers' chances for success depended upon the solidarity of their fellow workers. The Workingmen's Amalgamated Council, assuming direction of the strike, appointed a committee of fifteen—soon reducedto five—to negotiate a settlement. The committee's leaders included James Leonard, president of the Amalgamated Council and a leader of the Typographical Union; John Breen, president of the white screwmen; James Porter, secretary of the black longshoremen; John Callahan, state AFL organizer and president of the white cotton yardmen; and A.M. Kier of the boilermakers.[128] The Amalgamated Council's threat of a general strike brought the Board of Trade representatives to the bargaining table. But negotiations collapsed almost immediately, prompting a general strike on November 8. "The gauntlet has been thrown down by the employers that the laboring men have no rights that they are bound to respect," the committee declared, "and in our opinion the loss of this battle will affect each and every union man in the city." Electrical and gas workers walked off the job, plunging the city into darkness at night. Recently organized workers at the American Sugar Refinery quit in large numbers, as did clothing store clerks, musicians, car drivers, cab men, and paper hangers. Several thousand freight handlers on the Illinois Central,

Louisville and Nashville, Texas and Pacific, and Southern Pacific struck in sympathy, while adding their own demands for wage increases and union recognition. Historian Roger Shugg estimated that over 20,000 men from 42 unions joined the walkout. The *Louisiana Review* concluded that the strikers "have done nothing less than to order a war of classes—a war between the employed and the employers." Although employers succeeded in putting some non-union laborers to work, the strike seriously crippled the city's trade just as the busy commercial season had begun.[129]

As the sympathetic actions spread, the waterfront cotton unions were conspicuous by their absence. Although the screwmen's president John Breen, the white yardmen's president John Callahan, and black longshore leader James Porter numerically dominated the committee of five directing the strike, the dock unions did not participate in the action. "The stopping of the cotton trade would, in my opinion," Callahan argued, "in no manner, shape or form benefit those on a strike, as they handle freight which the cotton men are never called upon to handle. We can sympathize with and assist them financially, but we should not oppress our friends because our enemies oppress us."[130] Indeed, from the moment that cotton arrived in the city by rail or river until the moment it left on steamships for export, it followed a closed circuit of trade. The world of cotton transportation and processing involved a chain of labor separated from the rest of the city's internal commercial life. During the previous twelve years, the longshore and cotton unions already had won the battles now being fought by the new AFL unions; participation in the general strike would have jeopardized those earlier achievements. The "relations between the men and their employers was of the most friendly character," Callahan announced. The cotton press proprietors and cotton factors employed none but members of Cotton Yardmen's Benevolent Associations No. 1 and 2 and paid union wages, which had recently been increased, and while the yardmen were not required to handle any cotton touched by non-union labor, their three-year contract barred strikes for any other reason. Equally reluctant to rock the boat were waterfront employers, who lent no support to the 1892 open shop crusade. "We are not in this fight," explained Cotton Exchange secretary H.G. Hestor. "Everyone working for us is a union man and everything is working smoothly between the employes and employers."[131] For all their problems with waterfront unions and their concern with controlling their labor force, members of the Cotton Exchange had come to accept the union shop as a fact of life. Forged over a decade of conflict and sporadic cooperation, this relationship now produced the only significant arena of stability in the midst of labor's insurgency.

The threat of military repression ended the general strike on November 11, days after it began. While Mayor John Fitzpatrick had promised the Board of Trade that the city's police force of 250 men would keep the peace, employers recognized that local authorities had neither the resources nor the will to end the strike's "reign of anarchy." Governor Murphy Foster arrived in New Orleans on November 6 and met repeatedly with delegations from both the Workingmen's Amalgamated Council and the Board of Trade. Unlike Fitzpatrick, his sympathies lay with the merchants. On November 10, the governor assumed control of the city, denouncing the

> condition of affairs prevailing . . . during the past ten days, the danger to the peace and good order of this community arising from the paralysis of industry, trade and commerce, and from the suspension of the usual means of transportation, the insecurity of life and property caused by the perturbed state of the public mind.

Vowing to preserve the peace and protect lives and property, Foster placed several battalions of the state militia on alert. It was "well understood," the *Picayune* reported, "that many of the merchants and their clerks had joined the different commands and formed companies ready to be mustered in a moment's notice."[132]

Although the 5000 state soldiers and special deputies remained out of sight, Foster's threat produced its intended effect. The Amalgamated Council chose not to confront the armed power of the state; instead, it backed down, calling off the general strike. Under the final agreement, arbitration would settle the wage and hour issues; employers promised to take back the strikers without discrimination if such action did not "conflict with contracts already made," though they reserved the right "to deal directly with their men" without union interference. The following day, an arbitration board granted the Triple Alliance small wage increases and a reduction of hours, but the AFL unions failed to win their pivotal demand—the closed shop—and hundreds of union workers, particularly the freight handlers, streetcar drivers, and employees of Standard Oil Company, found that they had been replaced.[133]

New Orleans employers and commercial leaders emerged from the strike more conscious than ever of their class needs. The Amalgamated Council's promise to continue its fight at a later date, the *Times-Democrat* editorialized,

> shows the necessity of full preparation against the plan, of strengthening the militia so as to use it in the hour of need, and above all to get rid of the man, Mayor Fitzpatrick, whose official position is an encouragement to lawlessness and mob rule. If Mr. Fitzpatrick remains Mayor of New Orleans it means that the leaders of the late labor movement will try again what they failed at last week whenever the Chief Executive of the city can give them his aid and assistance. It is their belief that they can count on him that has made them announce their purpose to try it again at "another time." Mr. Fitzpatrick as Mayor is a constant menace to the peace, order and prosperity of New Orleans.

Fitzpatrick's handling of the May streetcar strike, the November general strike, and later the 1894–95 waterfront labor riots convinced commercial elites beyond any doubt of the importance of local political power and of the need to organize an effective challenge to the Ring in order to secure labor peace on their terms.[134] In the municipal elections of 1896, the commercial elite demonstrated that it had learned its lesson well when its new reform movement won control of the city's government.

Members of the city's merchants' associations also learned how valuable state officials—less beholden than local officials to New Orleans' working-class electorate—could be in the campaign against labor. Five years earlier, in 1887, elected officials at the state level had proven their worth to rural planters by brutally crushing the Knights of Labor strike in the sugar parishes. In 1892, urban merchants found a friend in Governor Foster, whose decisive intervention broke the general strike. Yet the state government's behavior in the two strikes could not have been more different. The *Mascot*'s earlier predictions of bloodshed, based on Governor McEnery's violent suppression of the sugar workers, proved wrong; the Amalgamated Council did not face the same fate as black Knights of Labor in rural Lafourche parish. For all of the commercial elite's animosity, the labor movement was too important in the politics of New Orleans, and hence Louisiana, to be suppressed so violently.

The Amalgamated Council "was the most ambitious labor movement of the kind ever attempted in this country and [it] very nearly succeeded," the *Savannah Tribune* concluded in its obituary of the strike. Perhaps it "would have done so but for the interference of Governor Foster." But the strike did collapse, as did the Amalgamated Council. The Mobile *Daily Register* only slightly exaggerated when it predicted that the "colossal failure of the strike means that the backbone of the unions is broken and the American Federation of Labor is out of business in this state." White and black discharged workers denounced the committee of five for its "treachery," and numerous unions withdrew from the Amalgamated Council over the next several months.[135]

An era of interracial cooperation and labor solidarity came to an end in 1892. The Amalgamated Council represented the culmination of a decade of experimentation in biracial and interracial unionism; in its brief lifetime, it united the city's older craft unionists, its unskilled workers, and black laborers. The failure to expand union power in the 1892 upsurge was a serious setback for both the labor movement and working-class race relations. If older craft unions survived its collapse, many unions of unskilled and black workers did not. An arena for interracial cooperation vanished and the cracks in the color line that extended into labor's ranks began to close. In subsequent years, labor's unity and strength, on the docks and throughout the city, confronted powerful challenges from two sources: the depression of the 1890s and the rise of white supremacy in political and civil society. Both temporarily crippled what was left of the labor movement and taught its leaders and members a number of conflicting lessons—lessons that would inform the character of the labor movement in vital ways after the turn of the century.

4

Turning Points: Biracial Unions in the Age of Segregation, 1893–1901

At the end of the 1880s, hundreds of black cotton screwmen and their families gathered to lay the cornerstone of a two-story union hall. The ceremony was an occasion for reflection as well as celebration. Following musical entertainment and opening prayers, several speakers assessed the progress of New Orleans' black workers and their prospects for the future. Despite "the fact that there has been much opposition and prejudice to be encountered," observed James Madison Vance, the union's attorney and Republican party activist, the black screwmen had made steady progress and now stood "at the head of the colored associations." Similarly, R. W. Gould found that although a strictly drawn "line between the races" had endured from the antebellum era to the present moment, the "condition of affairs has gradually been improved and better things are looked for by the colored race."[1] Infused with holiday oratory appropriate to such occasions, their portraits did not ignore the persistence of racism, but preferred to emphasize black labor's real accomplishments.

The optimism toward the future expressed by Vance and Gould drew upon black workers' experiences during the 1880s. Many no doubt believed that their associations might ultimately remove the color line in employment. The peculiar conditions of waterfront work—the largely unskilled labor, the potential of overcrowded labor markets, and the multiplicity of connected trades—had made biracial and inter-trade collaboration both possible and necessary for the advancement of all dock workers. Even though the Cotton Men's Executive Council was no longer the vehicle for sustained interracial and inter-trade cooperation after 1887, blacks and whites continued to labor in close proximity without overt hostility and dock unions maintained wage rates and work rules for the next seven years. And the New Orleans general strike of 1892 would demonstrate that the biracial impulse was not confined solely to the docks and cotton yards.

But if black workers' experiences in the 1880s underscored the New Orleans celebrants' portrayal of achievement, subsequent events would make a mockery of their predictions of future progress, at least in the short term. In

the interrelated realms of economics, politics, and race relations, blacks suffered severely. The century's worst economic crisis, from 1893 through 1897,
produced high unemployment and deep wage cuts for blacks and whites alike.
Corporate employers delivered staggering blows to the labor movement, spar-
ing neither white nor black union members. By the end of the century, the
political and social status of southern blacks had deteriorated further. The
fleeting interracialism of the Populist movement had collapsed under severe
governmental repression and internal strains. Throughout the South, state
after state had disfranchised its black voters as Democratic politicians consolidated a politics of white supremacy. Widespread segregation had found full
legal sanction, and white violence against blacks increased. At the same time,
white northerners had abandoned what little interest they had retained after
Reconstruction in the plight of southern blacks, embarking on their own
imperial ventures and subjugating "colored peoples" around the globe. The
Mississippi Plan, as C. Vann Woodward once noted of that state's efforts to
subordinate its black citizens, had become the American Way.[2]

New Orleans and its waterfront were deeply affected by the twin threats of
heightened racism and employer aggression. By 1893 the relative commercial
prosperity and earlier flexible racial codes had given way to a harsher reality
of economic distress and racial antagonism. Over the next two years, the
complex network of alliances sustaining the interracial and inter-trade cooperation of the 1880s disintegrated; in late 1894 and early 1895, the New
Orleans waterfront exploded in crisis. White screwmen and longshoremen,
seeking the total exclusion of black workers from their trades, utilized strikes
and violence to achieve their goals. Their repudiation of the interracial alliance led to a rapid decline of all union power, complete loss of control over
the labor supply, widespread elimination of work rules, and severe wage
reductions. By early 1895, the golden age of union power and the "era of good
feeling" between black and white dock workers had come to an end. "Looked
at from every standpoint," the black Indianapolis *Freeman* observed after the
New Orleans waterfront riots in 1895, "the condition of the Negro in America
is most deplorable and discouraging. It may well be asked, in what direction
shall we bend our vision to find the signs that promise release and freedom
from the NEW BONDAGE that has come to him within the years since he was
declared a free man and citizen?"[3]

Obituaries for New Orleans' biracial waterfront movement, however,
were premature. Their experiences after 1894 taught blacks and whites along
the docks that race relations, working conditions, and union influence were
bound inextricably together. But before dock workers could recoup their
losses, they would have to relearn the lessons of solidarity across trade and
racial lines. With the return of commercial prosperity at the turn of the
century, the alliance of black and white waterfront unions rose, phoenix-like,
from the ashes of the 1890s violence. Determined black unions, an improved
political environment for labor, the legacy of past collaboration and power,
and the specific character of dock work—all made possible the formation of
the new Dock and Cotton Council in 1901. The changes it effected were

indeed dramatic. Although "professional white politicians" sought "to make it understood that the races find it difficult to get along together in the South," the *New York Age,* a black newspaper, reported in 1913, "the best of feeling exists between the colored and white longshoremen [of New Orleans] . . . and a working agreement is in force between them which guarantees all a square deal."[4] Much the same could be said for other riverfront workers as well. While race and class dynamics on the New Orleans docks continued to defy any simple formula, the waterfront alliance challenged the strictures of white supremacy at the dawn of the Jim Crow era.

I

The collapse of the national economy in 1893 set the stage for the crisis on the New Orleans docks. The depression of 1893–97 produced business failures, massive wage cuts, and increased poverty across the nation; up to a fifth of the country's industrial labor force may have been unemployed in the winter of 1893–94. The depression hit the labor movement particularly hard. The craft unions that made up the young American Federation of Labor lost considerable ground, as employers successfully took the offensive against union work rules and wages. The Pullman Boycott, which had pitted the American Railway Union against both the nation's rail lines and the federal government, resulted in a major setback for the advocates of industrial unionism. In New Orleans, commerce suffered tremendously from "diminished trade and unsettled confidence," reported the Young Men's Business League in 1894. Statistics in "nearly every branch of trade reflect the depression and dullness which have prevailed." Although the city experienced relatively few business failures, shippers complained of low cotton prices, small crops, and non-existent profits.[5]

The depression of the 1890s, while of critical importance, was not alone responsible for New Orleans' waterfront crisis. Practices specific to the world of the city's docks also contributed to the breakdown of work-sharing agreements and interracial union structures. At the center of the controversy stood the so-called aristocrats of the levee, the cotton screwmen. White screwmen occupied the top of the port's employment hierarchy. As the most skilled laborers on the docks, they formed the port's strongest union, commanded the highest wages, and exercised the greatest control over the conditions of their labor. By the mid-1880s, they also received steady criticism from employers. Steamship agents and owners identified high labor costs in general, and the cost of the screwmen's services in particular, as "the backbone of all the excessive charges" facing the port. While few challenged the screwmen's wages of $6 a day for regular men and $7 a day for foremen, many singled out for attack the screwmen's 75-bale rule, adopted in 1878, which limited the amount of work performed. While nine hours ostensibly constituted a day's work, gangs of screwmen frequently quit early after loading their 75 bales.[6]

By the 1890s, the screwmen's work rules no longer affected all employers in the same way. Some ship operators remained highly dependent on the screwmen's skills. The careful, balanced storage of screwed cotton enabled smaller ships to sail more safely with far greater quantities of cotton than if the bales had been loaded loosely into their holds. "So closely do they pack the bales together," the *Picayune* remarked, "that when a ship is properly loaded, the cotton seems to be a solid mass, as if it were a part of the vessel, and so buoyant that a ship so loaded could scarcely be made to sink. . . ." Many tramp ships often arrived with the sole purpose of carrying away as much cotton as possible. With no strict sailing schedule, these vessels remained in port for as long as it took the screwmen to load them properly.[7]

Other employers were considerably less dependent upon the screwmen's services. New technological and economic imperatives drove them to search for alternatives to union power. Unlike smaller tramp ships, large transatlantic steamers followed specified sailing schedules. The screwmen's rule, one steamship manager complained, "gave ship owners very little opportunity to figure on a definite time for the departure of their vessels." More important, for many large steamships the careful screwing of cotton was far less important than speedy stowage and quick turn-around time in port. By 1894, the royal mail steamers of the West India and Pacific Line and the ships of Charles Stoddart relied little upon the screwmen's skills; similarly, the Cromwell Steamship Line carried only unscrewed cotton. Yet the screwmen's union required these companies to employ union men, observe union work rules, and pay union wages. The 75-bale limit "operates very unfairly . . . against the regular lines that make their living out of this port and come here regularly . . . [and whose] time is very valuable," complained agent Sanders of the West India and Pacific Line in 1886. By the early 1890s, a growing number of powerful, foreign-owned companies were objecting to the slow and expensive loading process that the screwmen's monopoly and work rules imposed upon them.[8]

Shipping agents' need to circumscribe the screwmen's work rules might have gone unaddressed had it not been for the rise of assertive black opposition to the white screwmen and the subsequent collapse of the unequal biracial agreements in that trade. The black critique of white labor focused on the general behavior of the national labor movement as well as the specific practices of New Orleans white screwmen. The case against the larger movement was articulated in the pages of the *Southwestern Christian Advocate*. The black weekly newspaper had always taken a tentative stance toward unions—praising local bodies that encouraged black and white cooperation, condemning those whose activities fostered violence or excluded blacks from membership. Though apprehensive at the unrest surrounding the Homestead strike in 1892, for example, it could still anticipate a better day when the labor movement welcomed "every honest loyal son of toil whether he be as black as ebony or as white as the parian marble" into its ranks. But by the summer of 1894, the *Advocate* had little good to say about organized labor. Surveying the outbreak of widespread class conflict across the nation, the paper concluded that "unions have failed, miserably failed thus far, in successfully dictating the terms by which our great interests have been managed." Strike-related violence and the considerable paralysis of the country's business during the nation-wide Pullman boycott signified that unionized workers represented an "irresponsible and unreasonable class of citizens." And not least was the indictment of national unions' racial practices. To the *Advocate,* the "existence of the color line in nearly all of the unions" and numerous strikes by whites against the employment of black labor had "impressed the colored laborer that he may expect but little sympathy" from white workers.[9]

But exclusionary practices, the *Advocate* suggested, were not an "unmixed evil." Systematic racial exclusion meant that few black wage earners participated in the violent strikes that obstructed the country's economic progress. "The Negro for the most part is a disinterested spectator," the paper observed during the May 1886 strikes for the eight-hour day, being "so situated that he cannot be seriously hurt or helped by any issue of the conflict."[10] If the *Advocate* called on white unions to admit blacks during the 1880s and early 1890s, its tone had changed sharply by 1894. "The well-known reliability of the colored man as wage worker," it now argued,

> should favorably commend him to every department of mechanical and industrial activity. He is thoroughly American; is not an anarchist or socialist; loves American institutions, and, with proper encouragement and protection, could be made a most efficient and valuable acquisition to our vast commercial, mechanical and industrial interests.
>
> We believe that the non-affiliation of the colored man into the labor unions of the day, places him in the forefront, among the most reliable wageworkers of the land, and will the more effectually solve some of the problems which are vexing the American capitalist.[11]

This prescription for black advancement resembled Booker T. Washington's philosophy on capital and labor. In his many speeches and writings, the Wizard of Tuskegee analyzed the situation of black Americans largely in economic terms, decrying their "agitation of questions of social equality" as extreme folly and cautioning blacks to abandon their interest in political and social rights. The development of a firm economic foundation took precedence over all else: "When it comes to business, pure and simple," Washington argued, "it is in the South that the Negro is given a man's chance in the commercial world." His hostility to organized labor led him to advise black workers to eschew strikes and unions and to stand by the better class of southern whites who offered their only hope of redemption. As anthropologist John Brown Childs has argued, Washington believed that "the cold unsentimental calculations of the capitalist would exert a tremendous corrosive power on the irrational racial sentiments of the agrarian South."[12] Black unionists in New Orleans, of course, hardly followed this advice to abandon their organizations. But as the relations between black and white workers deteriorated, black workers on the city's riverfront found it necessary to enter into uneasy and often unpleasant alliances with port employers in order to address long-standing and new problems.

If the *Advocate* framed its critique of white labor in broad terms, black dock workers had more specific objections to waterfront union practices, particularly the white domination of the screwmen's trade. Unlike longshore-men and yardmen, screwmen had no equitable work-sharing agreement. Rather, the white association imposed a strict twenty-gang, or 100 men, limit on their black counterparts. The black union's membership, however, ex-ceeded that number (in 1889, it boasted 500 members). In late 1892, the issue of the racial quota system split the black screwmen's ranks. A smaller group of roughly 100 men (which became the Screwmen's Benevolent Association No. 1) sought to preserve the status quo; a larger group (which became the Screwmen's Benevolent Association No. 2) rejected the twenty-gang limit altogether, announcing its intention to pursue an independent course.[13]

Black opposition to racial restrictions coincided with the shipping agents' efforts to circumscribe the screwmen's limit. During the early depression years, black screwmen of Association No. 2 secured work by undercutting union wages and dissolving long-standing union work rules. Utilizing their own black stevedore, they found a receptive agent in Charles Stoddart, who long had denounced high labor costs and the 75-bale limit. Stoddart found the Association's offer quite attractive: it charged 35 cents per bale of cotton, 15 cents per bale lower than the other white and black unions, and abolished the 75-bale limit. Unable to secure a promise of future labor peace from the white union, Stoddart turned permanently to the black union as "a guarantee against trouble" and of "peace and uninterrupted industry." The establish-ment of a low-wage black labor enclave had raised a serious challenge to white union power even before the depression began.[14]

The Cotton Men's Executive Council responded to these racial divisions not by addressing legitimate black complaints but by modifying slightly the status quo. Meeting in mid-January 1893, shortly after the split in the black screwmen's ranks, the Cotton Men's Executive Council expelled Association No. 2 for jeopardizing the screwmen's interests by working below the tariff rate. Agreeing to abide by the old twenty-gang limit, the black screwmen's Association No. 1 applied for—and eventually received—admission to the Council. Although both the white and smaller black screwmen's unions worked together in "harmony" under the old rules for the next year and a half, the earlier cooperative spirit had disappeared.[15]

By the opening years of the depression, black efforts to undermine on-erous racial restrictions and white screwmen's resistance to those efforts had produced a bifurcated labor market on New Orleans docks. The events of 1893 through 1895 represented a variation on the model of ethnic antagonism outlined by sociologist Edna Bonacich. A key source of antagonism between ethnic (in this case, racial) groups, Bonacich contends, lies in the emergence of a split labor market, in which different groups receive different wages for the same work. Historian George Frederickson, who has applied this analysis to industrialization in South Africa and the American South, has argued similarly that

> Racial or ethnic antagonism is thus aroused by a three-cornered struggle between capitalists desiring the cheapest possible labor, workers of the dominant ethnic group who resist being undercut or displaced by cheaper labor from a minority group or subordinate group, and the alien newcomers who are struggling to find a niche in the economy. The outcome of the conflict depends in theory on the extent to which the higher-priced workers can bring pressure to bear on the capitalist class to entrench their advantage either by excluding the lower-priced workers or by establishing some kind of industrial caste system which will allow them to monopolize the best jobs.[16]

New Orleans' black screwmen, of course, were by no means newcomers to the city or their trade, and at least a small group of blacks initially adhered to the restrictive union clauses. But the efforts of some black workers to shatter the industrial caste system and find a niche in New Orleans' commercial economy, as well as white resistance to those efforts, inaugurated a new era in labor and race relations on the waterfront—one that contrasted sharply with the experiences of the 1880s.

Two particular developments in the slow summer season of 1894 contributed to growing tensions between white and black screwmen. First, the press reported that shipping agents were searching for ways to eliminate the white screwmen's 75-bale limit. In July, one journalist alleged that at least one, and possibly both, associations of black screwmen were plotting to work for large shipping agents at a reduced tariff with no limit on the number of bales stowed. The consequences for white unionists would be severe: they would be "shut out from those ships or would have to work for the same [low] rate. They would make a hard fight against it, and a bitter factional and race contest might result." There is no evidence to confirm this particular rumor. But given the lack of coordination and communication between white and black unions, many whites probably believed it was true. The second development was the emergence of a black stevedoring firm, headed by Alcide Bessant. Black screwmen who had long complained that white stevedores discriminated against them now could turn to a black middleman for employment. Although Bessant failed to win any contracts outside of Stoddart's low-wage enclave, many white workers undoubtedly grew alarmed at the prospect of increased black competition.[17]

That alarm found expression in the increasingly harsh criticism white screwmen leveled at black screwmen during the late summer and early fall of 1894. While black Association No. 2 admittedly charged 15 cents a bale less than the white union, white screwmen falsely accused all blacks of tariff cutting. Whites also denounced alleged violations of the work-sharing agreement, renegotiated in January 1893, that allocated 20 gangs per day to members of the black Association No. 1. The whites charged the black Association No. 1 with dispatching up to 24 gangs per day—it had "insidiously started in to bolster itself up and reach out for more men and work than they had agreed to"—and admitting into its ranks green hands, unskilled men who had not served the two-year apprenticeship required by the white union. Underlying the white concern was a general fear that increased black competition would lead to the further degradation of their trade and higher white unemployment.[18]

White workers might have found a barely plausible basis for their fears in demographic trends. The city's population, which had grown at the rate of 13 percent in the decade of the 1870s and 12 percent in the 1880s, jumped by 19 percent in the 1890s. While blacks continued to constitute roughly a quarter of the city's population (27 percent) in 1900, the actual black growth rate rose from 12 percent from 1880 to 1890 to 21 percent from 1890 to 1900. Economic hardship and escalating white political violence drove many Louisiana and Mississippi blacks to New Orleans, a city with a reputation for relative racial liberalism and available work. Large numbers of blacks frequently were lured to New Orleans with the "promise of steady work, high wages and a good time," the *Southwestern Christian Advocate* warned in mid-1893. But these promises were "a mere subterfuge," and black migrants were compelled to return home or work for wages "which will barely give them support."[19]

White workers contended that the influx of blacks only exacerbated the unemployment and economic hardship experienced by whites. Blacks, one white worker complained in 1894, "had already all the steamboat and coal business. They got all the work on building street railroads and did all the carpenter work. . . . The result was that the white men must either keep their work on the levee or leave the city."[20] James Shaw, president of the white Screwmen's Benevolent Association (SBA), concurred. It had come to "such a condition in this city that you could go nowhere but the negroes were doing the work. On the streets, on the railroads and everywhere one is in darkest Africa," he complained that December. Plantation hands coming to the city "were constantly being used as a menace to labor." As for black screwmen, he denounced them for "gradually usurping the rights of whites." [21]

Black workers, of course, interpreted the situation very differently. L. J. Obert, president of the Colored Screwmen's Benevolent Association No. 1, acknowledged working over twenty gangs, but denied that blacks displaced whites. "The negroes were never given a job," he explained, "until all the white men were at work and there remained yet ships to be loaded." He also rejected other charges of bad faith. His members had always complied with the rules of their compact, had never worked below the regular tariff, nor loaded over 75 bales of cotton per gang per day. All members of Association No. 1 were residents of New Orleans, and were skilled, not incompetent, workers who had served a regular two-year apprenticeship; most of them were "sons of screwmen, who had been employed steadily since the war." Moreover, Association No. 1 men had stood steadfastly by the whites for two years and had "in no way aided or recognized the Colored Screwmen No. 2, parting from them on a question of principle, and remaining loyal to that principle." It was unjust, Olbert concluded, for the white screwmen "to throw them overboard now, because of some fancied or real injury sustained by other colored screwmen."[22]

Olbert's choice of words—"throw them overboard"—soon would serve as a description of reality, not merely as a figure of speech.

Tensions flared into the open by mid-October 1894. The SBA notified stevedores that, beginning on October 15, white union men would no longer work for any employer who hired black screwmen. In exchange for the total discharge of blacks, the white union promised to provide all necessary labor to handle every bale of cotton put on the docks. Agents and stevedores, fearing that a major strike in the early cotton season would cripple business, reluctantly complied with the whites' demand. Only agent Charles Stoddart continued to employ Association No. 2 blacks at the lower wage rate.[23]

The white screwmen's ultimatum marked the passing of an era. At the request of the black cotton teamsters and loaders, the Cotton Men's Executive Council—composed of only the white screwmen, the black screwmen of Association No. 1, and the cotton teamsters—met on October 17 for the last time. The expulsion of the black screwmen gave effective control of the organization to the whites, and the teamsters saw "no use in belonging to a council which was only maintained for the benefit of the white men." The resolution to dissolve the organization found approval almost without discussion. Council president James F. Breen, an employee in the city engineer's office and brother of John Breen, a former president of the white screwmen and now waterfront saloon operator, supported the measure. The Council's disbanding was hardly "a matter of any great importance," he noted. Since the "retirement of the Colored Screwmen No. 2 it had been merely a formality, the white screwmen having the entire control of everything." Breen was keenly aware that the black teamsters could do little "in the present difficulty . . . [I]f the [white] screwmen withdrew their support from them and agreed to work with other men as teamsters and loaders, the boss draymen could then reduce the wages or employ the other men." Almost a decade and a half after its formation, the Cotton Men's Executive Council finally crumbled; whatever remained of the initial spirit of interracial cooperation died with it.[24]

A period of conflict and violence quickly followed the Council's demise. Black workers attempted to regain their jobs, whites sought to preserve their dominant position, and stevedores and shipping agents used the crisis to manipulate racial tensions to their own advantage. Black screwmen met with stevedores and shipping agents to negotiate the conditions of their return to the levee. Toward the end of October, their efforts met with some success. Charles K. Lincoln, the stevedore for the firm of Ross, Howe and Merrow, had traditionally employed only white labor. The recent racial conflict, however, had eliminated black screwmen and increased the overall demand for white screwmen. As a result, he was unable to secure a sufficient number of whites to load his ships adequately. Under pressure from his firm, Lincoln broke ranks with his fellow stevedores on October 26 and hired a large number of black screwmen. This violation of the status quo raised the stakes dramatically. White screwmen immediately struck all ships in port, while the white longshoremen supported them by striking all of Ross, Howe and Merrow's ships. Their strike backfired, as some thirty-three gangs of black longshoremen and screwmen found immediate employment loading the cotton the whites refused to touch.[25]

White workers responded with unprecedented violence. That evening around nine o'clock, some 150 to 200 armed men, wearing false beards, handkerchiefs, and masks, gathered in the "shadow of the buildings" on Front Street. Their targets were six ships being unloaded by black screwmen. Creeping forward toward the river, they quickly occupied the levee from Second to Seventh streets, where four English ships being loaded by the firm of Ross, Howe and Merrow and two ships being loaded by the black stevedore Alcide Bessant for Stoddart and Company were docked. Posting sentinels, the crowd easily overpowered seven private watchmen. They first attacked the steamship *Constance*. "Swarming over the side of the vessel and overflowing the deck," the crowd broke open the hatches, seized the black screwmen's tools stored in the ship's hold, and threw them into the river. When the white men had completed their task, they moved on to the next ship. By one a.m., the raiders had boarded six ships and destroyed the black screwmen's equipment on each. "The crowd didn't do any damage to any of the freight and behaved all right so far as we were concerned," one ship captain told the police, "except they made prisoners of us."[26] And, he might have added, they destroyed roughly $4000 worth of the black screwmen's property.

White strikers renewed their attack the following day, Saturday, October 28. As many as fifty black screwmen and longshoremen were at work on several of Stoddart's steamships under the protection of a small number of police. At 2:30 that afternoon, at least one hundred whites rushed the docks from all directions. For the next two hours, the *Times-Democrat* reported, "the levee was in full charge of the white screwmen and their sympathizers, who terrorized every one there into submission." Whites fired pistols and Winchester rifles at the black workers who attempted to escape by jumping into the river, seeking refuge beneath the wharves, or fleeing down streets. As smaller crowds of armed whites pursued their black prey down the streets from the levee, others boarded ship after ship and, using steam derricks or sheer strength, hoisted the black screwmen's jackscrews and logs out of the ships' holds and into the river. By the end of the attack, two hundred shots had been fired; at least five people were shot and one black screwman, J. Gordon Taylor, drowned. The following day, white workers mistook the arrival of black screwmen to collect their pay as an effort to resume work. Immediately the whites mobilized, seized their weapons, and greeted the blacks with racial epithets and gunfire. "Upon the countenances of each one" of the white men, a reporter observed, "was an expression clearly indicative of their anger and hatred toward the negro."[27]

White longshoremen soon followed the white screwmen's example of black exclusion. On October 29, black longshoremen arrived for the evening shift only to learn that white longshoremen had voted to terminate the half-and-half arrangement and now refused to work with the blacks. With police patrolling in large numbers as the white longshoremen went to work, the levee from Jackson Avenue to Sixth Street resembled an "entrenchment behind cotton bales for the purpose of defending the city against some foreign foe."[28] The whites' action apparently took black longshoremen by surprise. "The colored longshoremen have been unjustly dealt with. It is an outrage the way they have been treated—without cause or provocation," black union leader Lafayette Tharpe complained.

> The colored and white longshoremen had no grievance. . . . Fourteen hundred colored longshoremen have been denied work to-day on the levee by the foremen, who claim that they received their instructions from their association to hire nothing but white men. We are sorry of this race issue, more especially when it comes unprovoked—without the slightest cause or provocation. We have a conference committee, twelve white men and twelve colored men, and we have worked on each and every ship, whether cargo or cotton, half and half. What induced them to break the conference rules and refuse to work with the colored longshoremen who have violated no rules nor regulations, we are at a loss to know.[29]

Another black longshoreman, Henry Crittenden, concurred: "White and colored have worked side by side for months and there have been no manifestations of enmity between the races," the *Times-Democrat* reported his saying. While it appeared that 1400 black longshoremen had been "driven off the levee for good," some now joined the black screwmen in Association No. 2, working on Stoddart's ships under the protection of fifty policemen. Barred from union work at union rates, the black longshoremen took refuge in the low wage, all-black enclave.[30]

In the short run, white crowd violence appeared effective. While the white SBA officially repudiated the riotous activities and insisted that no white screwmen were involved, its members clearly reaped the benefits of the disorder. Although the Cotton Exchange promised to protect black labor, its members capitulated to the white screwmen's demands, reaffirming the pre-riot but post-ultimatum situation. Following this agreement, for example, stevedore Charles Lincoln discharged his entire black labor force. Only Charles Stoddart continued to employ members of the black screwmen's Association No. 2, at the lower wage rate, through a black stevedore.[31]

But the pronounced racial tensions that produced a split labor market also gave steamship agents the opportunity to address their long-standing complaints about waterfront labor costs and the organization of work. The "whole thing in a nutshell," one prominent cotton businessman summarized, "is will the others who are employing white labor be able to stand around and see the ships loaded and unloaded at a less cost by negro labor while they are paying the regular rates to the whites?"[32] The temptation proved too great for M.J.Sanders, agent for the West India and Pacific line. With ships that followed a regular sailing schedule between the United States, Liverpool, and the West Indies and that required less care in the loading of cotton, Sanders seized the opportunity to reduce loading time by hiring black labor unencumbered by restrictive work rules. Denying that he had agreed to the original Cotton Exchange compromise, Sanders terminated his firm's twenty-year relationship with the white screwmen. The terms "dictated by the white screwmen were unfair and improper," Sanders complained, and the union's inability to provide him with a sufficient number of experienced men made the employment of "other labor" imperative. Asserting his right to hire any man he chose, he contracted with the new black stevedoring firm of Carey, Allen and Company to load his company's ships with the black screwmen of Association No. 1.

Individual efforts did not get agent Sanders very far. On November 4, a large cotton fire swept through the West India line wharves and freight sheds between St. James and Felicity streets, destroying an estimated 4000 bales of cotton, and large quantities of oil cake, cotton seed meal, staves, molasses, and other goods. Although the white screwmen's union denied any connection to the suspicious blaze, it maintained its vigilance against black competition. In response to false rumors that black workers upriver at Southport were attacking the white men's tools, an armed white force immediately mobilized. The "entire levee filled with [white] men, all armed and patrolling the levee in regular soldierly fashion," noted one journalist. The following day, November 6, an unknown attacker shot and wounded the black stevedore Carey near Southport.[33]

Neither arson nor armed patrols by white workers had the desired impact. Within days, the West India line and Charles Stoddart and Company put their black dock laborers—in Associations No. 1 and 2 respectively—back to work under the protection of a small army of police, both committing themselves to employing black labor on a permanent basis. Unsatisfied by guarantees of protection by city and state troops in the event of renewed violence, representatives of the West India and Pacific company discovered a powerful new weapon in the battle against white labor: the injunction. At the West India and Pacific company's request, a U.S. circuit court issued a restraining order barring members of the white screwmen's and longshoremen's associations from interfering in any way with black workers. The temporary order was made permanent in early December, and white workers reluctantly abided by its provisions. The violence temporarily subsided.[34]

Agents and stevedores sought an ally in the courts because municipal authorities and police offered them less than complete cooperation. Numerous witnesses testified that city policemen had stood idly by while armed whites seized the tools of the black union and assaulted its members. Police officers, black workers observed, "were like spectators upon a theatrical performance"; not only were the police in active sympathy with the whites, but many policemen also maintained membership in the white screwmen's union, owing "more allegiance to that association than they do to their oaths and the protection of life and property." The *Times-Democrat* concurred in a strong denunciation of municipal complicity in the riot: "All these riotous acts . . . were perpetrated in the full view of a police force which not only did nothing to interfere with them, but actually seemed to sympathize with the rioters."[35]

Not surprisingly, the strongest voices against the riots came from New Orleans' black community. On November 8, some 1500 people, including sixty preachers, gathered for a religious service and protest meeting at the Wesleyan chapel. The protesters adopted a series of resolutions condemning the behavior of the civil authorities and calling for greater Christian forbearance and faith. "At a time when the peace, prosperity, and even the very life, commercially speaking, of a great city like New Orleans, is endangered by the ruthless hand of the murderer, outlaw, assassin, and incendiary torch," they argued,

> when the blood of half a score of defenseless law-abiding citizens . . . [has] been slaughtered in cold blood, or wounded nigh unto death, by mob violence, in open day and in the presence of those whose duty it is to protect the defenseless and preserve the peace of the city; when the blood of these men is calling aloud for redress at the hands of injured justice and outraged law; when the only plea thus far given or hinted, by these insurgent violators of the peace and perpetrators of these deviltries, is the fact that those against whom these greatest of outrages are perpetrated are negroes and are incapable of resentment or redress; when the representatives of civil government, municipal and gubernatorial, seem to act with a measure of indifference to such condition of affairs, may it not well be said, as in the day of primitive Christian experience: "Lord, unto whom shall we go?"

Promising a firmer reliance upon God than ever before, the black protesters ended by appealing

> to all true Christians of every creed and faith to unite with us in asking God to turn and overturn, as in the days of old, until the arm of the oppressor shall be broken and the right of people of ever race and hue shall be respected and protected on every foot of American soil from east to west, and from north to south.[36]

If breaking the arm of the oppressor was God's work, black unions had the more difficult task of devising some strategy to cope with the new state of affairs. Sharp divisions immediately arose. A small number of black screwmen in Association No. 1 expressed their willingness to turn back the clock and accept the pre-ultimatum limit of 20 to 26 gangs, in exchange for racial peace. A majority of union members objected at a special meeting called to denounce the unauthorized offer by advocates of the racial quota system. "We deplore said unmanly actions," members declared in a public statement, and "can only characterize their actions as being menial and of the lowest and basest design, calculated to destroy both the veracity and stability of any organized body." The violence had destroyed the old foundation permanently. Some black unionists maintained:

> They had been driven off the levee and prevented from working and they would not now make any settlement that would not give all of them a chance

to go to work again. They would not agree to any contract which would give 100 work and leave 200 or more without it.[37]

H.J. Randle, vice president of Association No. 1, exclaimed that black workers would rather "go off the levee than make any such compromise." Principle and pride aside, black workers had a more pragmatic ground for opposing the restoration of the modified prior arrangement: they currently enjoyed more employment. A compromise move, one group of black screwmen observed, "would savor a great deal of imbecility, as they have all the work they can get now, whereas, under such an agreement as the one reported, they would have to turn three-fourths of their number adrift." For a majority of black workers, however, the situation was far more bleak. While black screwmen found work under Stoddart and Sanders, a far larger number of black longshoremen remained unemployed.[38]

The collapse of the Cotton Men's Executive Council, the demise of work-sharing agreements, and the refusal of white riverfront men to work with blacks, forced black workers to seek new allies in the struggle to secure employment. The thoughts of one union leader suggested both the indignation felt by black workers and the new directions that blacks might be forced to look. "We have been deprived of our implements of labor," he told the press in early November 1894,

> our wives, some of them, have been made widows; our children orphans, and our right to live denied. We can only rely upon those who seek to establish New Orleans as a great port of entry, who by their industry, their expenditure of time and money, have brought her to the height of her personal competition with other ports, who by years of toil and great loss have stood the ravages of time and the . . . sting of strikes, we ask them to aid us in the pursuit of a livelihood. . . .
>
> The strike is not settled. We are still unemployed. We have received no pay for our losses. . . . Of the Cotton Exchange and all other exchanges we now implore assistance in the reinstatement of affairs as they were prior to the strike. All we ask is a showing to make our living and to keep away the wolf from our doors. . . .
>
> To the charge made against us, that of "wage cutters," we simply say that prostitution of rights, threatening the continuance of our cotton trade and other trades, the denial of work and the assassination of our brothers, will force us to seek the arms of protection in the nearest way and under the most available terms.[39]

Led by former black union president E.S. Swann, a committee from the black longshoremen's association made one last appeal to the white union men in mid-December. When that move failed, black workers finally turned to their employers and the city's elite for assistance.

Black longshoremen resolved to break the deadlock and recapture their lost jobs. As "a matter of salvation for their very existence," they proposed to undercut the white longshoremen by reducing their tariff from 50 to 40 cents an hour for day work, from 75 to 60 cents for night work, and from $1 to 80 cents for Sunday work—in effect, restoring the pre-1885 wage standard. "We have done everything in our power," labor veteran Swann explained to his fellow longshoremen,

> but it was all in vain. . . . We did not like to reduce our rates without first trying to get them to divide the work with us, as has been done before, but as they would not, we are obliged to do something to get bread and butter. We have been trying to get them to come to some agreement with us for the past forty-eight days, but they have driven us off without cause or provocation, although we have always been loyal to them.
>
> We don't think it is right that we should have our men shot and killed on the levee like they were dogs, as they did with J. Gordon Taylor. . . . Now there are six white men working on the levee to every one colored man, and we have about 900 or 1000 who are on the eve of starvation. The only work we have is the little that is given to us by Messrs. M.J. Sanders and Chas. Stoddart. The colored Longshoremen's Association sincerely thanks these gentlemen for what they have done for us.[40]

Swann found a receptive audience among many individual stevedores. Immediately seizing the opportunity to slash wages for the first time since the depression of the 1870s, employers announced that they would employ any man, "irrespective of color," who agreed to work for 40 cents an hour, signaling, in the words of the *Times-Democrat,* that "none but colored union or white non-union longshoremen need apply for work . . . for no one expected the white longshoremen to meet the cut as a means of maintaining their monopoly." Anticipating a serious conflict, the contracting stevedores called on Mayor John Fitzpatrick to provide adequate protection along the entire riverfront.[41]

This time, white riverfront workers did not resort to violence. Nonetheless, they managed to halt most levee work as soon as stevedores cut wages. Although the white screwmen's union took no official action and disavowed responsibility for its members' actions, the white screwmen's tactics proved central. The men, one screwmen stated, had "come to the conclusion not to handle the cotton from negroes." Half of the white screwmen stayed home on account of "illness;" a few said they had to purchase Christmas turkeys; the rest claimed that there was simply no one (i.e. union white longshoremen) to bring cotton to the ships. Although the stevedores had plenty of black longshoremen on hand, there were too few experienced black screwmen available to carry on the work. Without the full number of white screwmen, there was little stevedores could do.[42]

The unsettled state of affairs troubled all parties involved. Whites feared wage cuts and loss of work; blacks dreaded renewed racial violence and sustained unemployment; employers faced financial losses due to the periodic cessation of work at the height of the cotton loading season. The first stage of the labor and race crisis ended on December 20, the second day of "absolute idleness" on the levee. A conference of black and white workers and cotton merchants agreed to return to the old state of affairs. Both black and white longshore workers would reestablish the pre-October arrangement, dividing existing work equally on all ships—except those of Stoddart and Sanders, who refused to participate—and restoring the 1885 wage rates and the old conference rules. As much as the stevedores desired a 20 percent wage reduction, they recognized that the old rules provided a basis for labor stability and rescinded the wage cut.[43]

The December 1894 compromise revealed both the precarious position of blacks in the depression-era New Orleans economy and the tensions within black labor's ranks regarding the new arrangements on the docks. While black and white longshoremen returned to an equal division of work, white screwmen reimposed the restrictive quota system on black screwmen in Association No. 1. But one fact had changed since the summer: two agents—Stoddart and Sanders of the West India and Pacific line—now fell completely outside these arrangements and employed only black workers. Tempted by stronger managerial control, lower wages, and fewer work rules, Stoddart and Sanders offered blacks a refuge in an increasingly hostile economic world. For their part, unemployed blacks and those union men bitter at the screwmen's racial quota system welcomed these employers' offer. But the new relationship being forged hardly represented employer paternalism and benevolence, on the one hand, or black acquiescence and passivity on the other. The bottom line for Stoddart and Sanders was economic; both kept their wage rates significantly below those of other agents and stevedores. Black workers' expressions of gratitude toward these companies during the fall (and later spring) crisis implied neither a wholesale adoption of Booker T. Washington's philosophy nor an abandonment of the principles of unionism. When Stoddart and Sanders refused to follow the port's other employers' example in restoring the wage cut in late December, the black longshoremen's union forbade its members still employed to work below the old association rate. And when black union longshoremen struck on December 27, Stoddart and Sanders quickly, and without interruption, filled their places temporarily with black non-union men.[44]

The end-of-the-year pact represented more of a truce than a solution to long-standing problems on the docks. It resolved neither the fundamental differences between black and white workers nor those between employers and unions. Many agents remained hostile to the white screwmen's union; labor solidarity across racial lines was all but dead on the waterfront; white screwmen still refused to share work equally with blacks; and the black screwmen's resentment of racial restrictions remained strong. Moreover, the December agreement contained the seeds of its own destruction, for it left intact—and indeed, legitimated—a racially split labor market on the docks. In the highly competitive world of transatlantic shipping, New Orleans employers constrained by union work rules and wage rates felt compelled to adopt measures enabling them to match the advantages of the all-black, low-wage waterfront enclave. In such a highly unstable and volatile environment, it did not take long for race and labor relations to deteriorate dramatically.

II

"This is an opportune moment to draw the attention of our great shipowners in England carrying on a large and regular trade with New Orleans," a British Consulate official remarked in a 1895 report, "to the fact that the fight against the exorbitant and ruinous charges of the screwmen and longshoremen must be met by concerted action on their part." By the start of the new year, shipping agents in New Orleans did not need to be convinced that the time for action was ripe. Their counterparts in Great Britain had already demonstrated that riverside and dock unions could be challenged successfully. Not only had the union upsurge on British docks, beginning with the London general waterside strike of 1889, been contained, but the Shipping Board and other employer associations had delivered a crippling blow to the young waterfront labor movement.[45] Much closer to home, the events of the fall of 1894 demonstrated clearly to the New Orleans agents of British shipping lines that the screwmen's control of the waterfront could be broken at last. They seized the initiative early in the new year.

The low-wage enclave of black labor proved too attractive a model for employers to resist. In February 1895, the Harrison Steamship Line of Liverpool ordered its agent, Alfred LeBlanc, to follow the example of competitors Stoddart and Sanders and cut the wage rates of the screwmen and longshoremen. Discharging his all-white labor force, LeBlanc contracted with a new black stevedore, George Geddes, and began loading his ships under heavy police protection with some 300 non-union black workers at 40 cents an hour.[46] In March, after experiencing lengthy delays in obtaining necessary labor, the Elder-Demster line also abandoned its long-standing preference for white workers. Rejecting demands by the newly organized Jefferson Association, composed of white unemployed Gretna men, to hire only local white Gretna residents, Elder-Demster's stevedores imported non-union black workers by tugboat. An attack on these new black workers by whites led the Elder-Demster line to obtain an injunction restraining white longshoremen from interfering with their employees. Each time a shipping firm switched to lower paid, all-black labor, stevedores and agents relied heavily on their allies in the judiciary. "In every instance the United States Court has been appealed to for protection by the agents of our regular liners," the British consular official noted approvingly, "and the loading continued in the presence of United States marshals." Backed by court injunctions, federal marshals, and occasionally squads of police, the agents and stevedores successfully rolled back waterfront wages.[47]

But New Orleans' men of commerce denied that their actions were motivated by "a question of wages or compensation"; instead, they asserted that managerial rights lay at the heart of the matter. In a statement adopted in March by representatives of fifteen businessmen's associations, many of which were not directly involved in river transportation, it was declared that the central issue was "simply and solely one of whether the merchants of New Orleans shall conduct their own business in their own way, or whether they shall be dictated to by a handful of employees." The only way to protect the commerce of the city, it was insisted, was to permit every man who wanted to perform "honest labor" to do so,

> regardless of race, color or previous condition: no man shall be interfered with in the pursuance of his daily avocation, and we insist on the right of every employer to hire whom he may choose, to have his work performed in such a manner as he may direct. We do not deny the right of the working men to combine together for mutual protection, or to stop work when they choose, but we deny their right to prevent others from working.[48]

Arguing that all labor should be "free and untrammeled" and that employers should retain non-union stevedores and foremen to direct the loading on their ships and ensure that laborers were performing a "commensurate amount of work," agents savored the opportunity to place the cotton screwmen at last "in the position of all other labor along the levee front," abolishing "anything like limitations" on the number of bales stowed per day. Given the high unemployment levels generated by the current depression, they believed that market forces would accomplish this task. "So many colored men were being employed and so many others were being brought in to the front," agent Ross told Governor Foster, that "the natural result will be the cutting of the prices of labor, and the questions will then settle themselves without interference."[49]

A managerial triumph on the docks, white longshoremen and screwmen countered, would have devastating consequences for their trades and their lives. Placing the agents' degradation of the screwmen's craft at the center of their analysis, president James Shaw and recording secretary John Davilla of the white SBA offered one interpretation of the recent crisis in a public notice:

> We saw an honorable occupation, which by dint of intelligent exercise had risen above the grade of mere physical force onto the plane of intelligence and skill, about to be dragged back down to the level of pure muscular exertion, where brawn without brains, and muscle without skill, would be all that were required.[50]

Longshoremen, with fewer skills to protect, emphasized less the preservation of their "honorable occupation" than the impact of job competition between rural and resident workers. Whites were "opposed to the employment of tramp labor secured from the plantations of other states," explained Henry O. Hassinger, general inspector of the white longshoremen's association and himself a foreman.

> The white men could not work for the prices paid these tramp negroes, for the season is a short one, and taking into consideration the many rainy days when it is impossible to work, the men do not make the large amounts they are supposed to earn. Then, too, they have to secure enough money to tide them over a long and dull summer.[51]

The employment of non-union labor meant that a class of unskilled, mostly black men would soon dominate the levee, Hassinger predicted. "They are unreliable and can be safely called a floating population."[52] Shaw and Davilla concurred, further warning that "it is . . . a folly for the merchants of New Orleans not to realize . . . that an irresponsible spendthrift labor population, impotent to perform its task, while temporarily and apparently a blessing, is eventually and really a curse."[53]

Were white screwmen and longshoremen protesting the increase in the number of inexperienced rural workers, black dock workers, or both? If the issue of race figured prominently in their discussions, they framed it in ambivalent ways. In practice, white strikers and rioters exhibited an intense hatred for their black counterparts, destroying their property, attacking and sometimes killing them. But in their public appeals, white union leaders were often unwilling to resort to the racist conventions of the day. According to Shaw and Davilla, it

> is folly for the public to remain under the impression that the problem on the levee involves social rights or race antipathies. There is no question of social equality nor any of race prejudice involved in the controversy. Any organization of skilled labor, finding themselves dragged down to be eventually classed as unskilled labor, would have made the issue thus thrust upon us.

Walking a very fine line, Shaw and Davilla justified their stance toward black workers not on the basis of their race—that is, not because they were *black* workers—but on the basis of their behavior. Black screwmen, they argued, had

> proved the truth of the assertion that he was not, and has never been worthy of trust. Every agreement we made with him, or compact to which he is a party with us, to adhere to rules which had been adopted, not for our individual, but for the aggregate, good, he violated. We declined and refused to further harbor him among us or to hold any contractual relations with him. He was an enemy to be feared, because he degraded the calling which he had adopted. He was and is the parasite of skilled labor.

Despite these disparaging remarks, white union leaders refused to acknowledge an explicitly *racial* foundation of their criticism. "If his skin had been such as would excite the envy of the fairest Caucasian," Shaw and Davilla said of the black worker,

> and his ancestry such as to fit him for association with the most exclusive circles, he would have received the same treatment at our hands. It is not a question of wealth, nor a question of color, nor a question of social rights. It is a question of preserving an honorable calling against the corruption to inevitably follow from dishonorable methods. We know the master and we know the puppet.[54]

Similarly, longshoreman Hassinger stressed that the current fight was "not a struggle of races, for the resident negroes are equally interested in seeing the work done by local men."[55] However true that latter claim was, white labor's behavior and its failure to appreciate the divisions within black labor's ranks belied its rhetoric.

Blacks could hardly ignore the racial dimension of the crisis as easily as whites. Yet differences of opinion existed as to how black workers should respond to the challenges or opportunities before them. Black Screwmen's Benevolent Association No. 2, the break-away group that sought greater employment for its members, offered one approach. Reiterating their hostility to the white unions, its members declared at a mass meeting that black screwmen and longshoremen

> do not intend to hurt the interests of the port by fooling with white screwmen. You have seen what they have done in the past. They make their application at reduced rates to get the colored screwmen off the levee, and just as soon as they get the colored screwmen and longshoremen off the levee, then they will find that the reduced rate is too low.

Moreover, they made clear their intention to "stand firm to the firms that stood by us in the time of trouble"—those of Ross (of Ross, Howe and Merrow), LeBlanc (Harrison line), and Sanders (Leyland line)—which now provided them with secure employment in the face of white threats.[56] Undoubtedly, it was this group's behavior, inspired by the white screwmen's racially exclusive practices, that most white workers found objectionable.

But since the crisis began the previous year, black workers had not acted as a unified group. In late December, the black longshoremen's association struck when agents Sanders and Stoddart refused to rescind their wage cut (white workers apparently ignored the short-lived and unsuccessful walkout). The following year saw continued tension between the two black screwmen's unions. If the Screwmen's Benevolent Association No. 2 accepted lower wages and undermined standards in exchange for a greater number of jobs and security, black screwmen in Association No. 1 attempted to maintain wages, work rules, and an alliance with whites, however distasteful they found the racial quota system imposed upon them. When the Harrison line reduced its employees' wages in early 1895, Association No. 1 rejected the company's job offer and forbade its members from accepting employment outside the West India and Pacific Steamship Company. Following the wage reduction, only Association No. 2 and black non-union men would work under Sanders, Stoddart, and the Harrison line.[57] The different approaches adopted by the two black screwmen's associations mattered little to whites. White workers' failure to recognize the important divisions in black labor's ranks demonstrated that, their denials notwithstanding, they viewed the crisis primarily through a racial lens.

As they confronted an expanding low-wage black enclave, white workers adopted a two-prong strategy. First, they declared that the widespread wage cuts constituted a lockout, and, accordingly, withdrew their labor from all ships in mid-February. Second, white screwmen offered a comprehensive plan that they hoped might "completely revolutionize the manner of conducting business along the river front." They targeted the contracting stevedores, "purely ornamental and unnecessary individuals," who made enormous profits but performed no useful work. The fee paid to stevedores, they declared, was "money practically thrown away," adding unnecessarily to already high port charges. Refusing to work any longer for middlemen, white screwmen announced that they would contract directly with steamship agents. To make their proposal attractive, they appointed a business agent, established an office, and, most important, reduced substantially their rates for stowing cotton. White longshoremen quickly following the screwmen's lead. (Cutting their charges on all articles of freight, they claimed that their rates "could not be met even by the negroes.")[58] In retaliation, black Association No. 2 screwmen lowered their wages even further—to 30 cents a bale on steamships and 35 cents on sailing vessels, five cents below the SBA's new rates. The "war in prices" kept relations between blacks and whites tense. While white screwmen and longshoremen found work on a few vessels, non-union and union blacks employed by stevedores and agents continued to receive most jobs.[59]

Furiously denying the charges of profiteering and angry at white union efforts to bypass them, stevedores precipitated yet another crisis by taking steps to replace permanently white longshoremen and screwmen who refused to work for them. They contracted with Norris Cuney, a black stevedore of Galveston, Texas, to provide them immediately with approximately 65 additional black workers. Since the Galveston shipping season was nearly over, as many as 500 black dock workers were potentially available. Mayor John Fitzpatrick denounced the importation of outside laborers as a breach of good faith that would foment "the ill-feeling and discord" which existed in the community. More important, he refused to provide additional police protection for the imported men. Nonetheless, on March 9, the Galveston crews arrived in the city on the Southern Pacific Railroad. Placed on a barge, they were kept away from local workmen and transferred directly to the ships of two stevedoring firms, A.K. Miller, Meletta and Company and Ross, Howe and Merrow. The arrival of the black Galveston workers, the *Picayune* observed, was "in itself enough to increase the already bitter feelings of the whites against the negro laborers."[60]

Importing outside blacks hardly solved the stevedores' problems. No sooner had they arrived than the new men demanded higher wages. Stevedores reluctantly agreed to pay them the white screwmen's old daily rate of $5. In addition, Galveston blacks claimed that New Orleans stevedores had misrepresented the local situation to them. Once aware of the real situation, one black Galveston foreman ordered his men to quit work while he negotiated with white screwmen. At a waterfront barroom on Jackson Avenue, white labor leaders James Shaw and John Breen related the history of recent events to the Galveston men, warning them of dire consequences if they continued to work. Convinced that the white men of New Orleans were "even worse than they were thought to be," many of the Galveston blacks accepted the white union's offer of a free return trip to Texas. Those that remained also quit work, adopting a wait-and-see attitude.[61]

Outright force soon replaced coercive persuasion. On the evening of March 9, between 300 and 500 armed white men forced their way into the

Morris Public Bathhouse at the head of St. Andrew Street, where the West India and Pacific Steamship Company stored the equipment of its black screwmen. Seizing 45 out of 90 sets of tools, the men threw them into the river. Two days later, on March 11, approximately 150 whites shifted the target of their anger from property to people, attacking black workers on the docks near Breen's saloon and the West India and Pacific company wharves. In the five-minute assault, whites fired at least twenty-five shots, hit two black men, and chased another group of black workers down the levee.[62]

A more deadly outbreak of violence occurred the following morning, when whites coordinated two armed attacks against blacks working on the levee. The first took place opposite the French Market between St. Anne and Dumaine streets. At 7 o'clock, several hundred whites made their way through a dense fog, taking cover behind Louisville and Nashville Railroad freight cars and piles of tarpaulin-covered cotton bales on the wharf. When a gang of twenty to thirty black workers began removing the covers from the forward hatches on the Harrison line steamer *Engineer,* the crowd of whites rushed the ship. "Bullets sang and whistled round the wharf like hail," as whites fired hundreds of shots at unarmed blacks dashing for cover. Not content with merely scaring off workers, one group of whites angrily boarded the ship and "emptied their pistols and guns at the unfortunate negroes," while another group "pursued with relentless pertinacity" blacks who sought escape on the wharves. Black workers "were given no quarter and were shot down like dogs," the *States* observed, and "[b]lood flowed like water." Another journalist noted that "shots seemed to come from doorways, windows, galleries, and street. Look where they would, the negroes saw pistols." Victims included black workers as well as blacks who happened to be working or passing through the area; by the end of the assault, three blacks had died. The handful of city police guarding the black workers stood by, taking no action.[63]

A second assault took place four miles uptown, where black longshoremen and screwmen worked on the ships of Ross, Howe and Merrow and the Elder-Demster line. At 7:30 a.m., between 200 and 500 men, hiding behind cotton bales and freight cars, "rose up out of the heavy mist like grim death spectres," the *Times-Democrat* reported, and attacked blacks working on the steamships *Merrimac* and *Niagara.* For more than an hour, one reporter observed, the whites "held a bloody carnival on the wharf." After killing one black and wounding another, the mob raced up and down the wharf, "terrorizing everyone and holding absolute sway" until the police drove them back. The two attacks left six dead and many more wounded.[64]

Military intervention followed the riverfront massacre, putting an end to the violence. Vowing to use the state's entire police power to uphold the law, Governor Murphy Foster on March 13 denounced the riots, barred assemblages near the riverfront, and placed the state militia on alert.[65] The following day, some black workers resumed their labors under police and state military protection. The initial procession of militiamen from the armory to the waterfront assumed theatrical proportions. As the soldiers left the Washington Artillery, they were "met along the line by enthusiastic and admiring crowds." The

St. Charles Street route—an upper-class area—took on "almost a Carnival appearance" as "bevies of fair girls . . . applauded them extravagantly." The "Cotton Exchange and other commercial bodies always received them with enthusiastic cheers." Yet as the soldiers moved into the working-class neighborhoods near the docks—"the district of saloons and tenement houses, the dark neighborhoods of alleyways and squalid houses and gutter-reveling children"—the contrast in reception could not have been more extreme:

> This was the land of the [white] screwmen and their sympathizers, of the longshoremen and their affiliates, and the women and children no sooner caught sight of the serried columns of militia than they made up their minds that woe unspeakable had come into their midst. . . .
>
> In the neighborhood of Rousseau and Jackson streets, a ragged, nondescript crowd of people, in which women largely predominated, jeered and hooted the troops as they marched by. A woman in a greasy, torn wrapper, cried out in a shrill voice: "Shame, shame! Ye're not statisfied wid takin' de bread outen our mouths. Ye want to shoot it out. Oh march away, and shame go wid ye." The men were silent. With set, dogged faces, they looked on and never said a word.

Troops cleared the docks of the crowds that had gathered; mounted police patrolled the wharves; a battalion from the Washington Artillery trained its howitzers on the screwmen's headquarters; and troops guarded black workers. With the riverfront under military control, union and non-union black screwmen and longshoremen gradually returned to work.[66]

Although an armed peace prevailed, the outcome of the "military protectorate" remained uncertain. After all, the militia could not occupy the levee indefinitely, and many businessmen believed that the "bloody business of breaking the law" would resume the moment the troops were withdrawn. While the commercial exchanges initially contributed $6000 for the maintenance of the soldiers, they expected the state government to assume the financial burden of keeping the militia on guard to protect their right to hire whomever they desired at the wages they set, for as long as necessary. Reluctant to draw too heavily upon the state's finances to support New Orleans' commercial elite, Governor Foster did not view military repression as a long-term solution designed to promote peace and prosperity. As the state's leading Democratic politician, he walked a fine line in trying to reconcile the opposing sides. "In the problem submitted for our solution," he explained in an interview with the *Times-Democrat,* "are two elements more difficult to handle than any that have ever tested political wisdom, to wit, the antagonism of race, and the conflicts of labor with capital. . . ." Claiming that his own power was "purely negative and preventative . . . with no legislative function," Foster declared himself "impotent to impose on any of the parties to this conflict any specific scheme of settlement." That task fell to the representatives of the city's organized commercial bodies and the white levee laborers. The presence of the troops served to preserve peace, to protect commerce, and to "allow the parties immediately involved in the labor difficulties an opportunity to come to some satisfactory adjustment." Given the passage of ample time in which the parties could have negotiated, and given the screwmen's assurances that they had no intention of indulging in additional violence, Foster withdrew most militiamen on March 25, ending their twelve-day occupation. The governor maintained some companies on alert, however, and local police continued to stand guard over the black workers. [67]

The March violence and subsequent military occupation of the waterfront dealt a severe blow to the white screwmen and longshoremen. In the riots' aftermath, white screwmen resolved to go it alone, promising to attend strictly to their own affairs. Some turned to other sources of employment, as they did every summer, journeying to the cod fisheries of Newfoundland or finding work in the lumber business in Scranton or Canada. But the majority who remained in New Orleans faced new challenges. Union leaders ordered their members to keep away from the waterfront, permitting only the union agency, the Excelsior Co-operative Association, to dispatch union men. At the same time, screwmen reiterated their earlier proposal to steamship agents to eliminate the "useless luxury" of the stevedore, and offered to sacrifice their 75-bale a day limit and traditional wage rate in exchange for work. A small number of white union dock workers found employment on the Hammond line ships through their hiring agent, and the screwmen's union sent recording secretary John Davilla on an unsuccessful business trip to Europe to bargain directly with ship owners. But in general, the white screwmen found little work. Most agents and stevedores continued to load their ships with black labor. [68]

White longshoremen were hit even harder. "It is a fact that white laborers have been almost entirely driven from the levee, and negroes employed in their places," the *Picayune* observed in May. In mid-April, a small but growing number of white longshoremen broke from the union and applied to stevedores and agents for employment at 40 cents an hour—10 cents below the union tariff. The union expelled one hundred such "scabs"—including longshore leader H.O. Hassinger, who stated with regret that it "may be a virtual confession of defeat to go back to work at the reduced rates, but what can we do? Starvation faces us." By October, the press noted that white labor presented "a pitiful picture of the suffering which has come to a once prosperous class of workmen through the demand for cheaper labor."[69]

Wracked by internal dissent, white screwmen admitted defeat. Union members held their leaders responsible for the Association's hard-line stance against working for stevedores. In late October, they disbanded their powerful executive committee of twenty-one—initially formed in 1882 and composed of the union's top five officers, five delegates to the now-dissolved Cotton Men's Executive Council, and eleven other members of the Association. Two weeks later, the SBA rescinded its order barring members from working for anyone except the Excelsior Co-operative Association and announced that any member could work for any stevedore. This action, the *Picayune* observed, "amounts to a complete abandonment of the contest on the part of the Screwmen's Association, and is the end of the long labor controversy over the loading of ships along the levee."[70]

If everyone recognized that the waterfront crisis of 1894–95 had brought to an end almost a decade and a half of union power in New Orleans, there was no consensus on the deeper meaning of the riots. All observers had witnessed the disintegration of the interracial alliance, increased white hostility toward blacks, and employer efforts to reduce wages and eliminate work rules. But opinion divided on the relative importance of racial versus class issues. Throughout the crisis, white screwmen denied an explicitly racial foundation to their attacks against blacks. Instead, they framed the matter around standards and prior pacts, holding blacks responsible for violating traditional rules (concerning wages, the bale restriction, and the quota system) and degrading a once honorable craft. Their employers adopted an economic interpretation that implicitly denied the claim that they were manipulating racial tensions to their own advantage. By enlarging the low-wage enclave of black labor, stevedores and agents were merely responding to market forces, which alone determined wage rates and conditions of employment. Race, color, or previous condition should play no role in the allocation of work, they held.

Many prominent white southern newspapers supported the contention that the waterfront crisis reflected fundamental economic issues. "The affair will be heralded as a race war [by a certain class of newspapers], and as a proof of the barbarity of the southern whites to the southern blacks," the *Florida Times-Union* contended following the March riots:

> Of course, thoughtful men know it was nothing of the kind. It was not a race war, but a war between strikers and men who had been employed to take their places. If these men had been white, the result would have been the same. Color had nothing to do with it. Just such riots have occurred at Pittsburgh, at Homestead, at Chicago, at Brooklyn, and in almost every northern city. . . .[71]

Less sympathetic to white workers though similarly stressing an economic analysis was the Baltimore *Sun*. The substitution of blacks for whites "may possibly have slightly intensified the bitterness of the strikers," its editors conceded. But

> the trouble was essentially industrial in character and did not originate in race prejudice. Many colored laborers regularly employed in New Orleans and other Southern cities along the water front and in the loading of vessels, and white men work amicably side by side with them all the year round. [The] riot was due to the same general causes that produced the labor disturbances in Chicago last year and in Brooklyn this year.[72]

The problem, according to the Mobile *Daily Register,* was rooted in economics—the "ability and willingness of the negroes to work for lower wages than the white laborers." Even if New Orleans whites killed or drove off the "first batch of cheap [black] laborers . . . there are more to come and come they will until they have absorbed all in New Orleans that is worth having." The law of economy, like the law of gravity, was irresistible, the *Register* lectured. "The side which acts according to that law will win. The opposition must lose." Only a commitment by white dock workers to lower their wages could secure their survival.[73] Of course, in denying the racial origins of the riots, southern white editors sought to shield their region from national criticism. After all, they implied, however unfortunate and unjust the waterfront riots, New Orleans workers were simply acting out a scenario first scripted in the industrial North.

Blacks, the riots' victims, were the one group of southerners who failed to dismiss the racial dimensions of the crisis. Deep racial animosity, reflected in the epithets hurled at them by union white workers—whatever the underlying cause—could hardly elude black riverfront union members. White screwmen denounced their Association No. 1 black allies—without cause—as rate busters and rule breakers, and lumped them with the Association No. 2 men. Drawing their own color line, white longshoremen ostracized and denied work to black longshoremen, for no other reason than to express solidarity with the powerful white screwmen. And for those blacks who had been shot, chased, or beaten by white rioters in October and November 1894 and March 1895, the reality of their attackers' racist anger was particularly poignant.

The national black press concurred. The *Washington Bee* blamed a number of white screwmen for letting their "race prejudice get the better of them" and "inaugurating a crusade against the colored brother" to drive him from the field of employment. The *Southwestern Christian Advocate* concluded that New Orleans' white laboring men discriminated against their fellow laborers for the "sole reason that they are colored."[74] The purpose of the white violence was clear:

> From observation and investigation we are of the opinion that there is a disposition and a determination on the part of the white laborers, skilled or unskilled to supplant Negro labor. . . . There is a combination of forces against the colored man that is very suggestive. Almost every other nationality enters into the alliance which threatens the utter elimination of the Afro American laborer as a part of the industrial system of the Crescent City.[75]

The Indianapolis *Freeman* saw the New Orleans experience not merely as an example of southern racial injustice but as a problem endemic to the nation. "Just now it is the white organized law-breakers and murderers of New Orleans we have to deal with," it noted. But "tomorrow the same condition of bloodshed and outrage is possible, and would be probable, in Indianapolis, Boston and Chicago, at the slightest encroachment of Negro labor on what are held the sacred rights of the province of white labor."[76]

In reality, it is impossible to separate the labor and race issues. Taken alone, neither can account fully for the violent drama of the early depression years. Undoubtedly, by the early 1890s New Orleans white screwmen and longshoremen had fused their long-standing consciousness of the rights of laboring men to a renewed belief in the superior rights of white men. For their part, black workers logically linked notions of laboring men's rights to the right to labor—that is, access to equal employment. The city's biracial union structure once had been predicated upon the assumption that all workers, black and white, gained more from unity than division; white workers eventually rejected that assumption. By the mid-1890s, white longshoremen and screwmen had come to privilege the "sacred rights of the province of white labor" over the rights of all labor. In doing so, they destroyed the entire foundation upon which the waterfront workers' power had rested.

III

From different perspectives, New Orleans businessmen and black residents saw in the Fitzpatrick administration's response to the 1894–95 riots the dangers of Ring rule in New Orleans. Long hostile to the labor movement and the pro-white labor machine, members of the commercial elite again focused their attention on the political arena, eventually taking control of the city government in 1896. Not surprisingly, their success severely circumscribed the political influence of white dock workers. Defeated by military occupation of the New Orleans waterfront, white levee laborers now suffered the depression's effects with few allies indeed. The change in municipal administrations did win the approval of some black workers, who blamed Mayor John Fitzpatrick for complicity in the racial violence. Attacked at the workplace in the early depression years, blacks now sought political solace in alliance with elite whites. But while they welcomed the elite's victory, they quickly learned that black support paid no political dividends in an age of white supremacy. By the end of the century, the position of blacks had deteriorated to a new low. Sharply at odds, black and white workers remained powerless to meet the economic and political challenges that confronted them.

While New Orleans' commercial elite often had denounced machine-controlled municipal politics, the Fitzpatrick administration's handling of both the 1892 general strike and the riverfront violence of 1894–95 gave it new cause for complaint. In each case, employers had turned to the state government and the courts for assistance when local officials failed to act. Following the screwmen's raid in March 1895, representatives of the city's commercial exchanges denounced what they saw as the white screwmen's "tyranny" and censured Fitzpatrick for failing to protect their property. Addressing fellow businessmen, reformer John M. Parker condemned Fitzpatrick's record on labor affairs. When the police proved "unable to prevent the lawlessness which then prevailed" during a streetcarmen's strike that preceded the 1892 general strike, the mayor had done "all he could to demoralize the community." In 1895, Parker held Fitzpatrick responsible for the riots, arguing that he had "encouraged the strikers to every act of violence by his apparent lack of sympathy with the class of men, employers and employees, who were engaged in the controversy with the screwmen." Shipping agent H. Meletta also condemned the "futile efforts of the police to curb the fury of the mobs, and the seeming disposition of Mayor Fitzpatrick to remain in an apathetic state while all this was going on." The only way, Parker argued, "to stop this business"—that is, the Ring's tolerance of labor's lawlessness—"was to go to the fountain head, which was in the city hall, and wipe that out of power."[77]

The commercial elite had not been politically inactive during the first half of the 1890s. In 1890, urban elite opponents of Louisiana's powerful Lottery formed an Anti-Lottery League, temporarily allying with rural Populists and, in one of the state's most bitter elections in years, defeated the multi-million dollar business at the polls in 1892. Critics of the Ring carried out thorough investigations of municipal corruption during Fitzpatrick's terms as mayor and pressed for his impeachment. When the Fitzpatrick administration awarded a franchise to the Illinois Central Railroad, permitting it to lay tracks through the upper-class Garden District, residents of that neighborhood protested by forming an uptown Citizens' Protective Association in May 1894. Successful in their efforts to stop the tracks, many of the Association's members went on to form a larger and more powerful Citizens' League in January 1896.[78]

The Citizens' League represented only the latest incarnation of the commercial elite's ongoing reform efforts. Like its predecessor, the Young Men's Democratic Association, it developed a well-organized and disciplined structure. Ostensibly non-partisan, the League denounced the "official venality and incapacity" which "have fretted public endurance almost to the point of revolution." Its platform, resembling earlier reform documents, called for a reduction in the size of the City Council, clean voter registration laws, a secret ballot, efficient administration of municipal affairs, an enlargement of the police force, and the elimination of "political loafers" from the city payrolls. Its mayoral candidate, Walter Flower, was a wealthy lawyer, cotton merchant, and former president of the Cotton Exchange. For the business elite, the campaign assumed a crusade-like quality. For black Republican leaders who endorsed the League, the campaign signaled their resentment of Fitzpatrick's support of white dock workers the previous year. Copying the tactics of its 1888 predecessor, the League dispatched some 1500 armed citizens to guard polling places and to protect voters and their ballots. Their efforts proved successful. The April 1896 municipal elections swept the League into power, delivering a decisive, if not mortal, blow to the Democratic machine.[79]

The Citizens' League worked quickly to implement structural reforms that would consolidate its hold over municipal affairs. Securing a new city charter from the state legislature, the Flower administration reorganized the municipal government to reduce the influence of the regular Democrats and enhance the reformers' position. The new charter attacked sources of Ring strength by drastically reducing the size of the City Council (whose members, elected by ward, tended to be dependent on the Ring), eliminating elective administrative offices within the executive branch, and granting broad appointive powers over city departments to the mayor. Finally, it created a board of civil service commissioners, appointed by the mayor, with overlapping terms to prevent political manipulation. However, while some of the changes enhanced business power in the long run, these reformers failed to solidify their newly won power. In 1897, defeated machine politicians reconstituted themselves as the Choctaw Club, rebuilt their ward organizations, endorsed state Democratic efforts to disfranchise black voters, and accepted state patronage from Democratic Governor Foster. In 1900, they staged a municipal come-back, putting their mayoral candidate, businessman Paul Capdevielle, in of-fice. With the assistance of state officials, Ring politicians softened civil-service regulations that restricted their patronage powers.[80]

Black voters, who backed the Citizens' League against the Ring, once more learned that elite reformers could be as hostile toward black interests as ma-chine officials were.[81] While their political experiences in the 1880s had driven that point home earlier, the consequences proved far more severe in the 1890s. In the aftermath of the 1896 statewide election, in which regular Democrats violently crushed the rural Populist insurgency, Governor Foster proposed to disfranchise the poor and uneducated of both races. By the end of the century, Louisiana's Democratic leaders had largely accomplished that task, ending an important source of potential political revolt. The elite Citizens' League, ignor-ing its black supporters, joined machine Democrats in endorsing these moves. (The "celerity with which the upper-class elements of that defunct coalition turned on their erstwhile allies and voted to disfranchise them," historian J. Morgan Kousser notes, "underlines the complete opportunism with which they solicited Populist and Republican votes. . . . [T]he New Orleans business-men-reformers' claim to favor 'honest elections' " was nothing more than "pure cant.") The consequences for blacks—and poor whites—were severe. In 1896, the state legislature "reformed" the election laws by barring officials from helping illiterate voters, requiring voters to re-register after January 1, 1897, and empowering registrars and representatives of political parties to purge the voting lists. According to Kousser, the registration act reduced the white elec-torate by more than one-half and the black electorate by 90 percent. The subsequent 1898 Constitutional Convention further restricted political partici-pation, requiring eligible voters to demonstrate literacy in their native tongue or to own property worth at least $300, and after 1900 to pay a poll tax. Although few Louisiana blacks could meet these stringent and discriminatory requirements, whites could take advantage of important loopholes. A grand-father clause permitted poor whites whose relatives had voted before 1867, when a Constitution enfranchising blacks had been adopted, to register within four months of the 1898 Convention's ending.[82]

Disfranchisement was but one, albeit very important, manifestation of blacks' deteriorating status in both New Orleans and Louisiana at the end of the nineteenth century. If the new constitution removed most black voters from the registration roles, it hardly removed the issue of race from the political agenda. In June 1900 the New Orleans City Council rejected efforts to revive the "star car" system of segregated streetcar seating. But only two years later, the white Louisiana state legislature overrode local officials, voting to require that streetcar companies either set aside separate cars or install screens to cordon off black riders from whites.[83] Politics affected blacks in other ways as well. Despite disfranchisement, the *Southwestern Christian Advocate* ob-served in October 1899, "there are few speeches being made during the present [city] campaign in which some one is not crying out to beware of the Negro." Choctaw Club Democrats promised, if elected, to employ only local white labor on municipal public works projects. While "reform" Mayor Flower advocated "home labor without regard to color," his administration's record on black employment inspired little confidence. The "great city of New Orleans gives nothing, absolutely nothing [to the black community] except a very few Negro policemen," complained the black press.[84] This effort to

> shut the Negro out, even from the most ordinary labor, seems to indicate a result, not to say one of the purposes of disfranchisement. He has no redress, his ballot is not feared. Disregard him if you wish, starve him, or strike him down, he is too weak to defend himself either by force or with the ballot.[85]

Over the two decades after Reconstruction's demise, blacks had voted regularly in municipal elections, a fact that Ring politicians and reformers alike had to take into electoral consideration. By the end of the century, disfranchisement had eliminated the need to make even a gesture toward accommodating black voters.

Another index of New Orleans blacks' deteriorating position was the increase in racially motivated violence. "There are few places in this country where the life of a Negro is held in less [sic] contempt than here," the *Southwestern Christian Advocate* concluded in July 1895, months after the waterfront riots. "In most of the instances the murders have been brutal and cold-blooded, the victims going to the death only because they were Negroes. . . . [T]he Negro of New Orleans has surely fallen upon evil times."[86] Racial violence, not merely a product of economic hardship and fierce competition for employment, survived the depression's end. In the first year of the new century, as historian Joel Williamson has written, white mobs in New Orleans unleashed "one of the most serious outbreaks of racial violence since Reconstruction." The focus of the upheaval was Robert Charles. Born and raised in rural Mississippi, Charles was one of thousands of rural, unskilled blacks to migrate to New Orleans during the 1890s and was a supporter of black nationalist Bishop Henry Turner. When white policemen harassed him one evening in late July 1900, he shot one of his assailants before escaping. For several days, Charles eluded vast numbers of police and white vigilantes. Once cornered, he remained defiant, fatally shooting seven whites who were firing at him, before being shot and killed himself. Seizing the opportunity this "outrage" presented, white mobs terrorized black neighborhoods, killing at least a dozen people.[87]

There was nothing new about racial or ethnic violence in New Orleans; whites had targeted blacks regularly during Reconstruction and the depression years in the 1870s and 1890s, and an elite-led mob had lynched several Italians in 1891. Given its severity, what did this particular explosion of white violence mean for New Orleans? William Ivy Hair, in his outstanding book on the episode, concludes that "white reaction to Charles vividly illustrated the hardening of racial attitudes which occurred around that time." Dale Somers, writing in the early 1970s, contended similarly that, by the end of the century, "blacks in the city knew Jim Crow and racial intimidation as intimately as blacks in the countryside." The Charles riot "established the pattern for Negro-white relations for the next half century."[88]

Undoubtedly, the Charles riot served as an important symbol of the changes under way in the Cresent City's racial codes and sensibilities. Racial hatred, discrimination, segregation, disfranchisement, and violence all worked to secure black subordination in the new racial order. But despite the rise of a virulent white supremacy and a general consolidation of a segregationist order across the South, the dynamics of race in New Orleans preserved something of the complexity that had marked the earlier era of the 1880s. Once again, that complexity was best reflected in the attitudes and behavior of the city's labor movement.

While the labor movement was by no means free from anti-black hostility during the 1880s and early 1890s, its racial codes revealed a remarkable flexibility. Often resistant to employers' efforts to separate black and white workers, the Cotton Men's Executive Council, the Central Trades and Labor Assembly, and the Workingmen's Amalgamated Council provided structured forums for mediating trade and racial issues. The interracial cooperation that marked the 1892 general strike, however, had been strained by the strike's defeat, and had vanished by the onset of the economic depression. Heightened white supremacy in politics and the rapid growth of white craft unions at the end of the century codified the change. Yet even in the age of Jim Crow, the parameters of action were not completely rigid. Different groups of workers drew different lessons from the economic crisis and the racial tensions that flourished in its wake.

IV

The great depression severely weakened, but did not destroy, the city's labor movement. "It is awful to think that in New Orleans, which once was one of the best organized labor cities, should now have but three live unions," AFL president Samuel Gompers wrote to organizer James Leonard in 1899. While Gompers underestimated the number of functioning unions, his general assesssment of the labor movement's decline was accurate. In its review of the 1897 commercial year, the *Picayune* observed that ever since the waterfront riots the labor movement had been in "a sadly demoralized condition as far as the wages of the laborer was concerned, as well as the demand for their services." The business depression, low cotton prices, and a relatively small harvest due to a severe drought, had produced a "great many more applicants for work than there was work to do." As a result, even the "most industrious and careful man virtually lived from hand to mouth." Lower wages, irregular work, and a flooded labor market often forced dock workers—whose "savings and earning proved inadequate"—to seek employment in other lines of work.[89]

The labor movement limped through the depression years. A short-lived United Labor Council succeeded the defunct Workingmen's Amalgamated Council and advocated "arbitration in all differences which may arise between employers and employes. . . . [S]trikes should only be resorted to after all other means of mutually satisfactory settlement have failed." The Council's Directory and Index of Unions in 1895 listed 54 locals affiliated either with the American Federation of Labor or the Knights of Labor. In an incomplete index two years later, the Trade Union and Knights of Labor Directory of the *Southern Economist and Trade Unionist,* a new weekly labor journal, listed 56 locals. In its first editorial, the journal predicted optimistically that in the near future, "trade unionism is certain to bound forward with tremendous strides, its future was never brighter than it is to-day, and it is our heart's desire to place this, our native city, not only in line with sister cities . . . but to make it in due time the brightest jewel in the glorious diadem of the American Federation of Labor." While excessive in its rhetoric, the *Southern Economist* was not far off the mark. As the depression gave way to renewed commercial prosperity in the late 1890s, the national labor movement, under the AFL's banner, expanded tremendously. Between 1897 and 1904, the AFL grew by over 360 percent, adding over one and a half million new members to its ranks. While much of that growth occurred outside the South, union membership in New Orleans followed the national pattern, as city workers rebuilt and extended the organizational structure of the pre-depression era. By 1903 the *Union Advocate,* a local labor paper, could boast that "there has sprung into existence so many new unions with such an overpowering membership . . . as to place New Orleans in the very front ranks of cities where organized labor is in the majority."[90]

Union organizing in the South, national AFL leaders believed, offered unique difficulties. Slavery, the " 'peculiar institution' which, until the commencement of the present generation, dominated the social and industrial life of the Southern States," the AFL's official journal, the *American Federationist,* argued, "has hitherto prevented any great expansion of the voluntary organization of labor, and has thereby imposed upon the American Federation of Labor today an immense and most difficult task, a task unparalleled in the history of the world." To take on this "herculean yet delicate mission," the AFL modestly appointed three general organizers—all "well acquainted with Southern conditions." William H. Winn, of the International Typographical Union and a general organizer; Prince W. Greene, president of the National Textile Workers' Union; and L.F. McGruder of the Iron Molders, journeyed across the South preaching their gospel of unionism, organizing new locals, and breathing new life into old ones. "Thus, step by step," one official explained, "do we plant outposts in the enemy's country."[91]

Organizer Winn found more friends than enemies in New Orleans, making his task far less difficult than anticipated. Throughout the summer and fall of 1899, Winn and AFL organizer James Leonard, one of the leaders of the 1892 general strike and subsequent member of the state's labor arbitration board, organized locals of boiler makers, carriage makers, broom makers, tobacco workers, freight handlers, and other workers. In June 1899, they announced plans to form a new Central Trades and Labor Council.[92] In the wake of the 1890s depression, labor's accumulated defeats, and the sharp rise in racial tension, the new federation heralded a spirit of harmony and conciliation between labor and capital. Addressing nearly thirty-five member unions in August 1899, Leonard declared that it was "time to do away with old methods which engendered violence, strife, dissatisfaction, bitterness and poverty, and which clog the wheels of industry and commerce." He called for the

> establishment of the reign of cooperation and good will and amicable discussion between labor and capital. . . . We will not resort to strikes or to any such violent measures. We will not countenance going into sympathetic strikes, but will invoke calm discussion with employers.

The Central Trades and Labor Council, which went into permanent organization on September 10, 1899, differed from the Central Trades and Labor Assembly of the 1880s, the Workingmen's Amalgamated Council of 1892, and a short-lived United Labor Council of the mid-1890s in another crucial respect. In an era of growing racial hostility, the new Council admitted into membership only white AFL-affiliated unions. Racial segregation, which increasingly permeated the city's political and social life, finally had infected its principal labor body.[93]

For James Leonard and the white craft unionists in the Central Trades and Labor Council, the crisis of the 1890s demonstrated that labor's success required abandoning interracial cooperation and adopting a white pro-segregationist stance. Indeed, it appears that Leonard's earlier, outspoken defense of a somewhat militant interracial alliance did little to win him support. For example, although he was president of the Workingmen's Amalgamated Council and a member of the Committee of Five that directed the 1892 general strike, a significant faction within his own Typographical Union Local 17 adopted a craft-first stance and strongly opposed any sympathetic action in that strike. Elected president of the local in December 1892, Leonard encountered growing criticism. In March 1893, union members refused to send him as a delegate to the International's convention, and when he resigned the presidency in May, the Typographical Union withdrew from the Amalgamated Council.[94]

A year and a half later, Leonard turned his energies toward the electoral sphere, spearheading an Independent Workingmen's Political Club. The Club denounced monopolies and trusts and called on working people to "purify the political atmosphere by electing good men to office" and "to place principle above party, patriotism above politics." An interracial movement, the Independent Workingmen offered an alternative to the increasing racism of white dock workers. While white screwmen and longshoremen were renouncing all interracial collaboration, some 25 blacks—including waterfront labor leaders Alexander Paul and Lafayette Tharpe—joined 225 whites at the Club's organizational meeting in September 1894. The black and white delegates nominated Leonard as their congressional candidate. Local machine politicians, however, denied the Workingmen physical representation at the polls, and Leonard lost overwhelmingly to the Democratic party candidate. After their defeat, white Independent Workingmen placed some of the blame on the "race question," which they believed had divided the white electorate unnecessarily; they seriously suggested that the elimination of black voters would enable the working class to further its—and blacks'—aims. The next year, an all-white Workingmen's Democratic Club of the Eleventh Ward advocated the employment of white labor in all city and state contracts. "Good government and white supremacy," the *Picayune* observed, "are the summing up of their tenets." As southern white workers exhibited a more strident racism, Leonard followed the path parallel to one trod by many white Populists during the same years, abandoning an earlier interracialism for a whites-only craft unionism.[95]

This was consistent with the AFL's official position toward black workers at the turn of the century. Federation leaders were, of course, aware of the dangers that a large, unorganized pool of black workers posed to their movement. The "wage workers ought to bear in mind that unless they organize the colored men, they will of necessity compete with the workmen and be antagonistic to them and their interests," Samuel Gompers wrote to John Callahan, a waterfront union leader and New Orleans AFL organizer in 1892. "The employers will certainly take advantage of this condition and do all they can to even stimulate the race prejudice."[96] At the same time, the AFL championed the craft union, whose exclusive structure aimed at restricting membership in the trade, as labor's most effective organizational building block. Adherence to a belief in craft union autonomy limited the AFL's ability to interfere in its affiliates' affairs. Over the course of the 1890s, earlier proclamations on the need to organize workers regardless of their race, creed, or sex, and the Federation's denial of membership to any international union which constitutionally excluded black workers from its ranks, had given way to a policy of tolerating all-white unions. Accepting the intractability of white workers' racism, the organizers of the AFL's southern drive did little to challenge the hardening racial lines. Discrimination and racism, one northern writer observed in 1898, had "entered into the very soul of the workday world, and infected even those workmen who are not organized . . . [W]herever the union develops effective strength the black workmen must put down the trowel and take up the tray."[97]

New Orleans black workers, however, neither took up the tray nor abandoned their unions when interracial cooperation fell victim to the white racism of the depression years. Nor did many heed the advice of religious leaders who counseled the pro-business perspective advocated by Booker T. Washington. In the aftermath of the Charles riot, for example, the *Southwestern Christian Advocate* reiterated its belief that the black worker's "only hope is to keep close to the best white citizens of the city. He has learned that these recognize his worth as a laborer and a well behaved citizen and will stand by him in the time of need."[98] Yet there is no evidence that black riverfront workers shifted their allegiance from their unions to employers in appreciation of the commercial elite's defense of its right to employ whomever it wished. Black waterfront unions survived the depression and attacks by whites, although they were in no position to challenge managerial authority. "[N]otwithstanding the many obstacles encountered," secretary James Porter reported at his union's anniversary celebration at Union Chapel in 1899, the longshoremen were in "splendid financial shape." Of the "many strong Negro trade unions in Louisiana," W. E. B. DuBois' 1902 study *The Negro Artisan* singled out black longshoremen, screwmen, cotton yardmen, teamsters and loaders, round freight teamsters, and Excelsior Freight Handlers. Paralleling the tremendous national growth of the AFL, black membership in New Orleans unions grew substantially after 1899. The DuBois study estimated that New Orleans had some 4000 black trade unionists in 1902; following a year of substantial growth, the Central Labor Union boasted that between 10,000 and 12,000 black members in some twenty unions had marched in its 1903 Labor Day parade. The *United Mine Workers' Journal* put the city's membership in some nineteen black unions at 11,000 in 1904.[99]

Working tirelessly, James Porter deserved much credit for the strength of New Orleans' black unions. As secretary of the black longshoremen's association, he attended the annual conventions of the International Longshoremen's Association (ILA), with which his union had affiliated. Delegates to the International's 1901 meeting in Toledo, Ohio, elected Porter to the post of ninth vice president. Upon his return to New Orleans, he began an ambitious unionization drive "with the purpose in view of bringing about a complete organization of all Water Front Workers of New Orleans and vicinity." Between the 1901 and 1902 conventions, Porter sent nine charter applications to the ILA and established several federal labor unions. Such "wonderful progress" had been made in the Crescent City that the ILA's eleventh annual convention acknowledged that it was "very much indebted" to Porter for the successes in that city.[100]

Porter and other black riverfront workers remained committed not only to their unions but to the national labor movement as well. Many black dock unions affiliated with the newly formed ILA early in the new century, often before their white counterparts.[101] Porter's commitment to the AFL also put him at odds with organizer James Leonard, a man with whom Porter had served on the Committee of Five that directed the 1892 general strike. When Leonard and white craft unionists barred blacks from the new Central Trades and Labor Council (CTLC), Porter spearheaded a new solution to the problems created by racial exclusion in the labor movement. If white locals refused to admit blacks into city-wide organizations composed of AFL-affiliated unions, then black AFL unions would form their own city-wide organizations. Shortly after the creation of the white CTLC in 1899, New Orleans black unionists pressed the AFL for official recognition of an all-black Central Labor Union.[102]

Porter's plan to confederate black unions encountered strong objections from white labor leaders. "The feeling here against a project of this kind is so great that I am afraid it would cause a great deal of trouble at this particular time," James Leonard wrote to Gompers.

> It is a very delicate question to handle, considering the prejudice that exists
> here against the negro. I thought at one time that it could be accomplished,
> but I am very much afraid that the chances are growing worse every
> day. . . . I am in hopes that some kind of agreement may be reached
> whereby both races will act harmoniously together, but as I said before the
> chances are very poor at present.[103]

That was precisely the problem. If white workers were unwilling to work
harmoniously with blacks, then blacks would have to go it alone. Writing to
Gompers in early summer of 1900, Porter complained:

> My members are very impatient and I cannot understand the remark you
> made in your letter . . . to me that there is no use kicking against the pricks,
> and we cannot overcome prejudice in a day. I did not understand that there
> is prejudice where the wages and interest are the same, and can only be
> upheld by concert[ed] action.[104]

The American Federation of Labor's own rules presented obstacles to Porter's demand for a dual central body. While the Central Trades and Labor
Council was free to exclude blacks, the national Federation had no power to
recognize an alternative city central. At the AFL's 1900 convention, however,
Gompers successfully advocated that the Federation issue certificates of affiliation to central bodies composed solely of black unions. New Orleans' black
Central Labor Union finally received an offical AFL charter in June 1901.[105]

By themselves, black unions could do little to secure employment for black
workers in clerical jobs or in the new, mechanized trades, but they did provide
substantial protection to members in at least two important industries—the
building trades and waterfront commerce. The 1902 DuBois study suggested
that New Orleans black artisans, many of whom were property holders, had
grown in number since the Civil War; they were "either gaining or at least not
losing" ground. One respondent noted that New Orleans black artisans who
"receive recognition in their respective trades, are widely employed and paid
remunerative wages."[106] A number of studies in the first decade of the twentieth century found that where blacks and whites performed the same work, they
received the same wages. While the unions could claim some credit, they were
only partially responsible for these blacks' success. In its 1911 survey of American cities, the British Board of Trade suggested that the types of industries
found in New Orleans also contributed. "It is probable that in New Orleans
there is a larger number of white and negro people in very much the same
economic position than in any other American city, or anywhere else in the
world," the British investigators noted, because New Orleans' industries "are
of a kind which employ mainly unskilled or semi-skilled labour, with the result
that both white men and negroes are found doing the same kind of work and
earning the same rates of pay."[107]

Yet such interracial tolerance—if it can be called that—in unskilled and
semi-skilled work had clear limits. "It must not be supposed," the Board of
Trade report concluded,

that social equality of the two races is recognized, even amongst the unskilled labouring population, for on the whole the "colour line" is drawn
with all the strictness common to the Southern States. The two races will
work side by side, but they will not play together, go to the same schools, or
sit together in tramway cars.[108]

Disfranchisement, legalized segregation, and increased racial violence were
certainly the hallmarks of the new racial order; in crucial ways, racism infused
and shaped the twentieth-century labor movement. Most whites repudiated
"social equality," adhered to a strict color line, and refused to identify with
blacks in any way (especially outside of the workplace). Their actions weakened the overall strength of the working class in the post-depression era.

While the white craft unions of the Central Trades and Labor Council
institutionalized segregation and white supremacy within the local labor movement, a very different dynamic operated on the riverfront. The most important challenge to Jim Crow in the city came from the ranks of organized dock
workers. Repudiating the violence of 1894–95, they built an interracial labor
movement that stood alone as an example of black and white solidarity. One
black longshoremen reported to W.E.B. DuBois in 1902 that "in New Orleans, we have been the means of unity of action among the longshoremen of
that port, both in regards to work, wages, and meeting" together. The secretary of the black Central Labor Union argued similarly that "by amalgamation
of organization and through international connections we expect to have the
color line in work removed." Although the color line would remain a permanent fixture on the docks, there was indeed cause for optimism. In 1908, a
commission investigating labor conflicts and port charges concluded that from
the perspective of waterfront employers, "one of the greatest drawbacks to
New Orleans is the working of the white and negro races on terms of equality."[109] Clearly, dock workers by 1908 had traveled a tremendous distance
since the crisis of the mid-1890s.

Two incidents that preceded the formation of the new biracial alliance—the
1900 Charles riot and the 1901 longshoremen's and trimmers' strike—cast
some light on the process of reconstitution. In the first, white riverfront workers apparently took little part in the widespread violence aimed at blacks in July
1900, in sharp contrast to their role in the 1894–95 riots. The *Times-Democrat*
reported only one incident implicating riverfront workers in anti-black violence. On July 27, a group of whites, whom the paper presumed to be longshoremen or screwmen, pursued and wounded a black man along the levee in the
French Quarter; shortly thereafter, a group of Italians, who had no affiliation to
the dock unions, killed the victim. Other "hoodlums"—not dock workers—
also seriously wounded two black riverfront workers on the levee near Jackson
Avenue. During the height of the crisis, two white longshore labor leaders
accompained the Commissioner of Police and Public Buildings on a tour of the
riverfront, where they found the men "busy and thinking little of mingling with
the mobs"; according to the *Picayune,* dock union leaders assured the mayor
that no union had participated in the ongoing "lawlessness."[110]

White longshoremen, demonstrating opposition to the mob violence, also quit work. At least half of all black dock workers remained at home during the worst rioting; those who went to work remained apprehensive, and eventually stopped work early. Although the white longshoremen "do not feel kindly towards the negroes," reported the *Picayune,*

> but for their own self-respect they assured the negroes working near them that they should not be molested. There has always been more or less feeling toward the negro on labor questions which have arisen between them, but the dominant spirits among the white longshoremen were busy yesterday impressing them with the fact that it would greatly damage their cause if they allowed the labor questions to influence them into any action with the disorderly hoodlum element making itself so obnoxious to all the good people by its outrageously wanton conduct.[111]

Based as it was on pragmatic, strategic reasoning, such behavior did not signify a renewed commitment to interracial collaboration or solidarity. It was, however, a first step, and perhaps laid some groundwork for subsequent negotiations. At the very least, it indicated that all white workers did not fully embrace the ethos of the new, increasingly hostile racial order.

The second step in the reconstitution of the waterfront alliance took place a little more than a year later. An improved economic climate after the turn of the century provided the context for this experiment in renewed cooperation.[112] In September 1901, white longshoremen struck to increase the wages of the approximately 250 grain trimmers in their union and to reestablish the position of the union foreman. The fact that trimmers' work was "very trying on the health of the men," the union explained, ought to be a factor in wage determinations. Moreover, trimmers

> frequently had to wait for five hours in idleness before getting a couple hours of work. At times, the men had to go from one point to another to load one vessel, and the time lost in going and coming was quite an item, involving considerable cost. Often the trimmers would start on a vessel at Chalmette, continue on it at Stuyvesant Docks and finish at Southport. [Thus] a week would be consumed in getting out three days' work. For the times lost, there was no compensation.[113]

The white longshoremen announced that they would handle no cargo of any stevedore or shipping agent who had not signed the new contract by September 12. Alfred LeBlanc, the Harrison line's agent, objected most strongly to the union foreman issue, insisting on his right to hire whomever he wanted, union or non-union. Leyland line agent Sanders protested the wage question, arguing that his firm, with a regular sailing schedule, could not afford to pay the same rates that ships coming irregularly did. In general, stevedores and agents feared that the trimmers' wage demand would trigger new demands by other longshoremen. Most rejected the union's ultimatum.[114]

Reminiscent of the 1880s, white longshoremen pledged to get "every class of laboring people into the Federation of Labor and [prevent] them from taking the places of the strikers." Establishing a close working relationship with black longshoremen was their first priority. When the black union agreed to participate fully in the strike, a joint committee assumed control over its direction and established its headquarters in the office of the all-black Central Labor Union. On September 12, some 1700 black and white longshoremen and 250 grain trimmers struck all employers who refused to sign the new wage scale. Longshoremen secured the support of other waterfront workers as well. Fearful that employers might utilize black roustabouts to load and unload ships, an interracial committee of longshoremen presented its case to the riverboat hands, who promised not to break the trimmers' strike. Both the black Central Labor Union and the white Central Trades and Labor Council endorsed the strike; president Robert E. Lee of the white CTLC promised the strikers moral and financial support, declaring that all of organized labor "would help to win this battle for the men." Union ship carpenters backed the strikers by withdrawing from the ships of two key companies, the Harrison and Leyland lines. While meeting in separate halls, both black and white screwmen also agreed to perform none of the longshoremen's work. By September 17, all of the port's small firms had signed the new contract.[115]

But the revival of interracial and inter-trade collaboration threatened to reverse the victories won by large employers during the 1890s, and several of the largest steamship companies—the Harrison and Leyland West Indian and Pacific lines—remained adamant. Agent Sanders resorted to strikebreakers to implement a new, innovative scheme to by-pass the longshore unions. Promising no discrimination against blacks or even union men, Sanders proposed to hire 250 men, and pay them between 30 cents and 40 cents an hour (well below union rates), but to guarantee them $50 a month—in place of the existing system of irregular longshore work. But getting men and getting them to work during a strike, Sanders discovered, were two different matters. Some 150 union strikers were on hand to greet twenty new laborers hired by Sanders, persuading them, as well as forty to fifty Harrison line strikebreakers, to abandon work and join the union. This "intimidation," as Sanders called it, continued unabated.[116]

Unconvinced by the agents' case, machine Mayor Paul Capdevielle refused to involve himself seriously in the matter. Indeed, his sympathies appeared to lay with union workers. From what he had seen and heard, he noted striking laborers "were trying to do just what was right." LeBlanc and Sanders, deserted by fellow employers and unable to hold onto their strikebreakers or secure adequate police protection, surrendered on September 22. The contract, signed before the joint executive committee of the black and white longshoremen, represented a complete union victory.[117]

Between 1000 and 2000 union men resumed work on the docks. "The conditions on the levee are now in the most peaceable condition that they have been for a long time," one labor leader observed.

> White and colored laborers are working in harmony. They realize that opposition to each other destroyed their power at the first big strike previous to the recent . . . trimmers' walkout, and they have recently been combining

their efforts. Every important action is indorsed by the coordinated organizations on both sides, and there are plans being formulated for an equitable distribution of the work, so that neither side shall have any complaint.[118]

A collaborative spirit extended even to the ranks of the cotton screwmen. Ever since the winter of 1895, black screwmen had performed most of the work for the Harrison line, monopolizing the work on the forward hatches of its ships. Toward the end of the 1901 strike, Harrison agent LeBlanc attempted to manipulate racial divisions once again by offering work on forward hatches to the white screwmen. Despite deep animosity between the two unions, the white workers rejected the company's offer, declaring that black screwmen were entitled to retain their positions. At a subsequent meeting, white screwmen agreed unanimously that "there would be nothing done that would cause bad feeling between the two races working on the levee." Had there not been "cool judgement," the press noted, "the incident might have opened up an avenue to trouble."[119]

Waterfront unions cemented their informal working alliance with the founding of the Dock and Cotton Council in early October 1901. Eight unions—the black and white cotton screwmen, black and white longshoremen, black teamsters and loaders, the black and white cotton yardmen, and the black coal wheelers—joined the new organization. With the exception of the coal wheelers, all of the Dock and Cotton Council's members had participated in the Cotton Men's Executive Council of the 1880s and early 1890s. As its first president, Dock and Cotton Council delegates elected Judge James Hughes of the white longshoremen; as first vice president, Isom G. Wynn of the black yardmen; and as secretary, veteran black labor leader James Porter. Under the Council's rules, six delegates from each constituent union attended Council meetings, and each union had equal vote in all decisions. While white workers always held the Council's presidency, blacks occupied important roles. It was not uncommon, for example, for a black officer to chair a Council meeting in the absence of the white president.[120]

Conditions unique to the world of waterfront labor help to explain the reconstruction of the biracial alliance in the age of Jim Crow.[121] The largely unskilled nature of the work made this recovery of the legacy of interracial collaboration imperative. While, in theory, familiarity with the techniques of loading, unloading, coaling, rolling, or carting cotton or round freight gave experienced workers important advantages over green hands, employers could easily replace these workers in practice. Most individual unions, representing specific groups of workers in a wide array of waterfront occupations, possessed insufficient power to control the labor supply or prevent unilateral employer actions (the white screwmen being an obvious exception). However, alliances among unions and, most important, between black and white unions, could reduce job competition, dramatically increase a union's resources, and tip the scale toward labor. The experience of the 1880s had illustrated this plainly. Pacts between black and white branches of the yardmen, longshoremen, and to a lesser degree screwmen, had mediated racial tensions; work-sharing agreements and joint rules had reduced the ability of employers to exploit racial tensions. The Cotton Men's Executive Council had institutionalized a solidarity linking all cotton waterfront workers in a powerful circuit. From 1880 to 1887, and to a lesser extent, to 1894, formal and informal alliances among groups of waterfront workers kept wages high and union work rules strong. The collapse of the inter-trade and interracial alliance after 1894 presaged the collapse of union power.

The strength of New Orleans black waterfront unions also appreciably narrowed the options available to white waterfront unions. White screwmen and longshoremen realized that, unlike whites in the skilled crafts, they were powerless to enforce a whites-only policy on their employers. The economic depression and the upheavals of 1894 and 1895 had underscored their vulnerability. Neither black unions nor aggressive employers had any intention of permitting whites to drive black workers from the riverfront. Moreover, from 1896 to 1900, the Flower administration was sympathetic to business interests and would not turn a blind eye to such attempts, as the earlier Fitzpatrick administration had done. Had whites again sought to eliminate black waterfront workers, they would have found the task impossible.

Dock leaders came to appreciate the real difference that the alliance made with regard to conditions and wage rates. As the rise of union power in the 1880s had depended in large part upon inter-trade and interracial alliances, the fall of waterfront union power in the 1890s had resulted from the collapse of those alliances. Shipping agents and contracting stevedores would later look back on the late 1890s as a golden age of high productivity and profits; black and white workers recalled it as a dark age of fierce competition, degraded working conditions, and low wages. In the early twentieth century, union leaders of both races pointed to the riots of 1894–95 as a turning point for the worse for all longshore laborers. The lessons of division were evident enough. With the return of commercial prosperity and the election of a machine mayor more sympathetic to white working-class voters, longshoremen initiated a process that quickly led to the reestablishment of joint work rules, unified wage rates, and interracial and inter-trade collaboration. At the start of the new century, waterfront workers would relearn the lessons of solidarity.

"In its heyday" in the early twentieth century, radical Covington Hall recalled to researcher Abram Harris in 1929, the Dock and Cotton Council was "one of the most powerful and efficient Labor Organizations I have ever known."[122] Its formation in 1901 resurrected biracial and inter-trade alliances in the age of segregation. Under its auspices in the ensuing years, riverfront workers eliminated the dangers of a racially segmented labor market and successfully met many new challenges posed by changing technologies and renewed employer assaults. More than ever, the issue of work rules and job control assumed a central place in the conflicts between workers and their employers. During the first decade of the new century, the docks of New Orleans constituted a battleground not between blacks and whites but between social classes with very different visions of the economic order and the role of unions and workers within it.

5

To Rule or Ruin:
The Struggle for Control in the Early
Twentieth Century

In early 1908, the Louisiana General Assembly launched an investigation into the manifold factors adversely affecting New Orleans' commercial prosperity. Initially unfamiliar with union practices on the city's waterfront, the state legislators on the Port Investigation Commission were disturbed by the evidence of interracial collaboration that they found. "[W]e are practically under negro government," exclaimed state senator and Tensas Parish cotton planter C.C. Cordill, upon learning of the power of the joint executive committees of both the black and white longshoremen and screwmen. At another hearing, he complained: "This is the worst nigger-ridden city in the South." Racial practices were not the only subject of condemnation. Echoing Cordill's theme, steamship agent Alfred LeBlanc denounced New Orleans as the "worst labor-ridden city in the country." Commission members had no difficulty discerning the connection between the two observations. In its final report to the Louisiana state legislature, the Port Investigation Commission concluded that racial equality on the docks was the "fruitful source of most of the trouble on the New Orleans levee" as well.[1]

If equality was largely responsible for "trouble," it was because a racially united labor movement was considerably stronger than a divided one. And it was the degree of union strength that determined whether riverfront workers could resist successfully employer efforts to intensify, or "rationalize," the labor process. In the opening years of the twentieth century, waterfront labor's most important conflicts centered not on race but on on-the-job control. Indeed, the docks of New Orleans resembled a battleground on which employers and workers continually struggled to define the character of labor relations. From 1901 to 1908 the port experienced repeated strikes by the most numerous and powerful riverfront workers, screwmen and longshoremen, that paralyzed the flow of commerce, while less dramatic (and less successful) walkouts, involving round freight teamsters, cotton teamsters and loaders, railroad freight handlers, and coal wheelers, tied up specific sectors on the docks. Issues ranged from the parameters of managerial authority, work rules that sharply limited exploitation, and wage rates, to the hours of work and meals. The contractual agreements that ended the waterfront strikes represented not substantive resolution of fundamental differences, but merely truces in the enduring struggles for control.

I

The conflict over work rules on the New Orleans docks was a local variant of a much broader phenomenon involving American workers and managers in the late nineteenth and early twentieth century. During the nineteenth century, skilled workers often exercised considerable collective control over the performance of their labor, based upon their irreplaceable skills and knowledge of the work process. Toward the end of the century, employers intensified their drive for dominance. Technological innovations that eliminated previously indispensable skills, the introduction of scientific management, and the homogenization of labor all offered the possibility of reducing or eliminating the power of the skilled craft worker and securing managerial authority for supervisors and owners. Skilled union craftsmen engaged in running battles with the managers over work rules defining the scope of labor's power, the nature of managerial authority, and the amount of work and the conditions under which it would be performed. Of particular concern to employers intent upon maximizing productivity was the restriction of output, or stint. Unions placed specific limits on the rate of production, or as labor economist John R. Commons defined it in 1904, the "quantity of output of the workman in a unit of time" affecting the "intensity of exertion while at work." Union craftsmen defended such restrictions as "protection against driving and killing a man," a necessity to prevent "demoralization" and "sweating" and a means to ensure health and dignity. In contrast, employers attacked such practices as inherently unfair, uneconomical, and unnecessarily restricting progress.[2]

Skilled workers were not alone in their concern over work rules and job control. As an examination of the New Orleans riverfront demonstrates, unskilled longshoremen and the more skilled cotton screwmen, like skilled craft workers, experienced much of what historian David Montgomery has called "functional autonomy" on the job. That is, they exercised considerable control over how they executed their assigned tasks, how much work they performed, and how many workers were assigned to each task. During periods of union strength—from 1880 to 1894 and from 1901 to 1923—union work rules very clearly delineated the conditions under which labor would or would not be performed. Not surprisingly, New Orleans waterfront workers, like their skilled counterparts, found themselves subject to employer efforts to cut production costs through a reorganization of the labor process after the turn of the century.[3] Key groups of dock workers not only resisted the assault on their traditional sources of power but also succeeded in expanding that power.

Reaching back to the early 1880s, the conflict over work rules on the New Orleans waterfront was rooted in employers' concerns over cost, efficiency, and managerial principles. During that decade, dock workers implemented strong conference rules limiting managers' authority; in the 1890s, employers defeated the unions, reversing their gains. By the turn of the century, the changing character of the shipping industry heightened employer concern with the costs and organization of labor. Competition between leading exporters intensified dramatically, undermining an earlier stability produced by economic collusion. During the 1890s, the three Liverpool firms that dominated the New Orleans trade—the Harrison line, the West India and Pacific Steamship Company, and the Leyland line—had abided by cargo allocation and freight rate fixing agreements. The 1900 merger of the Leyland and the West India and Pacific lines, and its 1901 purchase by U.S. railroad magnate J.P. Morgan, undermined the carefully engineered regulation of competition. Economic consolidation, then, increased competitive pressures after the turn of the century, forcing shipping agents to find new ways to increase volume and cut costs. More important, however, were longer-term technological and structural changes in the ships themselves. The new steam-powered ships were vastly larger than the older sailing vessels they replaced. With steel girders supporting upper decks, builders eliminated the rows of pillars in the ships' holds, permitting the storage of considerably more cargo.[4] The economic pressures generated by new shipping technologies, size, and competition levels meant that speed, not skill, was the most important variable in the economic equation.

Any attempt to reduce in-port turn-around time in New Orleans encountered one basic obstacle—the so-called aristocrats of the levee, the cotton screwmen. Under the methods that prevailed during the nineteenth century, jackscrew-wielding screwmen forced as many bales of cotton as possible into the holds of steamships. It was an essential function, for the careful stowing of cotton could mean the difference between an agent's turning a profit or a loss. Now, screwmen were an expensive, often dispensable group of workmen. These "floating warehouses now on the sea," one agent remarked of the new steamships, "require modern methods" of loading and unloading cargo. Many agents of large firms no longer had their cotton screwed at all. Rather, workers loaded the "fleecy staple" by hand. "For the past six or eight years," another agent explained in November 1902,

> during six months out of the twelve, the use of screws in the port of New Orleans has been discontinued on many steamers.
>
> The rates of freight during the last two years has reached such a low basis that cotton screwing by the regular liners has become practically a thing of the past. There is absolutely no doubt whatever that the days of screwing cotton into vessels are numbered, if not already in the past.[5]

Technological and engineering changes gave rise to simpler methods of loading and unloading cotton, but screwmen resisted all efforts to reduce them to the level of longshoremen. These men, agents complained in 1902,

> do not appear to recognize that this is an age of machinery. They do not appear to have sufficient foresight to see that the days of the screwmen are fast pulling away and that they should be thankful that they are alive at all.

In a few short years the work of the old time screwmen will have been extinct altogether. There will be no further need for his services. He will be a relic of the past. These men, in view of this fact, should be tickled to death with what they have and should try to foster those conditions . . . as long as possible, instead of trying to drive themselves out of business still faster by heaping impossible demands upon the men who conduct the commerce of this port.[6]

New Orleans agents did not have far to look for examples of higher productivity and stronger managerial control. In Galveston, Texas, most agents had abandoned cotton screwing and required their workers to stow the bales in the holds of ships as quickly as possible. "The stuff is rushed from the cars across the docks into the ships at a rate that would make an old-timer screwman's hair turn grey," the *Picayune* reported in October 1902. Galveston's screwmen were "disposed to complain that theirs is almost a lost occupation." New Orleans-based businessmen anxiously followed the notable increase in the shipment of coastwise cotton in that Texas city, even making "parity with Galveston" a rallying cry in their public relations campaign against the screwmen.[7]

To achieve that parity, New Orleans' four dominant shipping lines designed a technical solution to diminish the screwmen's strength. In 1902, they introduced the "shoot the chute" system of loading, designed to maximize speed and minimize skill. Under the "shoot the chute" system, screwmen calculated, a crew of four or five men would have to stow anywhere from 400 to 700 bales of cotton a day by dropping them into the ships' holds at a great speed. This dangerous process, they complained bitterly, "annihilated all regard for the number of bales to be stored," undermining their power and control over the performance of their work. The white screwmen reasoned accurately that they could neither abolish the "shoot the chute" system nor regulate the number of bales on their own. Employers could counter any protests by simply hiring more black screwmen. Black screwmen had an additional reason for welcoming an alliance with whites. Agents had failed to provide them with half of the available jobs, as had been promised. An alliance with whites would accomplish what an agreement with employers did not.[8] The key to success, for both black and white screwmen, lay in reducing racial divisions and preventing a revival of a split labor market.

Relations between the two screwmen's unions had remained strained since the crisis of 1894–95 destroyed white restrictions on black employment. For the rest of the decade, wage differentials remained a "bone of contention" between the organizations. The "white association held aloof from the colored, and all relations were of a distant nature." That policy, the *Picayune* observed, "was found to be a big mistake." Experience also had taught black workers the costs of division. Black screwmen were "tired of being used as an instrument to starve our brother workmen, the white men, and who have the same right to live that we have," explained A.J. Ellis, the black screwmen's delegate to the Dock and Cotton Council, in April 1903. "We stood for it for eight years; got nothing for it but abuse, and depreciation of manhood."[9]

Employers' efforts to abolish the screwmen's trade brought those black and white workers into a firm and enduring alliance. Following the successful trimmers' strike and the formation of the Dock and Cotton Council, the white and black screwmen agreed to abide by a uniform wage scale in April 1902.[10] This was only a beginning. Under the Dock and Cotton Council's auspices, white and black screwmen entered into an unprecedented agreement the following autumn. Under the final "amalgamation" compact (as they called it) adopted on October 29, 1902, they pledged both to act jointly with regard to all demands and to share equally all available work. Their purpose was not "to stop work and sit down in ships and draw money that was not earned," Ellis explained. "But we did . . . [so] to stop men from forcing us to starve other men and ourselves as well. . . . I would like the public to remember there is a spirit that exists to rule or ruin, but not among the Screwmen, white or black." To prevent employers from playing one race off against the other (a frequent union complaint), the new half-and-half compact provided for the integration of work crews, with each consisting of two blacks and two whites, headed by either a white or a black foreman. Both unions promised to enforce the new system strictly.[11]

Their alliance cemented, the screwmen next drew up demands. First, they insisted that foremen hire only half-and-half gangs of union men and distribute them evenly in every hatch abreast of each other. Second, gangs would take orders only from a designated walking foreman, under penalty of a $25 fine per gang for each offense. Third, the union screwmen would recognize no foreman's authority unless he was a member of either the black or white union. ("All the screwmen are asking for," explained Dave Taylor, the white chairman of the Joint Executive Committee of the two unions, "is a union foreman who knows his business. We have been subjected to the dictates. . . . of men who know nothing about the handling of cotton . . . who hardly know what a bale of cotton looks like, let alone how to handle and stow it.") Fourth, the members of the white Screwmen's Benevolent Association No. 1 would work on the starboard side of all ships arriving in port and the members of the black Screwmen's Benevolent Association No. 2 would work on the port side. Last, any gang dispatched to terminal points outside the city—such as Chalmette, Westwego, or Southport—which found that there was no work to do would receive one-quarter of a day's pay for the time lost, unless they were prevented from working because of rain.[12]

Stevedores, shipping agents, and the press immediately recognized the new rules' threat to employers' prerogatives and pocketbooks. The screwmen's high wages, they argued, attracted a "large surplus of labor to the river front," and the anticipated restriction of output would spread out the existing work at the employers' expense. "If [the union foreman] rule is to be strictly enforced to the letter with all the hidden things that it may conceal," one leading agent declared, "then we might as well close up the port of New Orleans and get off the map." Members of the New Orleans Steamship Conference, the principal association of the largest waterfront employers, shared that opinion, moving to oppose any attempt by their workers to regulate or reduce the number of bales handled.[13]

In early November 1902, the screwmen took action to implement their new rules. Their target was the four largest companies—the Harrison, the Elder-Demster, the Leyland West India and Pacific, and the Head lines—which together handled between 50 and 70 percent of the port's cotton. Least dependent upon the screwmen's skills, these firms had inaugurated the "shoot the chute" system, performed their own stevedoring, employed foremen of their own choosing, union or non-union, and had rejected their workers' demands. On November 3, the unions struck eleven of their ships currently in port.[14]

From the start it was clear that not all employers were affected equally by the strike and, in the end, the division between large and small employers proved critical to the outcome of this (and future) struggle. Operators of smaller firms reasoned correctly that the screwmen's goal was the reduction not of screwed cotton, which they relied upon, but of hand-stowed cotton, which affected only the large companies. For small firms still dependent upon the screwmen's skills, any attack on union rules would derive less from economic imperatives than from ideological ones. But at this point in the cotton season, they concluded that a full-scale anti-union assault would prove too costly an undertaking. Many smaller shipping lines agreed to the unions' demands, on the condition that workers make no attempt to reduce the number of bales screwed daily; in exchange, the screwmen continued to work on their ships during the strike. The strike "is an effort on the part of the four big companies to get the smaller companies to help fight their troubles," one stevedore explained, "but we were not going to do it." Thus, the Steamship Conference failed to secure the sustained support of the more numerous smaller lines (some of whom were not members of the Steamship Conference, while several others quit as a result of the October crisis). Without allies, and with large quantities of cotton rapidly piling up on the docks, the "big four" sacrificed principle to expediency and gave in.[15]

Although assured that the new rules involved no reduction in the amount of work to be performed, as soon as the screwmen resumed work, employers suspected a conspiracy to limit the amount of cotton handled. "Gangs do not handle precisely the same number of bales of cotton per day," observed one prominent stevedore, "and they vary it by a bale or two, so that it cannot be proven that there is a fixed limit. [But] the limit is there all the same." Unable to afford another work stoppage at the height of the cotton season, agents postponed punitive action. But they did not forget the issue and prepared to resolve it once and for all when the shipping season ended.[16]

New Orleans steamship agents and contracting stevedores mounted a co-ordinated attack on the screwmen when the cotton shipping season drew to a close in early April 1903. If the immediate source of their hostility was the restriction of output made possible by an interracial alliance, they also rearticulated a long-standing (if intermittently applied) belief in managerial prerogatives. The controversy centered squarely on the control over the labor process and the definition of "a fair day's work." In a statement of its principles, the Steamship Conference concluded:

> 1)That the employer always has the right to direct where the employees shall work.
>
> 2)That the employer's orders must be carried out, no matter whether the employer or his agent is a member of either association [union] or not.
>
> 3)That the Conference [members] are the only parties to determine the character of the stowage of the cotton, and that consequently, the employee never can have the right to say what shall be the character of the work done, and
>
> 4)That the employer has a right to expect for the wages agreed upon as much work as can be reasonably done. . . .[17]

Recognizing that the work-sharing agreement between black and white workers had shifted control decisively toward the unions, the Steamship Conference insisted upon a restoration of conditions that had prevailed before the November agreement. That is, it demanded that screwmen repeal their half-and-half compact, take orders from stevedores as well as foremen, accept the reinstatement of the "shoot the chute" system, impose no limit on the amount of work performed, and observe all other conditions of the old contract.

If the immediate cause of the anti-union offensive involved conditions specific to the docks, the agents' outlook also tapped a much broader challenge to the American labor movement by employers across the nation. The American Federation of Labor had grown dramatically in the years immediately following the depression of the 1890s, its membership rising from 447,000 in 1897 to a peak of 2,072,700 in 1904. Trade agreements and experiments in labor-capital cooperation had failed to stabilize labor relations, and by 1903, increasing numbers of employers were abandoning whatever tolerance they may have exhibited toward unions. Organizations such as the National Metal Trades' Association, the National Founders' Association, various employers' associations and citizens' alliances, the American Anti-Boycott Association, and the Citizens Industrial Association launched an open shop drive that halted the AFL's growth by 1904; for the rest of the decade, union membership levels remained stagnant. The ideological and organizational issues came into sharp focus when New Orleans hosted the 1903 convention of the National Association of Manufacturers (NAM), roughly one week after the Steamship Conference announced its April ultimatum.[18] The NAM's declaration of principles condemned labor boycotts, strikes, and "illegal acts of interference with the personal liberty of employer or employee," while Dayton, Ohio, anti-labor activist John Kirby denounced labor unions for abetting murderers and rioters. New Orleans employers did not take their anti-union animus that far. In their public insistence upon management's right to control, agents and stevedores demonstrated far more interest in increasing productivity than in destroying unions altogether.

The agents' April ultimatum elicited a hostile response from the screw-men. At a large joint meeting of the two unions, black and white workers found the pre-November conditions "so objectionable and so inimical" to the unions' best interests that it would be impossible to return to them. On April 6, agents and stevedores declared a lockout of the port's 2000 screwmen.[19] From the start, the city's labor movement extended support to the workers. "No body of self-respecting men would think of submitting to the arrogance of these self-constituted autocrats," declared the *Union Advocate,* the official organ of the Central Trades and Labor Council. The Dock and Cotton Coun-cil, the white Central Trades and Labor Council, and the black Central Labor Union all endorsed the screwmen's cause. Most important, though, was the direct support of the other levee unions, particularly the longshoremen. Three times that spring and summer the Steamship Conference offered the black and white longshoremen's unions the work of the cotton screwmen; three times the longshoremen rejected the offer.[20]

Over the slow summer months, a series of arbitration efforts failed to resolve the deepening conflict. In mid-April, both the Steamship Conference and screwmen's unions appeared before an advisory committee of the New Orleans Cotton Exchange, which had offered to mediate existing differences. On April 18, after four days of hearings, the Arbitration Committee issued its findings. Since "perfect harmony now exists between the members of the white and colored associations," the Committee accepted the equal division of work between the black and white unions and the integration of work crews. At the same time, it recommended that 1) a fair day's work for one gang consist of 85 bales tightly screwed, 95 bales snugly screwed, 175 bales hand stowed, and 225 loosely hand stowed; and 2) that the shipping agents, as employers, should have the right to employ union or non-union walking fore-men. The shipping agents accepted the compromise measure, while the screwmen rejected it, offering instead a counterproposal to stow 90 bales of cotton with screws and 150 bales by hand, all other rules remaining the same. The agents and stevedores rejected the counterproposal without consider-ation. Subsequent mediation efforts by the Cotton Exchange, the New Or-leans Board of Trade, and the stevedores failed to resolve the deadlock.[21]

The crisis came to a head in the fall of 1903. Following strikes in Septem-ber by longshoremen and Southern Pacific freight handlers, the Steamship Conference refocused its attention on the port's most skilled workers and the still-unresolved question of the definition of "a fair day's work."[22] There could be no permanent peace on the riverfront, the representative of the Elder-Demster line insisted, "so long as the members of the screwmen's associations are permitted to arrogate to themselves the right to select their work." In a declaration of war, Steamship Conference chairman Alfred LeBlanc an-nounced, "We have our backs against the wall. We propose to fight. . . . Not another man of them will ever work in our ships under those conditions." The shipping interests had finally realized, the *Times-Democrat* reported, that "unless the matter was fought to a finish now, the untold agitation and dis-putes and trouble would arise in the future."[23]

Approximately 400 screwmen reported to work on October 1, refused to increase their limit, and found themselves locked out. In their place, the Steamship Conference put sailors and twenty additional non-union men to work stowing the first bales of cotton on October 6. Two days later, 200 white strikebreakers arrived on special Illinois Central rail cars from St. Louis. The agents' plan involved the centralization of all cotton operations at several large cotton wharves. The Texas and Pacific, Illinois Central, and the Southern Pacific railroad lines all promised to work in close cooperation with the agents and stevedores. When negotiations in the office of New Orleans mayor Paul Capdevielle failed to resolve the crisis, employers ended their armed truce on the levee and, on October 10, put large numbers of strikebreakers to work on a Leyland line steamer. Taking few chances, Leyland managers secured a restraining order in U.S. District Court against the screwmen and their sympathizers. As added security, 165 U.S. marshals and New Orleans policemen were on hand to keep back the 3000 dock workers who congregated along the streets behind the wharves.[24]

Despite an initial show of strength, the Steamship Conference's strategy collapsed only one day after the strikebreakers began loading. The sharp division between large and smaller shipping lines, which had produced an early settlement in the 1902 screwmen's strike, reemerged. Not all agents favored the "big four's" leadership. Some smaller agents expressed satisfaction with the screwmen's proposal to screw 85 to 95 bales with jackscrews; the larger agents' sticking point—the number of hand-stowed bales—did not affect them significantly. The "best thing to do was to let labor alone"; one local steamship agent argued. The demands for the vast increase in hand-stowed cotton, he noted critically, were "merely the efforts of the merger companies and were intended to reduce levee laborers to $1.50 [a day] like railroad employees."[25]

Smaller agents had acquiesced reluctantly to the demands of the "big four," but they eventually broke ranks over the question of importing strikebreakers. In early October the large companies unilaterally moved to hire non-union men. Agents of the smaller firms began voicing new doubts by the second week of the lockout. Some declared that, despite their earlier endorsements, they would never employ the imported strikebreakers on their ships.[26] When a number of stevedores openly refused to employ the non-union, out-of-town men, the Steamship Conference expelled them from the organization. Nine smaller agents then bolted the Conference, as they had done the year before, leaving only the "big four" in that body. Informing the mayor that they would "no longer sustain the contentions" of the four larger lines, they contacted the screwmen and offered to settle on the basis of 90 bales screwed and 160 bales hand stowed. At their largest mass meeting in months, 1500 black and white screwmen debated, then accepted, the new limit. Isolated and lacking support in a city growing increasingly tired of the economic warfare on the docks, the four large shipping lines admitted defeat and signed the agreement. The following day, the levee bustled with activity, as the screwmen and their riverfront allies returned to work; only the "cliques, combines, the pools, [and] Napoleons of commerce" were out of sight. After a year of continuous struggle, the screwmen had won a decisive victory.[27]

The conflicts over work rules and managerial authority that raged from November 1902 to October 1903 and the renewed interracial collaboration between levee laborers marked a decisive turning point in the history of New Orleans docks. William J. Kearney, stevedore for the Harrison line, captured the transformation in waterfront power relations. "I don't consider that I am in any way responsible as an employer, because the men take everything out of my control," he testified before the Port Investigation Commission in March 1908. He complained that the stevedore "had absolutely no control over his men; that the foreman of a gang was simply a figurehead and that a conference committee, consisting of twelve whites and twelve negroes, had the absolute say as to what sort of work the men should do." Commission member Parkerson expressed incredulity: "These niggers and whites rule the whole roost."[28]

The events of 1902–03 changed not only relations between employers and workers but relations between black and white dock workers as well. Less than a decade before, New Orleans had experienced a racial and labor crisis on the docks. The riots of 1894–95 involved the repudiation of a fragile and unjust interracial structure, the destruction of complex work rules, and the plummeting of wage rates. Less than a decade after the violence, the unions entered into an unprecedented alliance that enabled them not only to resist employer efforts to reorganize the waterfront labor process and eliminate their trade, but to wrest control of the labor supply and process for themselves. There was little question that by late 1903, cotton screwmen had regained their position as aristocrats of the levee. The "amalgamation" of the white and black unions stood in stark relief against the consolidation of Jim Crow and the legacy of racial exclusion and violence in the screwmen's ranks. Once again, waterfront workers learned that their own well-being on the job was wholly dependent upon interracial collaboration.

II

Just as screwmen were not the only target of managerial efforts to weaken union control and enhance managerial authority, waterfront workers in less skilled jobs struggled over working conditions and work rules. But the degree of control they exercised corresponded closely to their position in the hierarchy of labor on the levee. With their skills still required by many smaller agents and stevedores, the cotton screwmen retained their privileged position. Longshoremen occupied the second rung on the occupational ladder. Through their conference rules and unique relationship with the screwmen—many screwmen were members of the longshoremen's union as well as their own—their wages and working conditions were the envy of New Orleans' unskilled laborers. Both screwmen and longshoremen controlled the supply of labor in their respective trades, and no employer could hire non-union men without incurring the wrath of both groups. Despite the general intensification of work and the stevedores' and agents' assault on their work rules in the early years of the century, screwmen and longshoremen successfully resisted a reorganization of the waterfront and preserved their relatively privileged positions.

The intensification of the labor process brought labor relations in the longshore trade to a crisis point by the fall of 1903. There "has been a disposition on the part of some stevedores and some agents," a special committee of black and white longshoremen complained in September 1903,

> through their foremen, to so reduce the working force on ships that it has become almost impossible for any man to stand a day's work on ships loading and unloading. . . . We do not feel that we should be reduced to a condition a little less than slavery. We are at a stage where the men individually cannot even complain of overwork for fear of being boycotted. It has come to such a stage that the handling of the commerce of this port has been reduced by nearly one third in cost since 1901 and at the price of suffering and enslaving of the longshoremen . . . [T]here seems to be a disposition on the part of the stevedores and foremen to work a man to death.

E.S. Swann, former black longshore president, put it succinctly: "Work in the penitentiary, under the guns, is no harder than what we do on the levee front." In response to the speedup, longshoremen formulated a new work rule in August 1903 which would require employers to assign four men to each side of a ship's hatch.[29]

Vowing to head off any efforts by longshoremen to imitate the screwmen's restriction of output, agents and stevedores not only rejected the eight-men-to-a-hatch clause but demanded an anti-limit, or rule change clause, barring workers from adopting any rule that could reduce the amount of work performed or add to the cost of working ships. In the strike that ensued in early September, the largest shipping lines began diverting their vessels to other Gulf ports and ordered their crews to unload and load those vessels already in port. The longshoremen backed down. At a joint mass meeting of over 2000 black and white workers on September 12—a session that black longshore president John B. Williams chaired according to the half-and-half rule whereby blacks and whites spoke alternately—the conservative faction of union leaders (including James Porter of the black union and Recorder James Hughes and Harry Keegan of the white union) successfully advocated eliminating the eight-men-to-a-hatch rule. The terms of the new three-year contract relegated claims of overwork to the grievance system. A standing arbitration committee—to be composed of three black and three white longshoremen, three agents and three stevedores (in case of a tie, the committee would choose a thirteenth arbitrator)—would judge all complaints. Agreeing to "do all river front work that may be required of them," the unions also accepted an anti-limit clause. In exchange for union concessions, employers granted the unions a veto power over the selection of foremen, and agreed to pay car and ferry fare for longshoremen ordered to Chalmette, Southport, or Westwego. Moreover, if the men found no work available when they arrived at these docks, they would receive wages for two and a half hours.[30]

Yet no sooner had the contract been signed when trouble flared up among the rank and file. To the dismay of both employers and union officials, longshoremen refused to handle goods on several coffee ships because of disagreements over the number of men assigned to a hatch. "There is no need to make contracts unless we are going to live up to them," black longshore president Williams told his men, "and I am very nearly worn out trying to bring order out of this chaos." Reasserting its authority, the unions' executive committee reaffirmed the contract, ordered the men back to work, and emphasized that only joint committee members had the power to interfere with work on the levee.[31]

The issue of overwork remained a source of bitter complaint among longshoremen. Union officials blamed stevedores for the the problem. As a result of underbidding for contracts in loading and unloading, which in turn was caused by intense competition, stevedores drove their employees at a pace "beyond human endurance," making each man do two men's work. "We have been in the power of the stevedores for the past three years," white longshore president Chris Scully complained in 1906. "The longshoremen will not continue to work two men to a side of a ship when by rights four should be employed to handle the work. We are not slaves owned by the stevedores." Not surprisingly, longshoremen made the elimination of overwork a central priority when their contract expired. To halt the stevedores' abuses, the unions insisted upon the right to change work rules during the term of the contract. The anti-rule change clause had to be eliminated.[32]

Negotiations in the City Hall office of Democratic party Mayor Martin Behrman produced a quick settlement to a one-day strike in September 1906.[33] While Behrman preferred to stand above the conflict, his desire for order and uninterrupted commercial activity led him to advocate moderation and compromise to both sides. The settlement he helped to engineer constituted a modest advance for the unions. While it did not implement the eight-men-to-a-hatch rule, the 1906 contract changed the formula governing pay for men dispatched to out-of-town docks; those who traveled to Chalmette, Westwego, or Southport and found no work would receive four hours' pay. More important, it altered the composition of the arbitration system to increase labor's representation. The new committee would be composed of two black and two white workers and three stevedores, who would judge individual overwork grievances and any other matters that might arise. Under the plan, work would stop while the Arbitration Committee conducted its investigation; if the Committee ruled in the workers' favor, labor would receive payment for the time not worked. On September 2, the dock workers unanimously accepted the contract. With the conflict apparently resolved, some predicted that "peace on the riverfront is practically assured for another year."[34]

Problems, however, were only beginning. For the next three weeks longshoremen repeatedly disputed the authority of stevedores and agents, unilaterally implemented important changes with regard to how they would perform their tasks, and frequently ignored the orders of union leaders demanding contract compliance. Rank-and-file longshoremen interpreted the contract in their own ways, insisting that their specific grievances be addressed through actions unsanctioned by the contract. While union leaders eventually re-imposed discipline and secured their members' compliance, longshoremen never abandoned the fundamental principle that they, not their employers, should determine how and under what conditions the work should be performed.

Immediately following the signing of the new contract, disputes arose on the steamship *Abessinia*. Previously workers had placed five sacks of beet sugar on each truck they carried; now they insisted that five sacks were too heavy and that, in the future, they would carry only four sacks. His orders ignored, the contracting stevedore followed the guidelines carefully laid out under the newly signed contract. But assembling the Arbitration Committee, the stevedore learned, was no easy matter. After "much loss of time and no end of annoyance," the Committee unanimously ruled in favor of the steve-dore, instructing the men to resume trucking five sacks of beet sugar in each load. Then, according to the *Picayune,* "the men—that is, the rank and file—told the Arbitration Committee, in effect at least, to seek a climate where the thermometers are made without zero points, and by tacit agreement contin-ued to truck only four sacks." Only stevedore John Honor's threat of mass dismissals convinced at least some workers to abide by the ruling.

Stevedore J. P. Florio similarly discovered that his men were "not an easy lot to get along with." Six longshoremen worked in one of his ship's holds, receiving freight brought to them by two trucks handled by four men. Com-plaining that the work was "too much for six men" they demanded two more, halting all work for several hours. In the third week of September, longshore-men ignored stevedores' orders, causing the steamship *City of Mexico* to lose an entire day in port. At the same time, grain trimmers on the steamship *Sardinia* also repudiated their employers' authority. Ordered to Chalmette from the city's riverfront, the ship required only three additional hours of work for eleven of the original twenty-two grain trimmers. However, the *Picayune* observed, the "whole twenty-two rose up as it were in rebellion," demanding that the stevedore "[t]ake us all or none of us can go." The *Sardinia*'s stevedore reluctantly surrendered.[35]

Overwork was not the only issue that longshoremen addressed by sponta-neous work stoppages. Approximately 150 longshoremen loading one of the largest freighters in port, the *Montauk Point,* objected to the physical condi-tion of the sugar wharves. Complaining that "they would require the strength of mules" to truck their loads through the muddy levee to the short planks by the side of the ship, the men struck, forcing agents and stevedores to negotiate for improvements with the railroad companies that operated the wharves. On two other occasions longshore workers gave William Kearney, president of the stevedores' association, cause for complaint. He ordered a gang to Westwego one morning, worked them for several hours and paid them, accord-ing to contract, for four hours time. That afternoon he sent them to Chal-mette, where they worked for three hours. When the men demanded four hours pay, Kearney objected on the grounds that it was unfair to pay four hours' time to the same gang twice in one day. He put the issue before the Arbitration Committee, which promised to consider it. Kearney recalled the Arbitration Committee when his longshoremen, complaining of too few men, quit work on the steamship *Baroda*. Following a nine-hour stoppage, the Committee upheld the longshoremen. Kearney obeyed its ruling, hiring extra help in the hold at a cost of $44. "Contrast this action to the action of the longshoremen on the Abessinia," the angry stevedore remarked, "and you will see which side is living up to the contract."[36]

Stevedores denounced what they saw as longshoremen's "spirit of indifference . . . their failure to live up to their contract, and their evident intention to throw as many barriers as possible in the way of the bosses," and vowed to "fight the battle out to the bitter end," in the *Picayune*'s words. While they praised white and black union presidents Chris Scully and E. S. Swann for doing "all in their power to conserve the interests of peace," their attorneys prepared to file suits for damages against both individual longshoremen and their unions in court. They delayed litigative action, however, when Scully, Swann, and the Executive Committee of twelve whites and twelve blacks assured them that they wanted no more trouble.[37] Union leaders again counseled their members to comply with the contract. "Now I want to say that we are prepared to live up to our contract and intend to do so," black longshore president Swann explained. "Complications may arise at times, but in my opinion these complications can be adjusted without a resort to extreme measures." At a conference attended by leading steamship agents, stevedores, and the longshoremen's Executive Committee, union leaders examined each dispute on a case-by-case basis, sometimes justifying, but often condemning their members' actions. Swann and black longshore secretary James Porter, for example, faulted the longshoremen on the *Sardinia*, *Abyssinia*, and *City of Mexico*, but stressed that the violations were committed by individuals, not by the associations. Before a joint mass meeting, Scully and Swann again explained the contract's details, emphasizing the need to adhere to all rules. When the men voted unanimously to abide by the contract, leaders distributed recently printed, hip-pocket sized books containing copies of the agreement.[38]

While labor leaders brought the rank and file under union control, agents and stevedores came to regret having accepted this institutional framework for waterfront peace in the first place. Most frustrating was the shift in the balance of power from employers to union leaders. Employers directed their fire against the 1906 contract's arbitration system and the foremen's lack of power in particular.[39] The arbitration system gave workers "all of the best of it," Steamship Conference leader W.P. Ross complained, putting employers "at the mercy of the frivolous objections of laborers." Any longshoreman complaining of overwork could halt all loading and unloading on the ship until the Arbitration Committee was assembled. "The complaint acts as a fine upon the ship, for it stops the work," Ross noted; it "penalizes the stevedore, for he must put on extra men, but [it does] . . . not give us the opportunity to punish the agitator when it is found there is no real cause for complaint." Under the existing

system, the union maintained responsibility for disciplining men for the violation of contract rules. Workers charged with making "frivolous" complaints
were tried by a union court composed of the longshoremen's joint Executive
Committee. Now the agents sought "some means of imposing a fine upon the
objectors when it was shown by the Arbitration Committee that the complaint
had no foundation." Equally important, they demanded that employers once
again be accorded equal representation on the Arbitration Committee.[40]

Employers identified the absence of managerial control in general and the
lack of power exercised by the union foremen in particular as primary sources
of difficulty. They found that workers "had no respect for the foremen and did
not heed their judgement." Agent Ross concluded that the longshoremen

> have practically taken out of the stevedore and his foremen's hands all the
> power to keep discipline among the men working under them, and they
> openly state that they intend to carry this still further through their system of
> fines and expulsion of foremen, who are compelled to be members of their
> organization.[41]

Stevedore Kearney concurred: The foremen "are jerked up at every meeting
of the Union and raked over the coals for some violation of the rules if they try
to have the work done, and the longshoremen come out boldly and tell us that
the individual members of the Union are the bosses of the foremen." The
workers' rules that provided for a number of fines against the foreman "practically puts him in the power of any man he has employed." There was considerable basis for the charge of union control over foremen. When employers
proposed to give foremen—even union foremen—the power to decide how
many men would work, longshore leaders rejected the suggestion because the
"foremen were not competent to judge on that point." Agents and stevedores
withdrew a further demand to empower foremen to impose on-the-spot disci-
plinary action, but they continued to insist (without effect) on the need to
keep work moving during any dispute.[42]

In September 1907, disagreements over work rules and managerial authority erupted into the open as stevedores and agents tried to dismantle the
system that had institutionalized the longshoremen's power. Approximately
3300 black and white longshoremen struck most stevedores and agents on
September 5. Within days, shipping agents ordered crews of the vessels in
port to begin loading cargo under the protection of a small squad of police.
The limits of this strategy immediately became evident, when Dock and Cotton Council members refused to handle cotton touched by or destined for
non-union workers. From the outset, union freight handlers at the Illinois
Central's Stuyvesant Docks refused to unload rail cars whose cargo would be
loaded by the crews of steamships that had not signed the longshoremen's
tariff; black teamsters and loaders refused to truck cotton to the levee; cotton
yardmen refused to process it; screwmen refused to screw or load it; and coal
wheelers promised to refuse to fuel any of those ships. Moreover, waterfront
unions in Galveston, Texas, assured New Orleans strikers that they too would

refuse to load cotton from any ship diverted from the Crescent City. These sympathy actions succeeded almost entirely in tying up the levee for days.[43]

In the end, the nine-day showdown produced few changes. Although the strike occurred weeks before the heavy commercial season got under way, fears of serious damage to the port's prosperity led Mayor Martin Behrman and representatives of the city's commercial exchanges into the negotiation process. A final agreement retained the 1906 arbitration system and payment rate for out-of-city work; in exchange, the longshoremen dropped a new demand to repair damaged cargo and sew torn sacks. Both parties agreed to a clause prohibiting employers and unions from making any changes for the contract's three-year duration. Overall, the balance of power remained more or less as it had been under the previous contract.[44] Withstanding their employers' latest assault, the longshoremen had retained their victory.

The longshoremen's and screwmen's repeated challenges to managerial authority and competence, their emphasis on direct action on the ships and docks, their tendency toward federation or organizational alliance, and their propensity to cross trade lines to engage in sympathetic actions suggest an orientation that transcended the narrow, conservative craft unionism generally associated with the AFL. In recent years, historians David Montgomery and Bruce Nelson have argued that such patterns of behavior, rather widespread in the first and particularly in the second decade of the twentieth century, revealed the prevalence of a syndicalist impulse—an impluse that gained far more adherents than the membership figures for the Industrial Workers of the World (IWW) or other anarcho-syndicalist organizations might indicate. Nelson defines this "mood of syndicalism" as

> first, the impulse toward workers' control of production; second, the belief that direct action at the point of production was the most effective means for the achievement of working-class objectives; third, the determination to cross traditional craft union barriers in order to build solidarity with other workers—ultimately, the impulse toward One Big Union; and finally, an apocalyptic dimension, a striving for fundamental social transformation embodied in the Wobblies' exhortation to "bring to birth a new world from the ashes of the old." . . . The ideological content of the New Unionism was vague and eclectic, but the impulse was nonetheless powerful.[45]

New Orleans dock workers exhibited little interest in the IWW, argued for no fundamental transformation of society, and invoked no apocalyptic imagery in their speeches or resolutions. Yet shorn of its political radicalism, the concept of the syndicalist impulse—with its emphasis on control, solidarity, direct action, alliance, and amalgamation—captures important elements of the experiences of New Orleans waterfront workers. The port's most powerfully placed workers—screwmen and longshoremen—achieved a significant degree of control over the labor process through direct action at the point of production. Their successes were made possible, in part, by the extension of solidarity across sectoral lines that had traditionally separated waterfront workers. While the Dock and Cotton Council, like its nineteenth-century predecessor, did not function as "One Big Union," it did ally workers across trade and racial lines in an impressive, and effective, manner.

If waterfront workers pressed the limits of a narrow AFL craft unionism in the age of Jim Crow, they nonetheless remained within its boundaries, as the following examples illustrate. The Dock and Cotton Council offered no larger vision of social, political, and economic change. Respect for the occupational hierarchy on the waterfront, the increasing importance of contracts, the use of injunctions and suits for contract violations, the chances of success—all of these factors shaped the dock labor federation's strategy in the new century.

III

Other groups of waterfront workers fared less well than the longshoremen when it came to wages, conditions, and work rules. While not unaffected by the impulse toward control that infused the struggles of the screwmen and long-shoremen, these men remained more concerned with the basic goals of raising wage rates and curbing employer abuses. Cotton teamsters and loaders, round freight teamsters, and freight handlers occupied vital, but subordinate positions on the riverfront. Unlike the longshoremen and screwmen, these men were unable to control the supply of labor available to their employers in periods of conflict, and hence remained more vulnerable to economic pressures. Their experiences both underscore how workers' position in the waterfront's multi-tiered occupational structure continued to influence strongly a group's chances for success, and illustrate the limits of labor solidarity after the turn of the century.

For over two decades, dock workers employed in New Orleans' railroad freight yards enjoyed few of the gains achieved by longshoremen or screwmen in the more competitive contracting sector. Regular dock workers confronted a multiplicity of employers whose differing levels of economic might and often conflicting economic needs undermined managerial unity and strengthened union leverage. Freight handlers, however, remained vulnerable to economically powerful, resourceful, and unified railroad corporations, which ignored employee efforts to impose work rules or raise low wages. Shortly after the screwmen successfully imposed new work rules and the half-and-half compact on agents and stevedores in November 1902, some 400 black and white freight handlers at the Stuyvesant Docks and Southport struck the Illinois Central Railroad.[46] The outcome revealed once again the freight handlers' relative powerlessness and highlighted the differences in their work situations with those of the port's other levee laborers.

The Orleans Freight Handlers Benevolent Association, ILA Local 293, served notice on the local Illinois Central management in August 1902 for a wage increase from 15 to 20 cents an hour. With endorsements from of all the major levee unions affiliated with the ILA, the Central Trades and Labor Council, and the Central Labor Union, the union hoped to avoid a strike. But

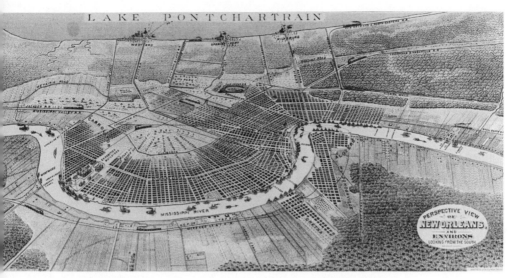

Map of New Orleans. James S. Zacharie, *New Orleans Guide,* 1885.

New Orleans levee, 1858. *The Illustrated London News,* June 5, 1858.

New Orleans levee, early 1870s. Drawing by J. Wells Champney. *Scribner's Monthly,* December 1873.

New Orleans levee, early 1880s. George Augustus Sala, *America Revisited,* 1883.

New Orleans levee, 1883. Courtesy of the Historic New Orleans Collection.

SCENE ON THE LEVEE, AT NEW ORLEANS.

Longshoremen in the late nineteenth century. By Bellew and Pierce. *Ballou's Pictoral Drawing Room Companion.* Courtesy of the Historic New Orleans Collection.

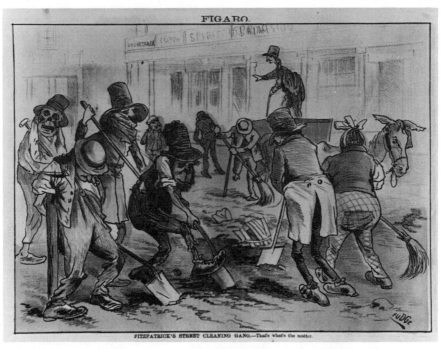

FIGARO.

FITZPATRICK'S STREET CLEANING GANG.—That's what's the matter.

Anti-Irish and anti-Ring cartoon. John Fitzpatrick, ward politician and Administrator of Public Works in the 1880s, received harsh criticism from businessmen and reformers for his liberal dispensation of patronage to immigrant voters before elections. *Figaro,* January 19, 1884. Courtesy of the Lousiana Collection, Howard-Tilton Memorial Library, Tulane University.

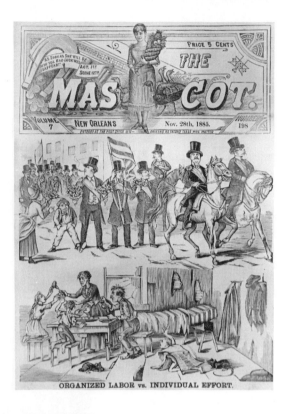

The cartoon credits the New Orleans labor movement—the Central Trades and Labor Assembly and the Cotton Men's Executive Council—with uplifting the city's white working class in the mid-1880s. *The Mascot,* November 28, 1885. Courtesy of Manuscripts, Rare Books and University Archives, Howard-Tilton Memorial Library, Tulane University.

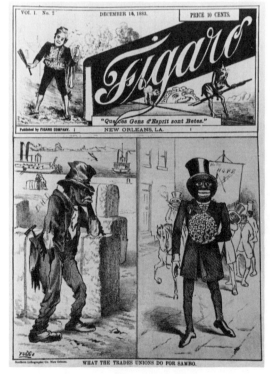

Anti-black cartoon. *Figaro* carried articles hostile to the city's black and Irish residents, the Democratic party machine, and the labor movement. *Figaro,* December 14, 1883. Courtesy of the Louisiana Collection, Howard-Tilton Memorial Library, Tulane University.

Shutting the Stable Door when the Horse is Gone.

New Orleans merchants constantly worried about the diversion of the city's cotton trade, which they blamed on high port charges, the extension of new railroad lines across the region, and the rise of commercial competitors in the South. *The Mascot* November 20, 1886. Courtesy of Manuscripts, Rare Books and University Archives, Howard-Tilton Memorial Library, Tulane University.

In late 1886, the Cotton Men's Executive Council admitted groups of employers into its ranks. The cartoon predicts the result of the admission of the Cotton Press Association. *The Mascot,* September 4, 1886. Courtesy of Manuscripts, Rare Books and University Archives, Howard-Tilton Memorial Library, Tulane University.

THE OFT' TOLD TALE.—The Big Shark Devours the Little Fishes.

The Cotton Men's Executive
Council split into warring
factions in 1886– 87.
The Mascot, April 2, 1887.
Courtesy of Manuscripts, Rare
Books and University Archives,
Howard-Tilton Memorial
Library, Tulane University.

The True Inwardness of the Latest Cotton-Handler's Strike.

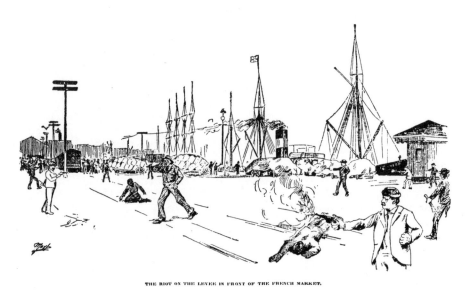

THE RIOT ON THE LEVEE IN FRONT OF THE FRENCH MARKET.

White cotton screwmen and longshoremen attacked black dock workers in a series
of violent confrontations in late 1894 and early 1895. *Times-Democrat,* March 13,
1895.

Mississippi River roustabouts, 1890s. *Harper's New Monthly Magazine,* January 1893.

Right: Thomas Agnew, president of the Screwmen's Benevolent Association in the 1880s, Administrator of Police and Public Buildings from 1888 to 1892, and a harbormaster from 1896 until his death in 1901. *Daily Picayune,* January 22, 1901.

Above: James E. Porter was active in the New Orleans labor movement from the 1870s to the early 1910s, serving as a leader in the Longshoremen's Protective Union Benevolent Association, the Cotton Men's Executive Council, the Central Labor Union, the International Longshoremen's Association, and the Dock and Cotton Council. *Daily Picayune,* September 2, 1903.

James Leonard, president of the Workingmen's Amalgamated Council in 1892, founder of the white Central Trades and Labor Council in 1899, and AFL organizer. *Daily Picayune,* September 2, 1902.

Thomas LeBlanc, president of the Central Labor Council, 1907. Courtesy of the Lousiana Collection, Howard-Tilton Memorial Library, Tulane University.

I. G. Wynn, president of the Cotton Yardmen's Benevolent Association No. 2, 1907. Courtesy of the Louisiana Collection, Howard-Tilton Memorial Library, Tulane University.

Thomas Harrison, a president of the Screwmen's Benevolent Association No. 1, and official of the Dock and Cotton Council, a Commissioner of Labor Statistics, and a vice president of the International Longshoremen's Assocation in the early twentieth century. *The Longshoreman,* August 1918.

Black delegates to the International Longshoremen's Association convention. Thomas P. Woodland was president of the Screwmen's Benevolent Association No. 2. *International Molders' Journal,* July 1918.

ATTENTION!

UNION MEN

of New Orleans

YOU MUST READ

the new Labor Weekly

"T^{HE} LABOR WORLD"

to know the true conditions of affairs
that now exist in New Orleans about

UNION MADE BEER.

YOU SHOULD KNOW

Fair Beer is sold for the same Price as Scab Beer,

AND IT IS

made. sold and delivered by Union men for benefit of Union men.

The following Breweries employ only Union Labor and are **FAIR** to the United Brewery Workers:

The Consumer's Brewing Co. of New Orleans, La.
The Ohio Union Brewing Co. of Cincinnati, Ohio.
The Moerlein   "   "   "   "
The Wiedemann   "   "   "   "
The Pabst Brewing Co. of Milwaukee, Wis.
The Miller   "   "   "   "
The Schlitz   "   "   "   "
The Blatz   "   "   "   "
The Anheuser Busch Brewing Co. of St. Louis, Mo.
The Cook Brewing Co. of Evansville, Ind.
W. J. Lemp Brewing Co. of St. Louis, Mo.

Send in your name and we will send you "The Labor World" each week and make you wise on labor conditions in New Orleans.

Phone Main 1523.

Address   116 EXCHANGE PLACE.
903 MAGAZINE ST.

A pro-Brewery Workmen advertisement during the factional disputes in the New Orleans labor movement during 1907–08. *The Labor World* was published by Oscar Ameringer. *United Labor Journal,* August 31, 1907. Courtesy of the Louisiana Collection, Howard-Tilton Memorial Library, Tulane University.

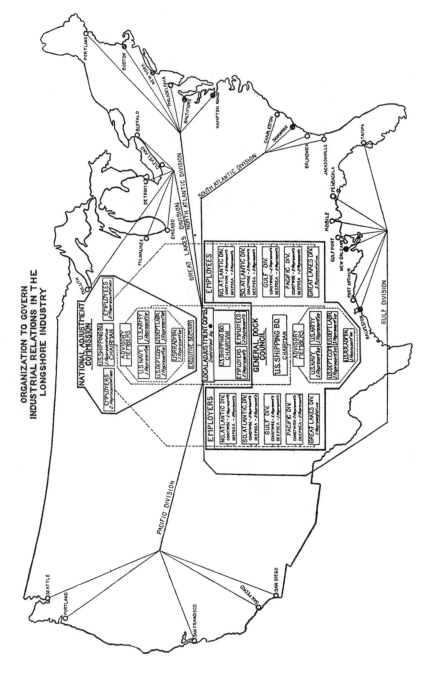

The National Adjustment Commission governed labor relations in the longshore industry during and shortly after World War I. *The Survey*, August 2, 1920.

local managers of the Illinois Central ignored the union's demands, increasing the hourly wages only slightly, to 16 cents. On December 9, some 400 freight handlers walked off the job, temporarily halting all commerce at the railroad's freight and handling departments at Stuyvesant Docks and Southport.[47] The Illinois Central easily deflected its workers' challenge. Managers procured 800 black and white strikebreakers from Chicago, St. Louis, Louisville, and other areas served by the railroad—all regular employees of the Illinois Central, not "professional strikebreakers"—and imported over a hundred agents from its secret service department to protect its property and employees. The Illinois Central was capable of holding out indefinitely. Behind its well-guarded fenced-in terminals, approximately 1000 men loaded and unloaded cargo for $1.60 a day; housed in the company's warehouses, they received not only three meals a day but nightly entertainment as well. The railroad delivered its ultimatum: unless the strikers returned to work immediately, the company would replace them permanently with between 200 and 300 new local men.[48] It was immediately apparent that the Orleans Freight Handlers had taken on more than they could handle. Backed by nothing more than the moral support of the screwmen and longshoremen, the freight handlers were little match for their employer.

With the strike effectively broken, local labor leaders advocated acceptance of the railroad's terms. The final agreement, worked out with several prominent Dock and Cotton Council leaders playing "conservative parts," provided wages of 16 cents an hour for a 10 hour day, a new grievance procedure, and, when possible, continuation of the union policy of dividing work equally between whites and blacks. The Orleans Freight Handlers ended its strike on December 23, two weeks after it began. The strikebreakers had cleaned up the docks so thoroughly, however, that initially there was not enough work for all the returning handlers to do. Only as freight accumulated did the company gradually rehire the men. Despite the Illinois Central's minor concessions, the railroad yards remained relatively immune to the collective efforts of their black and white workers for the next several decades.[49]

In the early twentieth century, black cotton teamsters and loaders in the Dock and Cotton Council fared only somewhat better than freight handlers and their counterparts in the sugar hauling trade. While sharing a general concern over work rules, each year they focused on working conditions and wages. Earning $3 for a 10-hour day, 300 black union teamsters and loaders struck in 1904, demanding that the grooming, feeding, and harnessing of the stock—which draymen required them to do before they began hauling—be treated as a part of their day's paid work. A settlement negotiated by a Dock and Cotton Council committee won them only a fixed dinner hour, set from noon to one. In September 1905, some 200 to 300 cotton teamsters struck again for a small wage increase. The Master Draymen's Association seized the opportunity to eliminate the fixed dinner hour, which it viewed as a serious inconvenience (since union teamsters would leave their wagons and teams of mules in the streets or on the levee while they ate), and insisted that the men report for work at 6:00 a.m., instead of 6:15. The master draymen, rejecting a teamster offer to resume work under the old tariff, began to hire strikebreakers. "We can not sign the tariff because it means the same thing as turning over our business to the teamsters and loaders," announced the president of the Master Draymen's Association.[50]

Within the Dock and Cotton Council, support for the teamsters followed racial lines. Some white delegates argued against a general walkout on the grounds that it would raise public opposition, given the epidemic of yellow fever then afflicting the city. Black workers strongly supported the teamsters. At a joint meeting of the black and white cotton yardmen, the whites advocated adherence to their contractual prohibitions against sympathy strikes, while the blacks, the *Picayune* noted, were "howling to support their black brothers, the teamsters," and insisted on striking. Black longshoremen and screwmen similarly pressed for sympathetic action. On September 11, the Dock and Cotton Council delegates voted against a general strike but ruled that union members would handle no cotton delivered by non-union float drivers. The result was that strikebreakers hauled their cotton to the press yards where they covered it with tarpaulins. But merchants interested in the steady flow of cotton devised novel ways around the Cotton Council's refusal to touch non-union cotton. They managed to bypass the drayage stage by switching bales arriving by rail at the terminals, sending them almost directly to the wharf's edge on cars. There, union freight handlers unloaded them, and once on the levee, longshoremen trucked the cotton to the side of the ships and handed it over to the screwmen on board.[51]

Arbitration finally ended the three-week cotton teamsters' strike. At the Dock and Cotton Council's request, Mayor Martin Behrman intervened and appointed cotton press manager Adam Lorsch as official arbitrator. Lorsch, siding with the master draymen, abolished the fixed dinner hour and instituted a "sliding," flexible hour. Now teamsters would have to deliver all cotton loaded to the presses before they quit work. Lorsch did create a grievance committee to handle future disputes, and decreed that the working day should begin fifteen minutes later. Thus, the final settlement shifted the balance of power somewhat toward the boss draymen, but the teamsters' union easily survived the strike intact. The Dock and Cotton Council's lack of commitment to the small black union—due to the reluctance of white unions to involve themselves too deeply—had left the teamsters largely to struggle on their own. But the mere possibility of a broader, sympathetic action on the part of the Cotton Council protected the teamsters from their employers' full wrath.[52]

Round freight teamsters—the men who handled the barrels, sacks of sugar, molasses, and rice—shared little of the success that characterized the longshore cotton unions. In 1901, 100 members of the small union struck unsuccessfully against the city's 30 to 40 boss draymen, receiving virtually no support from other riverfront unions.[53] The following year, they were barely more successful when they again struck for union recognition, overtime pay, and a union grievance committee. These labor leaders, the *Picayune* warned, were "determined to make this a union city from one end to the other." They failed. Dock and Cotton Council members, involved in their own struggles, took no sympathetic action. The screwmen had just concluded their fight with stevedores and agents over their new working rules, and union leaders had just barred sympathy actions and boycotts by longshoremen and screwmen during the Illinois Central freight handlers' strike. When teamsters successfully appealed to a group of longshoremen loading a record size rice cargo onto the steamship *Arcadia* to reject cargo delivered to them by non-union teamsters, officers of the longshoremen's union ordered the men back to work.[54]

While economic pressure compelled some of the smaller firms of boss draymen to sign the union tariff, the larger firms refused to negotiate, and slowly rebuilt their labor force by hiring strikebreakers. "There are plenty of men on the levee," one employer remarked, "who want to work . . . It demonstrates the folly of thinking that a handful of blacks can dictate to the commerce of this port. There might as well be no strike." Unable to control the supply of unskilled labor or secure the Dock and Cotton Council's active support, the round freight men submitted their case to the black Central Labor Union and agreed to abide by its decision. William Penn, acting chairman and president of the black longshoremen, declared the strike as good as lost and advised the men to return to work immediately. The following day, December 18, the union men "made a rush for the levee yard" where most of them received their old jobs back. The strike, the *New York Tribune* concluded, had proved an "utter defeat and disaster" to the union.[55]

In each of the episodes above, the relative weakness of the various black unions made the Dock and Cotton Council's stance critically important. The hierarchy of waterfront labor was particularly pronounced in the new federation; screwmen and, to a lesser extent, longshoremen, exercised a disproportionate share of power. Weaker black unions in the Council, like the cotton teamsters, and those outside its ranks, like round freight teamsters and freight handlers, could not count on automatic support from other riverfront workers. By the early twentieth century, the broad spirit of solidarity that had marked labor struggles in the early 1880s had given way to a more formal, calculated relationship between the various classes of labor on the waterfront.[56] While it shared with the Cotton Men's Executive Council of the early 1880s a commitment to interracial and inter-trade collaboration, the new dock union federation differed from its nineteenth-century predecessor in important ways.

Written contracts both codified the balance of power between workers and employers and placed limits on expressions of labor solidarity and sympathetic actions. Black longshore president William Penn explained the position of the Central Labor Union:

> We want to so keep and sustain our contracts that they will be regarded as binding and sacred by the business community. The more faith labor and

capital can inspire in each other the easier it will be to get contracts and to secure concessions to our demands.[57]

For unionized workers, the written contract legitimated the union's role as bargaining agent between employers and workers. For black workers in particular, the contract not only defined the terms of employment and limited the arbitrary power of the bosses, but it also reduced the ability of agents and stevedores to manipulate racial divisions. Given the intensification of labor conflict and the hostile racial climate, black union leaders appreciated the legal guarantees written agreements afforded them. At the same time, for labor, the risks associated with contract violations increased. Employers relied on court injunctions and the threat of legal action to compel their workers to adhere to their contracts. While the use of the injunction as a weapon against workers by no means eliminated sympathetic strikes or walkouts, it did create an environment in which solidarity became less routine and more calculated and strategic. Under such circumstances, union officials had to evaluate carefully the dangers or opportunities posed by taking sympathetic action on behalf of workers in other organizations. In many cases, they concluded that the dangers outweighed the opportunities.[58]

The case of the 1902 Illinois Central freight handlers strike illustrates the limits of labor solidarity after the turn of the century. Regular dock unions traditionally had offered little active support for strikes of freight handlers, and this case was no exception. Tempering avowed sympathy with a pragmatic, if self-interested, restraint, both longshoremen and screwmen voted to uphold their recently signed contracts. That meant that union longshoremen would offer the freight handlers moral and financial support, but would load goods placed on the wharves by non-union workers.[59] Screwmen, who had just completed their own negotiations with the steamship agents and stevedores over their half-and-half compact, similarly concluded that the moment was a "bad time for them to talk of trouble on the levee." One black union official explained why his members voted against a sympathy strike:

> The employers are beginning to have too little faith in contracts with labor unions. I feel that the Illinois Central fears that should it give in to the demands of these men they would have no assurances that there would not be more trouble. I do not believe the colored or the white longshoremen are in a position to strike under the circumstances. They have both declared, as have the screwmen, that it is their intention to do away with strikes in New Orleans, and that they would work for the upbuilding of the port. Both organizations are now bound by contracts that are binding so long as neither side violates [them].[60]

Leaders' stern declarations notwithstanding, many individual longshoremen working at Stuyvesant Docks refused to handle freight put on the wharves by the imported strikebreakers and walked off the job. Acknowledging that their contract prohibited them from striking, they insisted that their individual consciences prevented them from working with strikebreakers. In the end, officials in the joint conference of the black and white unions forced most longshoremen back to work, apologizing to stevedores and agents who threatened legal action against the unions. "It was the only thing for our associations to do," explained one official.

> We could not afford to go on record as violating a contract. Our contracts amount to something in the eyes of the public and to violate one means something to us. We would like to have extended this help to the Stuyvesant freight handlers but it could not be done under the circumstances.[61]

Union leaders' fear of the repercussions of breaching their contracts served to weaken the formal bonds between the waterfront unions. Protective of the institutional security of their own associations, union officials enforced contract adherence and policed their members' behavior. Despite the potential power of the Dock and Cotton Council, many secondary waterfront unions had to rely on little but their own, often inadequate, resources.

IV

The position of black workers in the New Orleans labor movement was the subject of considerable discussion in the opening years of the twentieth century. Black and white dock workers always defended their alliance in practical terms, as a means to end abusive treatment, stop competition for jobs, and reassert control over the labor process. Before the screwmen's amalgamation, black union leader E.S. Swann testified in 1908, "[i]t was a case of slavery," with foremen pitting black and white gangs against each other. "We saw that we were being used as a cat's-paw in the labor situation," recalled black longshore leader Alonzo Ellis.

> The steamship agents were making us the enemy of the white laborer, and at the same time robbing us of our earnings. . . . We didn't seek social equality with the whites. We just didn't want to keep up the strife and bitterness on the levee. We wanted peace. We started into the business as a question of meat and bread, and we did not want to be wedged between the white screwmen and the steamship agents, for we knew that if it ever did come to a pinch the white man would stand by his own color and we would get the worst of it.[62]

The effect of the amalgamation convinced even its white opponents that collaboration was preferable to competition. "I put up the most vigorous fight of my life in opposing amalgamation of our local with the colored men," Thomas A. Harrison, white screwmen's union delegate and former State Commissioner of Labor, recalled at the ILA's 1913 convention. "I am pleased to say today that the fight was of no avail and today I realize that my fight was a mistake." The reason was clear. The alliance which ended racial competition on the docks permitted the screwmen to reassert their control over the labor market and the conditions of their work.[63]

Few outside the dock unions' ranks shared this positive assessment of interracial collaboration. In May 1903, both the white *States* and *Times-*

Democrat concluded that the admission of blacks into the labor movement had "cast a shadow" over local trade unions, simply because the interracial alliance "has a distinct tendency to bring about social equality among the working classes of both races." The *Vicksburg Herald* went even further, denouncing the "unnatural mixup" as a "broad field for debasement and disorder in the screwmen's black-and-tan organization."[64] The Port Investigation Commission reported in 1908 that one "of the greatest drawbacks to New Orleans is the working of the white and negro races on terms of equality. It drags down the white man; it does not uplift the negro; and so we find white men working hopelessly for existence under these intolerable surroundings."[65] The equality of the races that "exists to-day on the Levee," the five white commissioners all agreed, "was a disgrace to a Southern city." State Senator C.C. Cordill, a large cotton planter from Tensas Parish, was most appalled by the interracial unity he found on the docks. "In this town there are five white men to one nigger, and THE NIGGER IS THE BOSS. It's the only town in the South where they'd stand for it."[66]

Such vehement condemnation, however, was only heard in periods of severe labor unrest. One such moment came during the Steamship Conference's lockout of screwmen in April 1903. The Cotton Exchange Mediation Committee made a last-minute effort to head off a general riverfront strike by proposing binding arbitration by the president of the International Association of Longshoremen, Marine and Transport Workers, Daniel J. Keefe. As a labor leader in the National Civic Federation, an organization which promoted class harmony over conflict, Keefe had a reputation in some business circles for fairness; he was not considered to be "a strike man" since he often "resorted to all sorts of arbitration to secure a settlement rather than allow a strike." The Mediation Committee reasoned that the screwmen could not refuse a decision by their international union's president. Both the Conference and the unions accepted the proposal and, pending the final verdict, the screwmen returned to work. But Keefe refused to become involved in the local dispute. Instead, he recommended as a substitute mediator James E. Porter, who was an ILA vice president as well as secretary of the city's black Central Labor Union, secretary of the Dock and Cotton Council, secretary of the joint conference of black and white longshoremen, an AFL organizer, and a participant in the current conflict. Members of the Cotton Exchange Mediation Committee were "indignant" at Keefe's insensitivity to the local situation and racial mores.[67] "The first and sufficient objection to Porter is that he's a negro," the *Times-Democrat* added.

> Now, we don't know a great deal about Porter. He may be a good man in the longshore, marine and transport business. . . . His line, however, is not arbitrating quarrels between white men, and the place for him to arbitrate white men's differences is certainly not New Orleans. . . . Keefe's decision . . . may be a composite of both viciousness and imbecility, it may be a cheap political expedient invented in order to postpone a settlement of the dispute. If imbecile, it is pitiable. If vicious, it is diabolical. If political, it is contemptible.[68]

If the press and the employers found Porter's nomination as mediator insulting because of his race, on numerous other occasions they accepted black participation on negotiation committees without comment. During conflicts involving round freight teamsters, freight handlers, longshoremen, screwmen, teamsters and loaders, and coal wheelers, black union leaders sat side by side with their white counterparts in presenting workers' positions to Mayor Paul Capdevielle or Mayor Martin Behrman and in negotiating with white agents, contracting stevedores, master draymen, or press owners. Employers seldom objected to black participation or their equal representation on union committees. For the most part, the press, politicians, and businessmen either tolerated or simply ignored the Dock and Cotton Council's interracial collaboration. In moments of serious labor unrest, despite their rhetoric abhorring alleged "social equality," it was less the interracial character of the collaboration that appalled them than the simple fact that interracial collaboration reduced employers' power.

The racial practices of the dock unions constituted the most conspicuous exception to segregation and racial exclusion in New Orleans. Elsewhere in the city's labor movement, black workers encountered varying degrees of white hostility. The dramatic growth of the city's labor movement following the depression of the 1890s had coincided with the rise of legalized segregation across the South. The Central Trades and Labor Council, the federation of white craft unions affiliated with the American Federation of Labor, had little to do with the black workers it excluded from its ranks. Despite their shared membership in the AFL, the white Council offered neither moral nor financial support to the black Central Labor Union. As an examination of Labor Day parades in the early twentieth century reveals, white craft unionists refused to accord black workers even symbolic inclusion in the white-dominated labor movement, save on terms of inferiority.

During the 1880s, Labor Day parades were celebrations of union workers' influence, power, and pride, demonstrating their commitment to the labor movement, and affirming their place as upright citizens in the commercial city. But after the turn of the century, the parades also demonstrated the strength of racism within labor's ranks and, thus, the limits of interracial labor collaboration in the age of segregation. Beginning in 1902, white workers affiliated with the Central Trades and Labor Council and blacks affiliated with the Central Labor Union marched separately, unlike the grand interracial marches of the 1880s. In 1903 the Dock and Cotton Council attempted to bridge the gap. During the height of the agents' and stevedores' spring 1903 assault on the screwmen, the waterfront Council announced that its black and white members would march together on Labor Day. The Central Trades and Labor Council rejected that decision as a "direct slap" in the face. "It has always been our policy to endeavor, so far as possible, to separate the white and colored unions," one white official declared. "If the Dock and Cotton Council wishes to parade on Labor Day with white and colored unions side by side, they then should feel equally at liberty to give joint picnics, etc."[69] Negotiations between white dock unions and the white CTLC produced a feeble compromise. White waterfront workers would march in the white federation's parade if the CTLC permitted black union men to participate as well. As a result, the white city central invited the Central Labor Union to "form a part" of one large, united labor parade. But "form a part" meant "fall in and follow the white parade," a position that the black workers rejected. "We offered to go into the parade," one black unionist stated,

> if they had offered to give us a show, but they wanted to make a rear guard of us and we objected. The Central Labor Union has been organized longer than the Building Trades [the association behind which the black workers would have to march] and is a stronger body. We thought we were entitled to second place, anyhow.

The black unions rejected the invitation, as they had rejected a similar offer the previous year. Other than their initial challenge, white dock unions made no further efforts to counter segregation in the ranks of the city's white AFL body. On September 6, 1903, thirty-seven white unions with 20,000 members and seventeen black unions with 10,000 members marched in their own Labor Day parades. Separate celebrations became an annual tradition.[70]

If impressive when compared with the racially exclusionary tendencies of the white CTLC, the interracialism of the Dock and Cotton Council also had clear limits. Indeed, the Council's white union members were themselves hardly untainted by the ideology of white supremacy. Despite the application of the principle of equal division to available work, the election of Council officers, and the composition of joint leadership conferences in some trades, there were limits beyond which many white workers were not willing to venture. Blacks served as vice presidents and secretaries of the Dock and Cotton Council, but whites always held the presidency. The city's press reported no black protests over this arrangement, but it did record another, perhaps more important, black challenge. White longshoremen apparently refused to extend the half-and-half principle to foremen, who were overwhelmingly white. This inequity provoked considerable conflict within the black union in the spring of 1904. Vowing to "make a firm stand for equality in everything," some black longshoremen threatened to break their compact with whites unless they received an equal share of foremen's positions. Embroiled in a factional leadership dispute, the black longshore union eventually dropped the issue in the interests of preserving the peace on the docks.[71]

But the issue did not disappear. In late August 1907, a faction within the black union resurrected the half-and-half foremen demand. One participant, frustrated by heightened segregation in the city, laid out the rationale. In the words of the *Picayune,* he declared that

> white people were screening the colored brother off to himself in the street cars; that they would not let him enter the saloons and drink there with them; shoved him closer to the sky in the theatre, and generally denied him his rights . . . [T]he colored brother is bound to have equality on the Levee if he can have it no place else.

This "equality," argued black longshore leader A.J. Ellis, did not constitute a challenge to white supremacy or a demand for social equality. Rather, the issue simply involved a "labor problem." The blacks sought equal representation in these higher paid positions to "conserve the peace of the port and bring added prosperity." After all, the black unionists argued, the longshoremen "could not effect the ends they seek without power over themselves," and it was precisely that power that the black workers demanded. A meeting of the black union men unanimously endorsed the proposal for an equal division of foremen, and union officials hinted that they might raise the issue directly with the shipping bosses.[72]

White longshoremen strongly objected to the proposal. Racial antagonism flared into the open at a crowded meeting of the two associations, jointly chaired by white union leader Chris Scully and black leader E.S. Swann, and held in the black longshoremen's hall. Outnumbered three to one at the meeting, the white longshoremen refused to submit the issue to a vote. "We'll have no nigger bossing us as foremen [sic]," declared one enraged white worker. But black longshoremen were equally vociferous in defense of their rights, and outside the hall a squad of police stood by ready to handle a potential riot. No physical clashes broke out, but prospects for continued cooperation between the two associations seemed bleak. It appeared to the *Picayune* that the half-and-half foremen principle would be "the rock upon which . . . the bark of peace would be wrecked and sent to the bottom of the sea of trouble."[73]

The anticipated break never developed. Likely recalling the riots of the 1890s and fearing the rising level of racial antipathy, more conservative blacks, led by President Swann, forced a reconsideration of the initial half-and-half foremen vote. This time, moderates defeated the so-called radical element and the union withdrew the controversial proposition. "The negro question on the river front is buried forever," the *Daily News* reported Swann as saying. "We have buried the color question, and it will not be dug up again." The move reestablished the racial status quo on the docks and secured a tenuous peace between the two unions. It was, however, a clear defeat for black unionists.[74]

But as infuriating as they found their inability to extend the equal division rule, New Orleans blacks were growing accustomed to defeat, suffering setback after setback in the public sphere in the early twentieth century. Most glaring was the black experience in politics. The Louisiana Constitution of 1898 disfranchised most black voters, but blacks continued to participate in the affairs of the small Louisiana Republican party, which remained a source of very limited black influence and federal patronage. By the start of the new century, even that participation came under direct attack from the Lily-Whites within the Republican party. Believing that the prominence of blacks in their organization prevented it from attracting southern white votes, the Lily-Whites sought to broaden their party's appeal by eliminating black Republicans from the party's councils.[75] On the verge of success in 1904, their project met fierce resistance from black Louisiana Republican party officials and the newly formed Equal Rights League, directed by New Orleans' black water-front union leaders.

The Equal Rights League, in cooperation with the black Central Labor Union, launched a series of mass meetings in February 1904 against the Louisiana Lily-Whites' efforts to send an all white delegation to the Republican national convention to be held in Chicago that year. The list of League speakers and leaders read like a *Who's Who* in the waterfront unions. Participants included I. G. Wynn, president of the black cotton yardmen; James Porter, J.B. Williams, and E.S. Swann of the longshoremen; T.R. Hickerson of the teamsters and loaders; Thomas P. Woodland and Alonzo J. Ellis of the screwmen; and J. Madison Vance, attorney for the black unions. "There is no politics in this movement," the *Picayune* described League president Wynn as saying, "but it is a strong and earnest protest against what the colored people consider a persecution, both in the North and in the South." Wynn promised that branch associations would be organized in every ward of Orleans Parish as well as in the rural parishes. That year, the League's efforts met with success. The Chicago convention rejected the Lily-Whites' delegation and instead accorded recognition to the "black and tan" delegation represented by black Republican leader Walter Cohen. But the impact of disfranchisement, the rise of Jim Crow, and the decreasing interest of northern Republicans in southern blacks was strengthening the hand of the Lily-Whites. In 1908, white Republicans at last achieved their party's official recognition as the legitimate Republican organization in their state. Although black political figures like Walter Cohen and J. Madison Vance remained active in party affairs, blacks were almost entirely excluded from the party councils.[76]

Effectively deprived of the vote and excluded from any influence in the relatively powerless Louisiana Republican party, large numbers of New Orleans blacks nonetheless continued to register to vote by paying their annual poll taxes. The *Southwestern Christian Advocate* advocated payment as the duty of citizens and a source of possible political influence. Moreover, the state applied the poll tax to the public school system, such as it was. James Porter was an active force in promoting poll tax payment. At his instigation, in December 1902, the black Screwmen's Benevolent Association voted to require its 700 members to pay the tax, "and thereby place themselves on record as favoring the educational fund and of making an effort to register and become citizens." Shortly thereafter, the Central Labor Union similarly ordered its constituent unions to follow the screwmen's example. Despite their lack of political rights and influence, black longshore workers continued to pay the tax well into the 1930s.[77]

Black challenges to the new racial order produced few successes. Take the case of the short-lived campaign against the new star-car system of segregated public transportation in 1902–03. In late July 1902, 200 representatives of 60 black organizations denounced the erection of screens and curtains designed to relegate them to the back of the cars. If whites insisted upon segregating blacks, the Reverend E. A. Higgins declared, then "why don't they give us separate cars, with negro conductors and motormen. . . . Before I would ride in a screened 'Jim Crow' car I would walk my feet off.'' Black religious leaders in the New Orleans Ministerial Alliance, with the full support of the *Southwestern Christian Advocate,* called upon blacks to cancel plans for anniversary celebrations, picnics, social excursions, and the like because of the injustice and inconvenience of segregated streetcars. Following the implementation of the car law, black patronage declined somewhat, as blacks choose to "stay at home," or to walk or resort to buggies and bicycles for local transportation. While the Ministerial Alliance's campaign emphasized the system's injustice, the participation of many black groups was based equally upon pragmatic concerns. In November 1902, for example, the black Longshoremen's Protective Union Benevolent Association cancelled its annual memorial service to honor its dead at the Carrollton burial grounds. "There was nothing else to do," President William Penn noted. "We could not get transportation service in the cars" because the streetcar company simply could not accommodate the estimated 3000 blacks who usually participated in the service. Unable to transport their members in a quick and efficient manner, other black groups, such as the Odd Fellows, also postponed or cancelled their balls, picnics, and excursions in 1903. But resistance to the law proved only temporary and ineffective. By and large, black riders obeyed the star-car law. As the *Advocate* observed in mid-1906: "While they detest it they make no open rebellion."[78]

On the issue of public transportation, black New Orleans apparently had retreated from public confrontation by the 1910s. In 1914, a large black mass meeting joined Booker T. Washington's campaign for better accommodations on southern railroads. Its efforts contrasted sharply with that of the Ministerial Alliance a decade before. "Let no one mistake the negro's motive in entering this protest," the participants declared. "He is not pleading for the repeal of the separate car laws, but for an honest enforcement of them." The meeting not only accepted segregated transportation but even held blacks partially responsible for it. "The negro is not without fault in this matter. His public manners, or rather, lack of public manners, make him a very uncongenial fellow-passenger. The Jim Crow car, as it is now run, seems to be designed to meet the needs of that noisy ill-mannered class." The movement's goal was simply to gain improvements in the segregated facilities. Among those chosen to petition various railroad officials were Walter Cohen, the city's black Republican party leader, black union president Albert Workman, and middle-class supporters of the black labor movement, the Reverend H. H. Dunn of Central Congregational Church and Dr. J. T. Newman.[79] Whatever their private feelings on the matter, black leaders seem to have concluded that, given the degree of white intransigence, they could achieve more tangible benefits by working within the system than against it.

While public protests against segregation diminished after the turn of the century, the extensive network of black unions, fraternal lodges, and social clubs continued to address the needs of their members and communities through private charitable activities. As the largest black union in both New Orleans and the South, the black longshoremen's association continued its nineteenth-century involvement in the broader affairs of the black community, requiring its members to pay poll taxes and contributing funds to black churches, schools, and medical institutions. Black union officials also participated in a wide range of activities addressing black concerns. In 1914, for example, longshore leaders Albert Workman and James Porter supported workmen's compensation laws and joined black lawyer J. Madison Vance in seeking more legislative support for Southern University. Another black dock leader, Alexander Paul, served as first vice president of the Emancipation Celebration League in 1912, while Porter acted as assistant secretary. During the First World War, the Central Labor Union assessed its constituent members an annual levy to support the new black Providence Hospital and Training School for Nurses; Albert Workman served as president of its administrative board.[80]

In the late 1910s and 1920s, black union leaders continued to cross the line between labor and political protest. During the summer of national racial violence in 1919, Alexander Paul, Thomas P. Woodland, Sylvester Peete, and other black delegates to the International Longshoremen's Association convention insisted that the ILA "enter its solemn protest against the burning and lynching [of blacks] . . . the greatest menace to civilization." "It is a daily occurrence in the southern states of this country, that negro men and women are being unfavorably and brutally murdered and burned at the stake by mobs and outraged in every conceivable and inhuman manner," they declared. The resolution, which decried "oppression and unjust persecution of the negro race" and proclaimed the "fundamental principles of this Government that every man, even the humblest citizen should have a fair and impartial trial," received approval without further discussion. In the 1920s, Alexander Paul remained active in black protest, serving as black labor's representative to the New Orleans branch of the National Association for the Advancement of Colored People during its successful drive against New Orleans' segregated housing statutes.[81]

The black retreat from open protest against Jim Crow and the heightened emphasis on charity, group improvement, and racial uplift before World War I was a response to the hardening of the new racial order across the South. At the same time, such trends did not indicate a renunciation of public and militant action in all spheres. The city's union movement remained a vehicle for struggle to define the place of both blacks and organized labor in the southern economy. While racial considerations were important in determining the contours of relations among different unions and between black and white workers, they were by no means the only factor. Two case studies from 1907–08 explore the sources of interracial collaboration during the climax of labor conflict in early twentieth-century New Orleans. The first is a jurisdictional dispute between the interracial and industrially organized brewery workers and the white, craft-oriented teamsters, that split apart the New Orleans labor movement. Sizable numbers of white unionists worked closely with blacks, not only to resist the white Central Trade and Labor Council's attempt to destroy an important New Orleans union, but to defend industrial unionism against the craft-dominated AFL: Interracial collaboration reflected both institutional imperatives and an ideological commitment to what black unionists called the "fundamental principles of unionism." In the second case, black and white dock workers reaffirmed their alliance, despite persistent racial tensions, and withstood the steamship agents' and stevedores' renewed drive to reorganize the waterfront labor process. Moreover, they resisted the state government's later efforts to separate the races on the levee. Although by no means free of the ideology of white supremacy, the labor movement contained important currents which ran counter to the dominant impulse toward racial exclusion and subordination in the early twentieth century.

V

Despite the triumph of "pure and simple" craft unionism at the end of the nineteenth century, important currents of oppositional strength persisted within the nation's labor movement. Socialists continued to operate within the AFL, articulating their positions and debating strategies for moving toward the "cooperative commonwealth." Industrial unionism flourished in radical unions like the United Brewery Workmen and in non-radical unions like the United Mine Workers and, to a degree, the International Longshoremen's Association. Outside of the AFL, the Industrial Workers of the World (IWW), formed in 1905, advocated not only industrial unionism but the organization of the unskilled and the overthrow of the capitalist economic order as well. Although the United Brewery Workmen rejected the Wobblies' antipolitical perspective and never joined their ranks, its commitment to industrial unionism and long-standing political radicalism earned it the enmity of the Federation's leaders. By the early twentieth century, the Brewery Workmen found itself at the center of a power struggle within the AFL over broad organizational and ideological issues, a struggle with important ramifications for the New Orleans labor movement.[82]

The unionization of New Orleans' breweries in the late nineteenth and early twentieth century had been a long and difficult process. Unsuccessful strikes for recognition in 1901 involved a city-wide boycott of local beer by the Central Trades and Labor Council. By 1904 the brewery proprietors finally signed three-year contracts with most of the industrial union's locals.[83] At the same time, however, the International Brotherhood of Teamsters, under the direction of organizer Patrick McGill, made important inroads in the breweries. In October 1905 a craft union of keg beer drivers, Local 701, secured a five-year contract, providing somewhat lower wages than the brewery workers' contract and requiring drivers to be members of the Teamsters' union. McGill failed to pressure the brewery workers into relinquishing control over their drivers, but the Teamsters made no further move to dislodge its rivals, waiting for a more propitious moment to assert its jurisdiction over all drivers.[84]

That moment came less than two years later. The AFL set the stage by endorsing the jurisdictional claims of various craft unions to the members of their trades working in breweries. At its November 1906 convention the Federation issued its Minneapolis decision, barring the Brewery Workmen from admitting any new engineers, firemen, or teamsters and allowing existing members to join their respective craft unions. But the industrial unionists overwhelmingly rejected the AFL order in a referendum vote; the beer drivers of New Orleans, Local 215, unanimously repudiated the directive. On June 1, 1907, the AFL Executive Council expelled the Brewery Workmen from the Federation. In New Orleans, this act precipitated a jurisdictional and ideological struggle between the Brewery Workers and the AFL craft locals, a struggle which deeply divided the city's labor movement from mid-1907 to mid-1908. The brewery workers' historian, Hermann Schlüter, summed up the episode in 1910, concluding that the "jurisdiction disputes [among unions] and the injurious manner in which they were often conducted fill one of the ugliest pages in the history of the labor movement."[85] The battles of 1907–08 bear out his assessment.

In late May 1907, New Orleans Teamsters fired the first volley by calling on brewery proprietors to enforce their contract and fire all non-Teamster drivers; most employers readily complied. The Brewery Workers retaliated by calling for a union boycott of those firms and declaring a general strike. On June 1, approximately 130 walked out of the American, Security, Standard, Pelican, and Louisiana brewing companies, temporarily tying up their operations.[86] Almost immediately, strikebreakers filled their places, and many of the companies, guarded by police, resumed business. Only at one firm did work remain at a standstill: the Louisiana branch of the New Orleans Brewing Association, located in a "rough area" adjacent to the docks, where sympathizers of the Brewery Workers held "the fort." Here, the threat of physical harm kept strikebreakers away.[87]

The jurisdictional crisis split the city's labor movement apart, inaugurating what participant Covington Hall called an "internecine labor war." The federation of white craft unions affiliated with the AFL, the Central Trades and Labor Council, faithfully implemented the Minneapolis decision, endorsed the Teamsters, and expelled the Brewery Workers. Denouncing the industrial union's boycott as "unjust and calculated to do an injustice to employers" who were willing to "abide by the decisions of the only recognized bona fide labor organization of the country"—the AFL—the Council called upon "all members of organized labor, who believe contracts or agreements made between labor and capital should be sacred and carried out, to assist in protecting" brewers who had supported the AFL and the Teamsters. One faction, led by machinist and Council president Robert E. Lee, adamantly supported the Teamsters. The *Picayune* reported that Lee had sent all the non-union men he could find to replace the strikers. Although Lee denied the specific charge, he admitted that he had directed unemployed union men to find work at one of the breweries. Perhaps most damaging was the CTLC's overtures to the brewery strikebreakers. At a meeting at Typographical Hall in July, AFL organizer James Leonard enrolled 57 strikebreakers into a new federal union.[88]

These actions generated tremendous controversy within the labor movement. Not only did individual unions censure Lee for allegedly directing strike-breakers to the breweries, but AFL organizer and longshoreman Rufus Ruiz and seven union presidents—including those of four black and white dock unions—telegraphed AFL president Samuel Gompers to insist that he block the formation of Leonard's federal union. Gompers ignored their protest, and Leonard proceeded with his plan. Anti-Leonard protestors, drawn from the ranks of the longshoremen, brewery workers, and their allies, found themselves dispersed by police when they physically demonstrated against the formation of the strikebreakers' union.

Waterfront laborers were perhaps the brewery workers' most important allies. Among the most powerful in the city, the dock unions had practical reasons for opposing the Teamsters' challenge. "The Longshoremen have taken sides with the Brewery Workers from the commencement of the controversy," observed organizer James Leonard in his correspondence to AFL officials,

> and the reason is their busy time is during the winter. In the summer they go into the breweries and work during their dull season. This move on the part of the Teamsters has caused some of these longshoremen to lose their jobs, because they would not join that union, and they are up in arms against the A.F. of L. for expelling the Brewery Workers . . .

The Teamsters' victory directly threatened this system of seasonal employment rotation, since the small craft union, apparently, proposed no reciprocal relationship with riverfront workers.[89] But since breweries offered only limited employment to seasonal waterfront workers, this factor should not be emphasized too strongly. More important was the larger issue of the meaning and nature of trade unionism itself. Throughout the jurisdictional conflict, waterfront workers defended the right of the Brewery Workers to organize along industrial lines. In both practice and ideology, waterfront workers were closer to the interracial industrial unionists than to the segregationist craft unionists. While the basis for waterfront unions was the class of work performed, the Dock and Cotton Council unified the various unions in a powerful coordinating body that enabled them to transcend craft lines. Unlike many of the white craft unions in the CTLC, the waterfront unions relied on sympathy strikes—or the threat of sympathetic action—to achieve their common goals. And upon numerous occasions, longshore workers' behavior affirmed the credo that "an injury to one is an injury to all."[90]

The black Central Labor Union was equally prominent among Brewery Workers' supporters. The issues for blacks were clear cut. The industrially organized Brewery Workers was an interracial union. In contrast, Teamsters' Local 701 would be a "white man's union," its president, N.H. Gross, had explained in May 1907; blacks would no longer drive beer wagons.[91] When Teamsters organizer Pat McGill announced that the Teamsters' Joint Council (embracing all New Orleans black and white teamsters unions) had endorsed Local 701, Central Labor Union (CLU) president T.R. LeBlanc, Dave Norcum of the black cotton yardmen, and James Porter, secretary of both the

CLU and the black longshoremen, publicly repudiated McGill for issuing false statements. The same black labor leaders signed the telegram to Gompers protesting James Leonard's proposed federal union of strikebreakers; after its formation, they condemned the AFL organizer and the white CTLC for violating the "fundamental principles of unionism." During the year-long battle to defend the Brewery Workers, blacks also worked closely with whites in the Dock and Cotton Council and pro-Brewery Workers craft unions. In July 1907, delegates from sixty black and white unions unanimously pledged to boycott all unfair breweries "til such time as justice is done." At the same time, interracial committees attempted unsuccessfully to negotiate a compromise settlement first with the CTLC and then with the brewery proprietors.[92]

With the failure of the interracial committees, in late July black labor leaders initiated their own discussions with members of the white Central Trades and Labor Council. For the first time in its nine-year history, the CTLC agreed to "recognize the colored brother to the extent of meeting him on terms of equality" by holding a joint session with the CLU. A large black delegation led by LeBlanc, Swann, and Norcum sought to convince the white delegates to "join hands" in an effort to settle the ongoing jurisdictional dispute. But too many differences separated the delegates, and the unprecedented conference between the black and white city centrals produced no constructive results. The black workers, the *Picayune* reported, "have espoused the brewery workers' quarrel and went to the meeting . . . to make a determined fight for the strikers" while white delegates remained solidly committed to the AFL's Minneapolis decision.[93]

From the start, the presence of IWW members among the Brewery Workers' allies put AFL organizers on guard. "There is no doubt in my mind," Leonard wrote in early June, that the Brewery Workers' strike "is a move made by the I.W.W. to create dissension in the ranks of labor in this city." Pat McGill similarly condemned the Brewery Workers for "using all of the unfair means at their command to destroy the whole of the trades union movement of this city by having a lot of wind-jammers from the rotten Industrial Workers of the World to address mass meetings" held under the Brewery Workers' auspices.[94] The public statements of the small Orleans Industrial Union No. 38, IWW, which had endorsed the strikers early in the summer, no doubt heightened the craft unionists' concern. Although the "I.W.W. is in no way affiliated with the Brewery Workers, nor they with us," the radicals declared that

> wheresoever we see a labor organization striving to perfect its power, to unify the workers, all our sympathy is with it, and we pledge ourselves to aid the Brewery Workers by every means in our power. Our voice, our pen and our last cent is theirs if they wish it. We are both against Gomperism and for Unionism.

Local Wobblies taunted their opponents mercilessly. On one occasion, they challenged the Central Trades and Labor Council to debate the question, "Which Organization, the I.W.W. or the A.F.L., will best Promote the Interests of the Working Class?" Later that year, a headline in the radical *Labor World* boldly declared: "The A.F.L. Must be Destroyed!"[95]

Leonard's and McGill's assessments of IWW involvement proved exaggerated, if not wholly unfounded. The jurisdictional dispute and the city-wide labor war it provoked were firmly rooted in factional splits and conflicting visions within the labor movement, not in the machinations of IWW activists. The commitment to industrial unionism—and, for some, to inter-trade solidarity and interracial collaboration—that the radicals advocated both existed prior to their arrival on the New Orleans labor scene and survived their departure. But if radicals, socialists, and Wobblies neither instigated the 1907 upheaval nor radicalized the city's labor movement, their efforts were not without impact. Delivering countless speeches and publishing commentaries on the summer's developments, they provided leadership, organizational and material support, and a political language for interpreting the crisis. Tapping into an important undercurrent of discontent within the New Orleans labor movement, they helped to mobilize labor dissidents into a temporary alliance opposed to the AFL's policies.

The radical leadership of the United Brewery Workmen enabled Wobbly sympathizers to assume prominent positions in directing the New Orleans strike. Socialist Joseph Proebstle, one of the International's three chief leaders, personally took charge of the local struggle. His "kitchen cabinet" consisted of a number of leading southern radicals. Covington Hall was a native of Louisiana, a member of the Socialist party and sometimes its candidate for public office in New Orleans, and a strong champion of the new IWW; in the 1910s, Hall would promote industrial, interracial unionism among Louisiana timber workers. Peter Molyneaux, twenty-four years old in 1907, was also a socialist and a Wobbly sympathizer. Initially a reporter for the *Daily States* and the *Daily News,* both owned by Eleventh Ward boss Robert Ewing, he became editor of the *Daily News.* Although the *News* was the official paper of the CTLC, the pro-Brewery Workers Molyneaux covered labor affairs extensively for the paper.[96] Molyneaux had company on the journalistic front in the person of socialist Oscar Ameringer, who moved his newspaper, the *Labor World,* from Columbus to New Orleans in August 1907. Combining satire and sarcasm with political and economic education in his paper, Ameringer distributed the *Labor World* free to strikers and Dock and Cotton Council members.[97] Radical strike leaders also capitalized on a considerable degree of working-class support for the Brewery Workers' cause. Taking their "crusade" to the streets throughout the summer and fall of 1907, the strikers appealed directly to the customers in local saloons, and held mass meetings, torch-lit parades, picnics, and lectures to sustain their members' spirits and preach industrial unionism to other workers. Waterfront workers frequently participated in these public displays of unity, and longshore leaders often joined the radicals in addressing large crowds.[98]

The radicals' influence was most visible when the labor war reached its climax in mid-August 1907. Meeting at Screwmen's Hall, delegates from twenty-five unions—including the Brewery Workers and most dock unions—

denounced the Central Trades and Labor Council as a fetter on union growth and established a rival, independent city union federation, the United Labor Council (ULC). Employing a sharp language of class conflict, the constitution and by-laws of the new United Labor Council, which appear to have been written by Ameringer, closely resembled the Brewery Workmen's constitution as well as the famous preamble of the Industrial Workers of the World. The document called on the working class to emancipate itself from the oppression of both capitalism and the state, insisting that only a league of all unions could effect such change. "There is no power on earth strong enough to thwart the will of such a majority conscious of itself," the document stated, promising the organization's support for a new moral order of community ownership of the means of production. Among its twenty-four articles were clauses advocating the industrial organization of labor, prohibiting contracts that barred sympathy strikes, pledging member unions to support morally and financially any organization in its ranks on strike, and giving the new Council the right to order sympathy or general strikes in support of any of its members.[99]

Covington Hall believed that this body "rapidly became the center of working-class power in New Orleans, and was so recognized (by all except the heads of the A.F. of L. Central Council). . . . This was not only because of its own militancy, but also because it had the steadfast backing of the colored Central Labor Union and the powerful Dock and Cotton Council."[100] But the predictions—and hopes—of radicals notwithstanding, the United Labor Council assumed no such prominent role as an independent radical federation. The new body did little more than attend to "routine" business at bimonthly meetings, sponsor a lecture series, and establish a committee to help the Brewery Workers. In effect, after bursting on the scene, it slowly faded away. At least three reasons for the Council's failure stand out. First, on an ideological level, nothing in their previous or subsequent experiences suggests that dock workers, or other members, subscribed en masse to the anticapitalist doctrines promulgated by the radical federation. The ULC's very existence was a tribute to the radicals' power of inspiration and a testament to the depth of discontent with the AFL, not a sign of deeply held political beliefs.[101] Second, on a practical level, the Dock and Cotton Council shifted its energy to a prolonged fight with the port's steamship agents and stevedores immediately following the ULC's establishment. Dock workers' own struggle for control over the waterfront, and perhaps for the very survival of the screwmen's unions, took precedence over building an alternative to the city's white AFL central body.

Most important in undercutting the new organization's potential were various national developments during the Brewery Workers' strike. From the outset, the United Brewery Workmen had sought not independent radicalism but rather readmission, on its own terms, into the AFL. Ideologically, it had consistently opposed dual unionism. As a practical matter, it remained dependent upon AFL unions' support for boycotting beer produced by recalcitrant brewery owners. The Federation too had reasons for seeking reconciliation. Allies of the Brewery Workmen within the AFL were joined by some Federation leaders, who viewed readmission as a way of keeping this important union out of the ranks of the IWW. The AFL voted to restore the Brewery Workmen's charter at its November 1907 Norfolk convention; early the next year the Federation formally readmitted the union.[102] In New Orleans, healing the sharp divisions within the labor movement proved difficult. The Brewery Workers' strike—which lasted one year and four days—ended on June 4, 1908. The Central Trades and Labor Council readmitted the delegates from the Brewery Workers' locals of inside workers, beer bottlers, and box repairers on the condition that they withdraw from the United Labor Council, an action which spelled the death of the rival city central. The CTLC also recognized beer drivers' Local 215's delegates, voting in favor of the honoring of their contracts and the reinstatement of all men who quit work or were discharged during the previous year's strike. But the ultimate question of jurisdiction, it ruled, would be decided by a referendum vote by members of locals 701 and 215, when certain contracts expired in November 1908. After an entire year, the Brewery Workers' emerged weakened but apparently victorious.[103]

In the end, its victory proved elusive. While the union's strong left-wing leadership moved on, Teamsters organizer Patrick McGill continued his local efforts. One by one, the *Picayune* reported two years later, virtually all beer drivers had signed up with Local 701. McGill finally cracked the last two "citadels" of Brewery Workers' strength—the Consumers and Dixie breweries—when he signed a contract with those brewery bosses on August 1909, forcing loyal Brewery Workers' drivers to join the Teamsters or lose their jobs. Two and a half years after the outbreak of the labor war, every keg and bottled beer driver in New Orleans belonged to Teamsters' Local 701. Organizer McGill, boasting to the journal of his international, claimed not inaccurately that the brewery workers "have received the worst licking they have ever been favored with in the history of their organization." In stark contrast to the 1907–08 upheaval, no recorded protests, from industrial unionists or black labor leaders, greeted the Teamsters' final success.[104]

The jurisdictional dispute brought to the surface sharply contrasting visions of the "fundamental principles of unionism." In their protests, black workers opposed both the Teamsters' racially exclusionary practices as well as the white CTLC's advocacy of a narrow craft unionism and its behavior toward the Brewery Workers and its allies. Although the Central Trades and Labor Council had promoted a whites-only craft unionism for almost a decade, substantial numbers of white craft unions were willing to defy Jim Crow and make common cause with unskilled black workers to defend the principle of industrial unionism. Aided by socialists and radicals, the alliance of black and white sympathizers of the industrial unionists demonstrated that interracial collaboration was indeed possible in the age of segregation. But this alliance failed to sustain itself beyond the resolution of the jurisdictional crisis; the white craft unions had made their peace with the AFL city federation by mid-1908, and the broader labor movement resumed its development along racially distinct trajectories. The waterfront, however, remained an important exception to this trend. During the height of the "internecine labor war," the interracial labor alliance on the docks withstood the most serious employer assault of the decade.

VI

A strike wave in the fall of 1907 marked the climax of the struggle for control over the waterfront labor process. Long frustrated by their contracts with longshoremen and screwmen, shipping agents and contracting stevedores launched an offensive to strengthen managerial authority, weaken union work rules, and increase work loads. Wages were not the issue. Rather, it was the fact that dock workers had put steamship agents in "such a position . . . so as to be almost controlled by the laborers." As we have seen, the first showdown with longshoremen in September 1907 ended in a stalemate, with workers retaining much of their recently won power. But however great their frustration with the longshoremen, it was nothing compared with how employers felt about screwmen. The 1903 contract, which placed a 160-bale daily limit on hand-stowed cotton per gang, "maintains and increases the hardships, delay, expense and difficulty under which cotton and tobacco are loaded at this port," the agents complained. The port's commerce "is simply being strangled to death." Since their defeat four years before, waterfront employers had lived unhappily with the port's most powerful unions. With the expiration of the screwmen's contract on September 1, 1907, they renewed their war for control over union rules restricting the amount of work performed. Initially demanding that screwmen stow 200 bales per gang per day, employers soon vowed to settle for nothing less than full parity with Galveston, which they defined as the daily loading of between 250 to 300 bales.[105]

It was a "case of now or never," the employers concluded, a "fight to the finish" with their most skilled workers. Few in the business community doubted that the outcome would determine conditions on the docks for the foreseeable future. A committee representing the city's numerous commercial exchanges threw its full weight behind the agents; its resolution endorsing the use of non-union labor, the *Times-Democrat* noted, was "tantamount to a declaration of war by the business men upon the labor organizations, and it was the consensus of opinion that it was better to fight the matter out now and retain the maritime interests of the port than to defer action with the chance of having all ships diverted to Galveston."[106] Determined that their ships would not remain idle, employers ordered crews to load the cotton on the docks; by the second week of October, strikebreakers from labor agencies in St. Louis and Chicago began arriving by rail. The vast Stuyvesant Docks controlled by the Illinois Central Railroad were converted into a "massive fortress." Officials placed watchtowers around the property and ordered armed guards to patrol the grounds. Railroads brought in almost 400 strikebreakers, housing them on a steamship anchored in the Mississippi River off the Stuyvesant Docks. The state militia stood ready to move into New Orleans at a moment's notice. With their own armed forces and the power of the state government firmly behind them, the agents announced that they would resume negotiations only when the screwmen agreed that the settlement would place the port on parity with Galveston—that is, when the screwmen surrendered.[107]

The crisis provided employers with their first opportunity in four years to restructure the organization of waterfront labor, and it stimulated creative thinking. One proposal floating in commercial circles suggested a piecework system, under which screwmen would be paid by the bale: the men would "receive pay for the work they did, and more pay for more work. All the questions of rules would be eliminated and the labor situation would be placed upon a simple and common sense basis." By October, rhetoric had hardened considerably. Agents called openly for the abolition of the screwmen's trade and the creation of a single system of general dock labor.[108] The agents, the *Times-Democrat* explained, believed that

> they have the strike of the dock laborers well in hand, [and] steamship agents and stevedores are making their plans for the future management of labor on the levee. This embraces a general association of laborers, combining in one body freight handlers, longshoremen, screwmen, and men of kindred trades. It is claimed that the main reason for this is that the present labor trouble is the outgrowth of dissatisfaction by a class of labor which in reality has no right to exist. The clamor now among the employers is for the elimination of the screwmen's unions and for the consolidation of the three classes of dock labor into one.
>
> In this way one contract will have to serve for all, and there will be no such thing as fighting over a bonafide contract as a means of showing sympathy for another union.

The employers' importation of strikebreakers and their rejection of arbitration offers suggested to many observers that their "real object" was "not so much a just settlement of the pending difficulties as it is the destruction of the Screwmen's Union."[109]

Workers were convinced that the outright assault on the screwmen was "only the forerunner of a war of destruction upon all the other labor unions" on the riverfront. Individuals and unions took action even before the Dock and Cotton Council announced its position. On the Leyland line wharves, union officers ordered the longshoremen to knock off when sailors began handling cotton. Black cotton teamsters refused to deliver loads to ships at those wharves; the Illinois Central freight handlers refused to unload cars with cargo bound for the Leyland or Austro-American lines; and 200 Morgan Steamship Line freight handlers similarly struck in sympathy. On October 4, the Dock and Cotton Council finally ordered a general strike. Over 8000 workers struck every shipping agent and stevedore that had not signed the screwmen's tariff.[110]

Despite the bitter factional disputes during the previous spring and summer, all of New Orleans' organized workers offered at least moral support to the levee laborers. As expected, the United Labor Council, in which the dock workers were a strong force, endorsed the strike. More surprising was the action taken by Patrick McGill, the southern Teamsters' representative and a key figure in the AFL war on the Brewery Workers, who called out an estimated 175 black and 165 white stave cart drivers and 160 black round freight teamsters in sympathy. Central Trades and Labor Council president Robert E. Lee, no friend of the dock workers, also expressed his sympathy. "No foreigners can come here and dictate the terms and conditions under which free-born Americans are to labor," he declared, referring to the position of the agents of the European shipping lines. With little prospect of settlement, the city's press reported that the Central Trades and Labor Council, the Central Labor Union, and the United Labor Council were considering a city-wide general strike to force the agents to their collective knees by tying up every union shop in the city. Only once before, in 1892, had New Orleans businessmen faced the united power of most unionized workers. What might have been the most significant development in the city's labor movement in a decade and a half, however, never materialized. "Sympathetic strikes are antiquated," explained CTLC president Lee. "[W]e win fights these days with moral support and money. No matter how serious the situation becomes, the unions affiliated with the Central Trades and Labor Council will stand by all fair employers." Instead of declaring a general strike, the white AFL central body offered moral and financial aid.[111]

Pressures for arbitration grew as the strike dragged on. The screwmen's willingness to make modest concessions impressed Mayor Martin Behrman, and meetings between the mayor and union officials produced a proposal that finally ended the strike. Under the compromise plan, the screwmen agreed to return to work and load 180 bales a day pending binding arbitration. But the Dock and Cotton Council insisted that the screwmen not become scapegoats for all the port's problems. Dock workers, supported by the mayor, contended that since there were other factors endangering the port's trade besides "alleged labor oppression," a full investigation was required before any final decisions were made.[112]

The agents' intransigence elicited considerable criticism in many circles and produced divisions within their own ranks. The faction blocking the settlement, holding out for a 200-bale limit, included the port's largest shipping agents—Sanders of the Leyland line, LeBlanc of the Harrison line, and Ross, of Ross, Merrow and Howe—the same men who led the assault on the screwmen in 1902–03. Once again, a split between large and small employers undermined the anti-labor offensive. "It has been known for some time that things have not been altogether rosy in the ranks of the ship agents, and now they are on the verge of a split," the *News* suggested.

> Certain of the ship agents, becoming tired of the present condition of affairs . . . decided that it would be best to accept the proposition of the screwmen. . . . Others, however, were of a different opinion and argued in favor of standing pat. . . . The agents who oppose the settlement on that basis, it is said, control more ships coming into this port than all the others put together.[113]

Confronted by a growing rebellion from smaller agents and stevedores and criticism from the mayor and some commercial exchanges, the agents reluctantly accepted the 180-bale proposal as a basis of settlement. The Dock and Cotton Council next called off its three-week general strike, and the 8000 waterfront workers returned to work. Once again, the agents had failed to destroy the screwmen.[114]

Behrman's compromise proposal, which called for the creation of an arbitration committee to investigate the port's overall condition, immediately encountered difficulties. Holding fast to their half-and-half principle, screwmen appointed two black representatives to the investigative body. The representatives chosen by the steamship agents and stevedores—businessmen not from their immediate ranks—protested this arrangement. "The committee will have many important things to look into," one prominent businessmen on the committee declared, "and I, for one, can't see how a negro belongs on it." Oscar Ameringer described the scene as he remembered it in 1940:

> The hue and cry that followed the announcement of the make-up of the workers' delegation came near to bringing the stars in their courses to fall on Louisiana.
>
> "What! Meet with niggers in the same room, around the same table, discussing a problem concerning the superior race exclusively?" Was it not terrible enough to meet common dock wallopers, water rats, white trash, in the same room, around the same table, to discuss as equals—well, almost equals—the weal and the woes of an industry in which the workers had not invested a red cent? Was it not terrible enough that men could no longer run their own business as they saw fit?

The purpose of this turmoil, Ameringer argued, was to "destroy the solidarity of the two races." A more likely reason, though, was the lack of experience these business representatives had in dealing with interracial unionism or in negotiating with black workers. Unfamiliar with the long history of black participation in labor affairs on the waterfront, they refused to serve on any committee with black members.[115]

When the appointment of black representatives threatened the arbitration committee's formation and a renewal of the labor conflict, Behrman again intervened. Upon receiving the list of labor delegates to the committee, he insisted that the messenger, black screwmen president Nelse Shepherd, withdraw them and "let the members know that interests demand that there shall be no colored men on the committee." In their place, Behrman suggested that the black screwmen elect two whites to represent them. This the black unionists refused to do. Behrman personally appeared before a Dock and Cotton Council meeting and asked the black union members, Oscar Ameringer recalled, "to sacrifice their representation temporarily, in order that the peace, tranquility and prosperity of New Orleans might be restored." In the discussion that followed, "every white speaker declared himself in opposition to the withdrawal of the four Negroes" and the Council almost unanimously de-

feated the motion for reconsideration. "I tried to make these people understand that they would display bad judgement should they insist on colored
men being on the committee," Behrman later informed the *News*.

> I told them very plainly that the sentiment of the community is [against]
> having colored men figure so prominently in public matters. I suggested that
> surely there must be some white men in whose hands they might entrust
> their case. It did not matter whether the white men whom they might select
> would be screwmen, longshoremen, cotton yardmen or men from any other
> walk of life. Despite all that I said to them, however, they have persisted in
> being represented by men of their own race. Of course, under the terms of
> the agreement for the investigation, they cannot be denied this representa-
> tion. My sole purpose was to try to have them appreciate the sentiment of
> this community on a question of this kind.[116]

In the end, repeated failures to select a neutral umpire and the rejection of
the black delegates by the employers' representatives led to a new approach
to securing levee labor peace. Although the strikes were over for the moment,
Behrman explained, "there is a reasonable ground for the apprehension that
the port of New Orleans will continue to deteriorate until proper steps are
taken to remedy all of the abuses which are alleged to exist." At the mayor's
request, Governor Newton Blanchard called on the state General Assembly
to establish an impartial commission with full power to investigate all port
charges and take remedial action.[117] Thus, with the aim of "ascertaining
wherein lie the differences" in all port charges between New Orleans and its
southern rivals, on November 20, the Assembly created a five-person committee to

> examine into pilotage charges, towage charges, demurrage charges, dock
> and wharf charges, railroad and steamship freight rates, charges and regula-
> tions, labor charges and regulations, and all other charges and regulations
> affecting the cost and moving of merchandise and import and export busi-
> ness at the port of New Orleans, whether made or imposed by persons,
> firms, associations or corporations.

The Assembly-appointed committee—composed of two state senators and
three state representatives—began its labors in January 1908, conducting a
thorough investigation of conditions in Galveston, Savannah, Pensacola, and
Mobile. From mid-February through early May, the Port Investigation Commission focused its full attention on the numerous factors influencing New
Orleans' high charges.[118]

"Can you explain why we had a strike here last year, and Galveston has
not had one since 1881?" State Senator C.C. Cordill asked Cotton Council
president James Byrnes during the Commission's hearings. While Cordill
was wrong about the history of waterfront labor conflicts in Galveston, the
Port Investigation Commission claimed to find at the time "no hostility
between employer and employed at Galveston; none at Savannah; none at
Pensacola, and none at Mobile. At each port capital and labor were in
accord, working for the upbuilding of the port and common welfare of

both." The contrast with the Crescent City was stark. In its final report, the Commission found in New Orleans "a condition of hostility, distrust and suspicion existed which created and maintained an open rupture. Each side demanded, neither conceded; each accused and recriminated." Addressing his remarks to representatives of capital and labor, one legislator maintained that "you were both to blame and both were crippling the community to further your selfish ends. We must not mince words about it; both sides are to be condemned."[119]

But Commission members were as obsessed with the influence of black workers as with the broader issue of class conflict on the docks. Interracial collaboration was a special target of attack. In their eyes, the screwmen's amalgamation through their half-and-half compact, in which black and white screwmen worked abreast of each in every hatch, had created a "bad condition" closely resembling "social equality." Senator C.C. Cordill—described by Ameringer as the "composite portrait of the Kentucky colonel seen in whiskey advertisements"—put the question directly to Dock and Cotton Council president James Byrnes: "Do you think it fair for 600 men to enter into an alliance with a band of negroes to bring about a disastrous strike, and cause the men who grow cotton to lose $15 on each bale?" The senator answered his own question: "No six hundred men can in justice amalgamate with negroes and tie up this port." Cordill was similarly shocked when he learned how the dock unions actually functioned. Steamship agents testified that the joint Executive Committee of twelve black and twelve white screwmen exercised an unreasonable power on the waterfront. "Do I understand you to say that twelve white men and twelve negroes dominate the commerce of this port?" Cordill asked. "Well, we are practically under negro government." Another state senator concurred, warning of the "dangers of black supremacy in the dock labor" which "if allowed to continue in the amalgamation basis with negroes overriding whites future generations would suffer."[120]

The question "who Africanized the labor on the river front" received a good deal of attention. Small, local employers of waterfront labor had little good to say about the large firms that dominated the export of cotton, blaming British ship agents for the state of racial affairs. "I'm a Southern man . . . [and] I believe in white supremacy," stevedore John B. Honor testified. "Remove the cause, and we won't have the whites and the blacks together." That "cause" was M. J. Sanders and his interference in the local labor movement. "There'd be no necessity" for the amalgamation of the races, he concluded, "without the Leyland and Harrison Lines." Their manipulation of racial tensions had exacerbated the crisis of 1894–95; in subsequent years, deteriorating conditions drove blacks and whites to "amalgamate" in self-defense. Thus the largest companies involved in the 1907 strike were responsible for creating the very racial conditions the Port Investigation Commission found so distasteful. If the "Africanization of the waterfront" could not be reversed, the "amalgamation" of the races—the system of blacks and whites working side by side— was not beyond challenge. In their discussions of "the Negro Question," the Commissioners repeatedly called for the separation of the races on the levee for "sociological reasons." Waterfront unions, however, simply ignored their charges, and the Commission took no action to implement its proposal.[121]

The Commission's final report dealt with a wide range of issues, including pilotage, towage, demurrage charges, dock and wharf charges, railroad and steamship freight rates and regulations, and labor costs. Adopting an optimistic tone on the subject of New Orleans' future, the Commission recommended the elimination of both the Harrison and Leyland lines' monopoly of the steamship trade and illegal railroad charges on export cotton, the encouragement of tramp liners, and a thorough overhaul of the Maritime Branch of the Board of Trade's costly inspection system. The Commission's intervention also addressed the conflict between agents, stevedores, and screwmen. Its final report declared that labor in New Orleans was not overpaid, "nor has it found that it is any better or any worse than it is at other places." Empowered by the Louisiana General Assembly to resolve the outstanding differences between waterfront employers and workers, the Commission ruled that hand-stowing 187 bales of loose cotton per gang—up from the original 160 bales and the compromise of 180 bales—would constitute a fair day's work. That decision increased the screwmen's current limit by seven bales, but fell far short of the agents' and stevedores' initial demand that their workers stow 200 bales and their later demand that they stow 240 bales. Despite the opposition of a considerable number of workers, the "hatchet was buried" at long last in early May when agents, stevedores, and screwmen signed an unprecedented five-year contract. Pleased with the outcome, the Port Investigation Commission concluded that, in the face of a "satisfactory contract . . . between employers and employees, antagonisms heretofore engendered cannot live, and strikes and lockouts cannot occur. No further damage or danger to our commerce need be apprehended."[122]

How little did the Commissioners understand the dynamics of the waterfront. No sooner had the contract establishing the new screwmen's limit been signed than other waterfront workers disrupted the new levee peace. One leading employer pessimistically concluded that New Orleans was "really no better off today than it was this time last year, as far as the labor question is concerned." In early July, longshoremen struck for an extra man per gang in order to compensate for the screwmen's increased limit. Agents conceded only when E. F. Kohnke, chairman of the old commercial exchange committee which had endorsed the Steamship Conference's war against the screwmen, personally sided with the workers, arguing that employers technically had violated the "rule clause" in their contract which prohibited both employers and workers from changing the working rules in any way.[123] A strike by Illinois Central freight handlers in early July for a 40 percent wage increase posed a more serious threat to the new levee peace. The railroad, invoking its traditional strategy, immediately procured strikebreakers. But after almost a year of conflict, unions in the Dock and Cotton Council were reluctant to renew labor warfare on the waterfront and called for the strike's termination. After two weeks, the Council withdrew its support from what it considered to be the strikers' unrealistic demands; it ordered the freight handlers to accept a five-year contract based on the old conditions and rates of pay. When the strikers failed to obey, the Dock and Cotton Council threatened to expel them. Faced with the prospect of continuing their confrontation with the powerful railroad without the support of organized labor, the freight handlers yielded, signed a five-year contract, and returned to work.[124]

Despite their vast economic resources and determination, steamship agents and stevedores failed to alter the labor process on the New Orleans levee in any significant way, the small increase in the screwmen's limit notwithstanding. The Dock and Cotton Council held firm during the crisis of 1907–08; supported by the range of riverfront unions and a city government committed to compromise, arbitration, and the port's prosperity, longshore laborers and cotton screwmen emerged from the crisis relatively unscathed. In the struggle for control over the labor process, screwmen and longshoremen had shifted the balance of power decisively in their own direction.

At the same time, they successfully resisted pressure from local and state politicians to modify or renounce the interracial alliance that lay at the core of their power. "We have had white supremacy and we have had black supremacy on the levee," testified black longshore president E. S. Swann, referring to the racial crisis on the docks in the 1890s, "and there was trouble in each case. Now, we have amalgamation and freedom and we are getting along all right."[125] Swann was exaggerating, for on numerous other occasions, he and his fellow black union leaders protested racial injustice at the workplace, in the union hall, and at party conventions. But whatever its limits, the interracial collaboration fostered by the Dock and Cotton Council still violated the "sentiment of the community." The segregationist impulse that infused growing sectors of southern life, the labor movement included, encountered sharp, if partial, resistance from blacks and whites on the New Orleans waterfront. If the sources of that resistance lay not in some broader vision of working-class interracialism but rather in economic necessity and the historical circumstances of the New Orleans waterfront, it proved sufficiently strong to guarantee at least some protection and power for its members through the first two decades of the twentieth century.

6

The Search for Stability:
The Crisis of Labor Relations in the
Era of the First World War

The Port Investigation Commission's resolution of the waterfront crisis in 1908 ushered in a more stable era for both unions and employers. In sharp contrast to the class warfare of the opening years of the twentieth century, manifested in the bitter struggles over work rules and the extent of managerial authority, the post-1908 period was characterized by limited cooperation between the Dock and Cotton Council, steamship agents, and stevedores. This relationship remained intact for nearly a decade, eroded only under the domestic impact of the First World War. During the war years, the federal government became a party to the system of labor relations; the mediation machinery it imposed upon workers and employers constituted a new arena of struggle. The postwar era witnessed no return to the earlier, cooperative framework. Increasingly powerful employer associations, now backed by pro-business state and municipal governments, prepared for a final assault to eliminate union power. The New Orleans Steamship Association's victory in 1923 dealt a crippling blow to the black and white workers whose organizations had exercised formidable power, with one exception, for four decades.

I

The five-year contract lay at the heart of the post-1908 framework of labor relations. Although the Dock and Cotton Council initially opposed longer-running contracts, labor leaders quickly came to appreciate the security and power they offered. Thomas Harrison, president of the white screwmen, boasted to the 1913 convention of the International Longshoremen's Association that his union had "the best contract existing anywhere, because it is the only contract which permits us to say just how much cotton we will stow in one day and when we have done that we do no more."[1] Two years later, black screwmen's president Thomas P. Woodland described the New Orleans situation in similar terms:

> We have a five-year contract with our employers, and then we have a clause in that contract to the effect that nothing in the agreement will prevent the

local from going out on a sympathetic strike if a sister local connected with the International is in trouble and we are authorized by the I.L.A. to assist them. We have tried to work in New Orleans under a one-year contract, but it worked a hardship, hence we succeeded in having a five-year contract signed. We are not tied down in New Orleans . . .[2]

Employers of dock labor too had learned the benefits of the new system of labor relations. Abandoning the rhetoric of the employers' right to control and the principles of managerial authority, members of the New Orleans Steamship Association found that their workers' institutions could promote discipline and profitability. If waterfront employers had once expressed considerable irritation over the grievance process, they now encouraged their workers to utilize it to adjust any complaints that might arise. The bottom line, they learned, was that industrial peace could pay. Agents and longshoremen and screwmen's unions renewed their five-year contracts without hesitation in 1913. To encourage port development, the screwmen's unions agreed to minor increases in the amount of tobacco stowed each day; in exchange, the Association conceded on several working rules. "The screwmen met us in a spirit of fairness at all times," noted the head of the Steamship Association.[3] So strong was the commitment to the new framework of industrial relations that the agents and stevedores stood firmly by the longshoremen in 1914 when the unions rejected the municipal Dock Board's program of experimenting with labor-saving machinery. Exhibiting this new cooperative spirit, employers and unions had come a long way since the continuous struggle over work rules and the amount of work performed that marked the twentieth century's early years.[4]

The new system of labor relations extended to other cotton trades as well. Although cotton yardmen and the press owners still signed one- or two-year contracts after the 1908 settlement, their relationship also was marked by cooperation. Following a poor year for foreign trade in 1909, cotton factors, press owners, and cotton yardmen inaugurated a "general movement" in 1910 to rejuvenate the cotton trade by reducing the costs for storage and handling. Adopting what the press viewed as a "patriotic" stance, the yardmen "voluntarily consented" to a tariff with slight wage cuts, which they renewed without complaint in 1911, 1912, and 1913. Also contributing to the civic effort to promote local commerce, black teamsters and loaders voluntarily reduced their wage tariff slightly in 1911.[5]

Such collaboration between old enemies generated harsh criticism from one local radical, the socialist and Wobbly Covington Hall. The radical movement that Hall had promoted in 1907 and 1908 collapsed following the Brewery Workers' reconciliation with the Central Trades and Labor Council, and the IWW made little headway on the New Orleans docks in the ensuing years. Only on foreign ships did the Wobblies attract a small following among disgruntled seamen. Hall had little hand in even that, having moved on to organize rural Louisiana's black and white lumber workers in the IWW's Brotherhood of Timber Workers. After the crushing of that interracial labor movement in northern Louisiana in 1913, Hall returned to New Orleans and took a moment to reflect upon the history of the Dock and Cotton Council since his departure. Writing in *Voice of the People,* an IWW newspaper he edited, Hall condemned the Council's leaders as men who "were courting the favor . . . [of labor's] enemies, the shipping trusts." Collaboration with men of commerce was bad, but worse yet were the "treasonable" five-year contracts. The "worthless" agreements, Hall argued, "completely destroy the unity of the New Orleans Dock and Cotton Council which was only a few years ago practically an Industrial Union and the strongest and most militant labor organization in the Gulf States." The Wobbly indictment of the Council was savage, and Hall wasted no opportunity in condemning his old allies. "Where is the Dock and Cotton Council of the olden days," he asked his readers?

> The same old D. and C.C. that raised wages and made working conditions along the river front in New Orleans as good if not better than any port along the Atlantic and Gulf Coasts? Does it still exist? Yes, but sad to say, in name only . . . [I]ts members are slaving along the front, bound hands and feet, by contracts that were not of their making, and working under conditions which in the palmy days of old, would have caused them to turn the whole damn front topsy turvy at the very thought of it.[6]

In numerous articles, the *Voice* painted a dismal picture of working conditions on the docks: wage rates were low; union foremen competed for their employers' favor by driving union members "as if they were convicts on a chain gang"; blacks received less than their share of jobs; few men could get "more than three or four days of work a week"; half the workers on the front remained "unorganized and the old time organizations are making no effort to organize them"; and vast disparities in wage rates existed between different sectors of waterfront work. Once an aggressive body engaged in a continuous struggle for the workers' betterment, the waterfront unions had become benevolent societies dispensing sick and death benefits. So "ends the once powerful and militant Dock and Cotton Council of the Port of New Orleans," Hall concluded.[7]

Hall's portrait was only partially accurate. Irregular work had long characterized the longshoremen's trade not only in New Orleans but around the world, and disparities in wage rates and conditions persisted in different sectors of waterfront work. But while the rest of Hall's critique may have applied to the docks of United Fruit or Southern Pacific, where workers exercised no union power at all, there is no evidence that the principal Dock and Cotton Council unions labored under "chain-gang"-like conditions. The wages, working conditions, and work rules of its dominant members—the longshoremen and screwmen—compared very favorably to other ports in and beyond the South. The *New York Age,* a black newspaper, described the black Longshoremen's Protective Union Benevolent Association in 1911 as one of the "strongest and most influential labor organizations in the Gulf States." Before the conventions of both the Gulf Coast and South Atlantic District of the ILA and the International union, New Orleans labor leaders, black and white, praised their system of organization. "The conditions in my craft are second to none," boasted black screwmen's president and ILA vice president Thomas P. Woodland to the 1914 ILA convention. Moreover, Woodland attributed his union's success to the "industrial equality" involved in the equal division of work and the "amalgamation" of the black and white locals in a biracial alliance. As ILA organizers and as union officials in New Orleans, Woodland and Thomas Harrison assisted unions in other southern ports and at times lobbied them, with varying degrees of success, to model their organizing efforts along the lines established in New Orleans. Even Booker T. Washington, no friend of organized labor, offered guarded praise for the Dock and Cotton Council in his 1913 assessment of the "Negro and the Labor Movement."[8]

Hall not only exaggerated the Council's weakness, but he romanticized its history as well. Throughout its entire existence, the Council (like its nineteenth-century predecessor, the Cotton Men's Executive Council) had done nothing to challenge the pre-existing hierarchy of waterfront occupations. Cotton screwmen continued to occupy the highest position, followed by longshoremen and yardmen; beneath them stood various teamsters, freight handlers, and coal wheelers. The impressive solidarity of the early twentieth century notwithstanding, the Council, dominated by screwmen and longshoremen, often neglected the struggles of its weakest members. Even at the height of strike activity in 1907—Hall's "palmy days of old"— the Cotton Council did not behave like a good Wobbly union. Solidarity across racial and trade lines characterized the general dock strike that fall, but sectoral divisions and tensions persisted. As soon as longshoremen and screwmen signed their contracts in 1908, they turned their backs on the Illinois Central freight handlers, ordering them to end their strike. Primarily an organization of cotton handlers, the Council was most "aggressive" when its dominant members faced challenges; only then would it turn "the whole damn front topsy turvy." That aggressiveness never extended to serious support for the unorganized. Hall was correct to point out that the Council made no sincere effort to organize the rest of the waterfront; he was wrong to assume that things had ever been all that different.[9]

Hall was on firmer ground when he observed that the Council now courted the favor of the steamship interests. Such action was less an expression of opportunism than of a deeply felt conservative impulse. Black waterfront leaders had long counseled patience and the strict observation of the contract. In an uncertain economy, they felt, their members' security depended upon the stability that came from observing the letter of the contractual law. After 1908, Thomas Harrison, former state commissioner of labor, former figurehead of the radical United Labor Council, and longtime Cotton Council officer, converted to this philosophy. "[A]s long as I can speak in favor of this system [of industrial relations on the New Orleans docks] I shall do so," he told delegates to the ILA's 1912 convention. At the same time, he stressed labor's own responsibility: "I believe as long as a contract is entered into, whether it be good or bad, we should religiously live up to it."[10] This was a stance waterfront employers could easily endorse, and steamship agents and stevedores now relied upon waterfront labor leaders as indispensable allies in their quest for uninterrupted expansion. Businessmen applauded the Council's new-found patriotic spirit, its willingness to place the port's interest before individual interest. As a result, one Cotton Exchange official noted in 1911 that "there was a better feeling existing between the cotton interests and the levee workers to-day than ever before."[11] In a context defined by civic boosterism and concrete measures designed to make New Orleans a more competitive port, the business elite found it convenient to grant to the waterfront unions the status of junior partners in progress.

But along with status came responsibility, and in one particular moment of potential economic crisis, the city's commercial elite did not hesitate to solicit labor's help. The responses of businessmen, waterfront unions, and rank-and-file workers to the Illinois Central Railroad strike revealed in sharp relief the contours—the extent and limits—of the new system of industrial relations. In the summer of 1911, ten railroad craft unions of the Illinois Central formally joined together into a system federation and demanded that railroad managers bargain with them. The Illinois Central refused to recognize the union federation, and the chance of a strike involving between 12,000 and 20,000 men loomed large by late summer. Given the centrality of the railroad to the economic well-being of the New Orleans cotton trade, the city's men of commerce greeted the news of the impending tie-up with alarm. In late August, the Cotton Exchange established a committee designed to avert the Illinois Central walkout and prevent the "untold damage" that it would inflict upon the port by paralyzing its business. Joining delegates from the Cotton Exchange, Board of Trade, Sugar Exchange, Auction Exchange, and the Progressive Union were the leaders of the Dock and Cotton Council's affiliated unions, black and white. Labor leader Thomas Harrison assured the business community that the Council would do "everything in its power to prevent a strike."[12]

These efforts failed. In late September, Illinois Central workers, including hundreds in New Orleans, walked off their jobs. For the commercial elite, the rail strike was bad enough; worse yet was the threat of a sympathetic walkout by union freight handlers, which theoretically could drag the Dock and Cotton Council into the fray on the strikers' side. But when the strike did spread to the freight handlers' ranks, the outcome proved to businessmen that collaborating with labor leaders could pay off in times of crisis. When ninety white freight handlers unexpectedly joined the striking clerks and car repairers in sympathy on September 28, their actions met with stern disapproval from the ranks of waterfront labor's leadership. The Dock and Cotton Council made clear that while it would support the shopmen's strike, it would take no action in the conflict nor offer sanction to the freight handlers' unauthorized strike. When gangs of screwmen working two ships at Stuyvesant Docks refused to accept cotton delivered by non-union handlers, the screwmen's union actually replaced the recalcitrant men.[13]

Racial dynamics were evident as the local strike unfolded. As the white handlers marched to strike headquarters that first afternoon, they were greeted by a sign that read "Welcome Brother Carmen, Freight Handlers, White and Colored." Some individual black unionists "seemed infected with the strike germ," and a number of them attending an impromptu interracial meeting at Stuyvesant Docks had agreed to join the whites' sympathetic strike. Other black Illinois Central freight handlers at the Poydras terminals arrived at a similar conclusion earlier that day, joining the strikers. Yet the overwhelming majority of black workers remained on the job, and those who struck quickly returned. Black union leaders played an important role in stemming the strike's spread, repeatedly imploring their members to remain at work until the Dock and Cotton Council formally ordered them out— which the Council refused to do. In a public statement, the leaders of the black freight handlers criticized their white counterparts for striking without the approval of either the Council or the ILA. The white union, they complained, "did not even call our committee to accompany them" to their employers. They

> called their men off the docks, and expected our men to follow, which was against all the laws of reason and propriety. With these facts the public will readily see that the colored freight handlers are not "scabbing," as has been charged, nor are they doing anything a fair and unprejudiced mind would condemn. We therefore contend that we are right in staying at our work and living up to our contract.

The attitude of the white freight handlers gave the blacks little reason to cooperate: when blacks hesitated or refused to join them, white freight handlers "cussed" them out for being "a poor union lot." One white workman snapped: "All right, keep at work, nigger, keep at work right along, and bring your things down here and sleep. When we win this strike, we'll kick you out of the Dock and Cotton Council and be rid of you."[14]

That did not happen. While the New Orleans labor movement marched and petitioned on behalf of the striking shopmen, the waterfront federation offered little more than moral support. The national strike of the system federation actually lasted until 1915, when it finally collapsed. In New Orleans, defeat came earlier. The sympathy strike devastated the white freight handlers' union, and many members left the city in search of employment elsewhere. In December 1912, over a year after the walkout, hundreds of white freight handlers petitioned the Illinois Central management for reinstatement, only to be refused. Non-union men worked the Illinois Central's Stuyvesant Docks, while black union freight handlers found themselves concentrated on the road's Poydras and Levee yards. The strike in sympathy with men outside of the circuit of cotton unions violated the new, carefully constructed framework of industrial relations on the waterfront; unable to control the white strikers' actions, the Dock and Cotton Council did nothing to prevent the union's destruction. Black unionists who adhered to the letter and spirit of the contract with the Council's approval were the prime beneficiaries. For the next several years, work at the Illinois Central's Stuyvesant Docks and levee yards would be performed only by blacks.[15]

The changes produced by the 1908 settlement, then, were unmistakable. Employers retreated from their earlier confrontational positions, attempting instead to profit by the new framework. For its part, the Dock and Cotton Council consolidated its strength on the waterfront. The new system of labor relations enabled the stronger unions of longshoremen and screwmen to codify their work rules in five-year contracts, putting an end to the annual bouts of capital-labor conflict that had characterized the early years of the new century. While the weaker yardmen and teamsters continued to sign yearly tariffs, few complaints were heard from their ranks regarding the terms of agreement.

Yet even at the height of their power and influence, waterfront union men had reason to be anxious about the future. Within several years, the war in Europe would generate new strains between workers and employers and almost completely disrupt the collaborative labor relations system. But even before the U.S. entry into the war, the dramatic expansion of the port's facilities laid a basis for the long-term erosion of the Dock and Cotton Council's power. Indeed, the fate of New Orleans' waterfront unions was tied inextricably to the modernization of the port. And it was clear from the start that the city's elite, and not its workers, controlled the modernization process.

II

"Imagine an American city of three hundred and fifty thousand inhabitants and nearly two hundred years old that never had a cellar nor a single foot of underground sewer," remarked writer Frank Putnam in 1907. Or a city built upon marshy ground at the river's edge, that took its whole drinking-water supply directly from the rain. That was "New Orleans, three years ago." In a short time, however, the city's administrators had gone far in "laying permanently the foundations for a city of a million inhabitants," by constructing new drainage and water-supply systems as well as a complete network of underground sewers. Nowhere else in America, Putnam exclaimed, "has any generation been privileged to witness a transformation so complete and extraordinary as that which is now in process in the metropolis of the southern half of the Mississippi Valley. It is a transformation from mediaeval conditions to the standards of the twentieth century." The editor of the International Longshoremen's Association official journal agreed. In "every avenue of civic life [New Orleans] is undergoing a metamorphosis that promises one day to make her one of the mightiest cities of the world," *The Longshoreman* noted in 1917. Few cities "have undergone such revolutionary changes as has the romantic old city of New Orleans."[16]

While Putnam's metropolis remained the South's largest city in the first two decades of the twentieth century, its population increase was anything but impressive by New South standards. In the first decade, New Orleans' population grew by 18.1 percent, from 287,104 at the turn of the century to 339,262 in 1910, and by 14.2 percent the following decade, reaching 387,219 by 1920. The city's racial composition changed little during this period, with blacks constituting roughly one quarter of the population, the same percentage as at the end of the Civil War.[17] But the transformation from medieval to modern age embraced more than sewers, drainage systems, and growth rates. Despite regular, annual fluctuations, New Orleans' commerce generally flourished during the second decade of the twentieth century, as trade with Cuba, Puerto Rico, and Central and Latin America expanded tremendously. While the city's economic foundation remained overwhelmingly commercial, the new century witnessed a diversification of the export-import trade. Cotton maintained its leading role in the nation's second largest port, but by 1920 sugar, lumber, oil, rice, and grain had assumed a growing importance as export items, while coffee, bananas, nitrates, oil, sugar, and molasses formed the basis of the expanding import trade.[18]

This diversification of commerce depended upon the modernization of the port's infrastructure, which was itself the product of an alliance between Mayor Martin Behrman and the city's commercial elite. The leader of the Democratic party's Choctaw Club, Behrman dominated New Orleans politics without interruption from 1904 to 1920. The Choctaw Club differed from its late nineteenth-century predecessor in its solicitation of business support and its enthusiastic embrace of municipally run economic development. The Behrman regime consolidated much of the city's debt, inaugurated a vast street paving and repair program, and greatly expanded the city's drainage and sewage system.[19] Most important for the city's economy was the construction of a municipally owned and operated Public Belt Railroad connecting larger railroad trunk lines to the waterfront terminal facilities (and later to public warehouses). While appointments to the board administering the municipal railway remained in the mayor's hands, the commercial associations dominated the nomination process. The Cotton Exchange, the Association of Commerce, and the Board of Trade were each guaranteed two representatives on the Belt Line Board.[20]

The municipal and state governments' aggressive role in promoting both the development of new storage and processing facilities and a close working relationship with the city's leading men of commerce made possible New Orleans' new economic diversification. Although several railroad companies maintained private wharf facilities in and outside the city, most of the city's docks were publicly owned, with the state government exercising considerable control over riverfront resources in the early twentieth century. The Dock Board, created by the state legislature as the Board of Commissioners of the Port of New Orleans, had assumed jurisdiction over the port's water frontage in 1901. Over the next two decades, amendments to the State Constitution empowered the Dock Board to expropriate property, undertake vast construction projects, and operate large, modern facilities. Raising funds through the collection of wharfage, shed, harbor, and other charges, and through the floating of public bonds, the Dock Board initiated a vast modernization and construction program. Within several years it had rebuilt almost the entire wharf system, replacing with steel sheds most of the rather temporary structures along the city's waterfront.[21]

The commercial elite turned its full attention to the port's structural problems following the resolution of the 1907–08 strike wave. The hearings of the Port Investigation Commission on conditions and charges spurred creative and bold economic thinking. Commissioner of Public Utilities W.B. Thompson and other men of commerce now identified the lack of storage and warehousing facilities as the chief obstacle to the port's growth. In the second decade of the twentieth century, the Dock Board undertook the construction of huge riverfront warehouses to encourage cotton shippers to use the city as a market of deposit, instead of as a "through" port. With the approval and assistance of the Cotton Exchange, the Board constructed the first units of its Public Cotton Warehouse and Wharf Terminal in 1915 at a cost of $3.5 million. Two years later, the Board oversaw the completion of a public grain elevator and terminal capable of handling 2.6 million bushels—more than any comparable elevator along the Atlantic or Gulf coast. In 1918 it initiated construction of a five-and-a-half-mile long Inner Harbor Navigation Canal linking the Mississippi River with Lake Pontchartrain. By 1919, the 4.85 miles of waterfront under direct public control were capable of berthing eighty-four vessels lying alongside the docks. "When one looks back over the years to the dilapidated, dangerous and devious ways of the river front, and from that picture turns to the new, highly organized, clean, sound, economic construction of today,"journalist Flo Field wrote in 1917, "it is simply a great modern fairy tale." Local publicist Julia Truitt Bishop sounded a similarly enthusiastic note in 1915. "If you would really learn anything about the Port of New Orleans you must see it for yourself. No matter how eloquent the narrator, he cannot convey to your mind one-tenth of what has been done, much less what will be done, to make this one of the best and busiest and most economical ports in the entire country."[22] The commercial elite had good reason to feel confident by the second decade of the twentieth century.

Wrapping its port modernization program in the rhetoric of progress, businessmen linked the city's future prosperity to grand development plans. Along with enthusiasm for new systems of storage and financing came a strong commitment to applied technology and the mechanization of the loading, unloading, and processing of goods on the waterfront. The prospect of labor-saving machinery fired the imagination of state officials, the press, and the commercial elite. Proponents of change argued forcefully that New Orleans should not "attempt to compete for twentieth century trade with a nineteenth century outfit." Only "the obstructionists are striving to hold New Orleans back to nineteenth-century methods and conditions while its rival ports are keeping step with the twentieth century." A modernized, mechanized waterfront promised more than to increase the city's trade. According to Cotton Exchange president W. B. Thompson, it would "minimize the expense of handling and would relieve the laborer of the killing physical strain of piling one huge bale upon another with his hands and arms, and pulling it down and piling it up again many times over. . . . Rational methods would cut out all unnecessary labor." Yet increased efficiency, he argued, would not result in loss of jobs: "If by reducing the expenses of handling cotton we can increase the amount of cotton handled to the extent I am sure we can, then there would be remunerative wages for five men where now there is employment at barely living wages for only one."[23]

But the promise of job creation through commercial expansion was hardly sufficient to induce the key longshore unions to join the bandwagon in favor of mechanical devices that threatened their strength. The adoption of movable freight platforms, conveyer belts, tiltable storage cars, automatic scales, and the like introduced elements of uncertainty, and at times, physical danger, into the world of work. Working conditions, in fact, seemed worst and unions weakest where employers had free rein to mechanize aspects of loading and unloading. Dock and Cotton Council men had only to look as far as the rapidly expanding fruit trade for an example. The United Fruit Lines employed the latest machinery that had "revolutionized the labor conditions surrounding the discharging" of ships. A continuous moving belt carried baskets of bananas out of the ships and onto the wharves, where workers transferred the fruit onto another conveyer belt that carried it to the rail cars which would take it to northern markets. United Fruit paid its labor force—largely black and Sicilian by the 1910s—some of the lowest wages on the waterfront and easily resisted its workers' attempts to organize and strike for improvements.[24] The Philadelphia Steamship line, which began operating in New Orleans in June 1909, offered another example. Although it initially employed union longshoremen at union wages, it rejected the terms of the 1908 waterfront labor settlement and refused to sign the long contract demanded by the unions. The new company, as a coastwise shipping firm linking New Orleans with Philadelphia, claimed it needed the same privileges as its leading rival, the Southern Pacific Steamship line, to be competitive. In September 1909, it discharged its entire labor force, replacing it with sixty black and white non-union freight hnadlers paid at a rate of 30 cents an hour—10 cents less than union longshoremen. Longshoremen feared that the "rebellion" of the Philadelphia line might mark the end of the peace established in 1908. It did not, but the Dock and Cotton Council was forced to accept the addition of yet another lower-wage enclave on the riverfront when it failed to secure the support of the Behrman administration. Moreover, the Philadelphia and Gulf line wasted no time in dismantling union work rules and adopting new loading methods. While union longshoremen and screwmen only loaded ships "overall"—that is, over the sides of the ship using large ropes and steel cranes—the new non-union freight handlers were required to load a vessel through side ports, or openings, using an adjustable and sometimes movable platform, passing the freight into the ship's holds on "slides," or "shoot-the-chutes"— the method barred by the screwmen's unions in 1902 because of the danger it involved.[25]

Waterfront unionists, of course, were not opposed to port improvements per se. "You see," explained cotton yardmen leader John Callahan, "this is our town, our port, and we have always felt it a duty to assist in its upbuilding." Union leaders constantly complained of inadequate facilities that led to both poor working conditions and congestion on the docks that hurt the port's reputation. What troubled them was the character and trajectory of the commercial elite's modernization program; progress, they argued, should not come at workers' expense. There "has been a tendency to disregard labor and the labor element is never consulted in matters like building warehouses, etc.," Callahan explained in 1915. "We are Louisianians and want to see dear old New Orleans the first port in the nation, but we cannot sacrifice the livelihood of ourselves and dear ones to make her so."[26]

The behavior of municipal officials who were overseeing the commercial expansion process gave unions compelling reasons to be suspicious of the modernizers' intentions. From experience, organized labor had learned that close collaboration between government bodies and the commercial elite meant that public ownership could work to labor's detriment. Early in 1913, for example, switchmen employed by the Public Belt Railroad established a Benevolent Protective Association which denounced poor working conditions and demanded a wage increase. The Public Belt Commission, which administered the railroad, declined to recognize the new association or to meet with its members. When eleven switchmen refused to attend a new standard examination test required by the railroad, Commissioner of Public Utilities and Acting President of the Belt Railroad W. B. Thompson discharged "the malcontents" and an additional twenty-five sympathy strikers for insubordination. As president of the Cotton Exchange in 1911, Thompson had praised the Dock and Cotton Council for its conservatism and cooperation in seeking to prevent the Illinois Central strike. But the terms of the post-1908 settlement applied only to duly-constituted waterfront unions. Employers might tolerate unions where they already existed, but under no circumstances would they permit organized labor's influence to spread. Thompson adopted the position that city employees could not organize as "agents against the city." Since the "Public Belt Railroad is not a railroad in the sense of being a corporate entity [and] . . . is simply the instrumentality through which the city discharges this particular function of government," the current strike constituted "a menace to the public welfare" and an "open revolt against the authority of the Civil Government," a "revolt against the sovereign" that went "to the very foundation of government." Such a rationale left no room for compromise. A meeting between Thompson and a delegation of labor leaders arranged by Mayor Martin Behrman at organized labor's request came to naught, and Thompson stuck to his ruling that the Belt Railroad would retain all strikebreakers and would reinstate no man discharged for insubordination. An important and dangerous precedent against union labor had been established.[27]

State governmental bodies adopted a similar stance toward cotton labor unions. Shortly before the opening of the $3.5 million Cotton Warehouse in 1915, the Board of Port Commissioners made it clear to the Dock and Cotton Council that as a state institution, the Dock Board was barred by law from signing contracts with labor unions. While the Board promised to give preference in hiring to union members—because they were skilled workers—the proposed wage scale disappointed union men. "The scale submitted by the Dock Board no doubt may prove a money maker for the warehouse and the board," complained John Callahan of the white cotton yardmen, "but I cannot figure a living for a cotton yard man and his family. Why, in 1873, when but fourteen years of age, I was paid $12 a week as a loose picker in the Commercial Cotton Press. Today, as a cotton yard man, under the scale submitted, I would not receive that much."[28]

Longshore workers greeted the Dock Board's 1914 experiments with machinery with stiff resistance. In March of that year, the Board granted permission to the Automatic Transportation Company to initiate experiments with cranes, electric trucks, conveyer belts, and other labor-saving devices. Longshoremen successfully halted the first demonstration of an electric truck unloading coffee at the Poydras Street wharf in late April. The unions' walking delegate notified the stevedore in charge that the demonstration constituted a violation of the rule clause in their contract, a clause employers had inserted in 1903. A second effort to test the trucks days later met with the same response. "We were given no consideration. We did not know what was behind it," explained white longshore president Harry Keegan. Instead of experimenting with mechanical devices, he argued, the Dock Board should repair the terrible conditions of the wharves that had led to deplorable working conditions. But the crux of the matter lay less in the specifics of the Board's move than in its implication. Keegan noted: "We figured it was going to change our whole relations with the stevedores." From the unions' perspective, the Board's attempt at unilaterally overriding carefully balanced contract provisions would set a dangerous precedent. Conferences between union representatives, the Dock Board, the commercial exchanges, and the mayor failed to change the unions' position. The unions would interpret any attempt to install the machinery as a contract violation; they promised to strike any agent and stevedore who violated the rule.[29]

From the pages of the *Voice of the People,* the IWW noted the futility of efforts to resist machinery and instead advocated a different program: "Tell them straight out that we ain't opposed at all, at all to 'labor saving' machinery," Covington Hall argued,

> but that from this hour on *we mean to have some of the labor saved by these machines for ourselves,* that we are damned tired of them getting all the "saving," handing us the comb and keeping all the honey. In fact, let's *demand* the use of "labor saving" machines all over the wharves and seas, but let's so organize as to *save the savings* for *ourselves*—organize to *control* the machines, then it wont make much difference who *owns* them. . . . *Control,* that's the secret of *power* in modern industry.

The only measure possible was to "organize so strong that whenever there is a new machine put in to take your work you can raise your wages and shorten the hours of your drudgery."[30] But Hall's program required the repudiation of the collaborative framework of labor relations, the unionization of anti-union enclaves on the docks, and a renewal of class warfare—steps the Dock and

Cotton Council was unwilling to consider. Despite Hall's constant pleas to riverfront workers to abandon their unions and join the One Big Union under the IWW's auspices, the Dock and Cotton Council retained the allegiance of its members.

The longshore unions stood on firm contractual ground in their refusal to participate in experiments designed to bring about their own elimination. Their five-year contract, signed the previous year, included a rule clause barring both workers and employers from altering the contract in any way during its lifetime. This provision, by freezing the status quo and guaranteeing stability, constituted the core of the new, post-1907 system of labor relations. Confronted by their workers' rejection of the Dock Board's experiments, the steamship agents and stevedores refused to cooperate with the Board's demands without the unions' approval. The specter of 1907 clearly loomed large in their minds. One agent explained: "Both ship agents and laborers agree that this contract has brought peace to the river front. Previous to it there was war on every two or three years about the beginning of September, when the new commercial year began. It is this condition of peace which has made the ship agents chary of any interference with the contract whatever. Their business has been running smoothly under the contract, and they say this smoothness is essential to the satisfactory conduct of business. They are equally opposed with the longshoremen to any injection of any matter whatever that will disturb the agreement." Argued agent Alfred LeBlanc: "I am in favor of the installation of any machinery that will facilitate the prompt and less expensive handling of cargo. At the same time I want to live up to my contract with my labor. In fact, I cannot do anything that would violate that contract." Agents protested the "onerous restrictions" the Board placed on them by making them partisans against their will. A spokesman for the coffee importers complained: "In an attempt to force us into a position we cannot take, you have required that we either violate our contract or send our ships to an out-of-the-way uncovered wharf. The situation is intolerable."[31]

Party to no union agreement, the Dock Board viewed its task not as the short-term maintenance of labor peace but rather as the preparation for the port's long-term prosperity. Toward this end, it restricted access to the centrally located Poydras Street coffee wharves, assigning recalcitrant employers to inconvenient dock space much farther down the river.[32] At one point, it even threatened to extend its authority over the entire port and order all ships using its facilities to use mechanical devices in loading and unloading. But such a drastic step was unnecessary. The Board's own experiments with mechanical equipment proved mixed—the coffee conveyer failed completely, while an electric truck and a portable conveyer for handling and stacking rice functioned adequately. At the same time, the outbreak of the European war temporarily threw local commerce into utter chaos. Perhaps concluding that the new equipment posed little immediate threat to their livelihoods or perhaps fearing a war-induced depression, black and white longshoremen quickly came to terms with the Dock Board, voting to give the machines a trial run. In exchange, the Dock Board promised to "do whatever it could to further the interests of organized labor, when the vital interests of the board were not at stake."[33]

Modernization and mechanization of the New Orleans docks in the 1910s by no means eliminated raw muscle power and workers' skill in cargo handling. In many cases longshoremen adjusted without the serious loss of union control. In other cases, the reorganization of the labor process and the introduction of machinery enabled companies like Southern Pacific to achieve greater control over their freight handlers. More important than the actual use of new machinery were the social relations that prevailed where the machinery was used. So long as steamship agents and stevedores adhered to a framework that respected union work rules and power, the negative impact of technology and the reorganization of the work process on longshore laborers' lives could be minimized.

But the modernization program also gave rise to a vastly expanded state and municipal sector that oversaw development of the cotton warehouse, grain elevator, and public belt railroad. These operations remained firmly in the hands of the commercial elite and municipal officials who were unwilling to recognize public sector unions, much less adopt the collaborative framework embraced by the steamship agents and stevedores. Given the Dock and Cotton Council's history of ignoring workers outside of its immediate ranks, it is not surprising that the Council made little effort to support the unionization of workers in the public sector in the ensuing years of war and reconversion. In the aftermath of the war, steamship agents and stevedores would find powerful allies on the Dock Board and other public agencies. The Council's failure to extend fully the network of solidarity into the warehouses and elevators that were the hallmark of the city's commercial progress not only hurt the workers there, but eventually weakened the edifice of Cotton Council power as well.

III

The entry of the United States into the First World War prompted new federal intervention into all aspects of American life. To secure popular support and compliance with emergency regulations, ensure the efficient mobilization of military manpower, and manage a war economy, the federal government created a vast array of voluntary and often not so voluntary bodies. Wartime agencies assumed control over food distribution; adjustment bodies tried to eliminate strikes and labor conflict through arbitration; and federal monies poured into shipbuilding and construction. The U.S. Shipping Board altered trade routes and diverted to New Orleans a considerable grain trade. New Orleans' experience in this period largely conformed to the national pattern. The Louisiana branch of the U.S. Employment Service, for example, monitored the labor supply, attempted to channel workers into war industries, and saw to it that local police enforced anti-idleness laws. "Four-Minute Men" preached patriotism to local theater audiences; government representatives and volunteers visited factories, stores, offices and meeting halls, targeting working people in general and labor unions in particular for subscriptions to war savings stamps and liberty bonds. Military intelligence officers closely monitored the city for signs of "subversive" activities, even before the declaration of war. In late March 1917, seven hundred local state militiamen were assigned to active duty along the riverfront, patrolling the long stretch of wharves in Algiers, Gretna, and New Orleans. The Dock Board undertook regular inspections as well, dispatching its patrol boats along the river to protect its property. By the time the nation entered the war, six companies of the First Infantry of the Louisiana National Guard, quartered in the sheds of Illinois Central's Stuyvesant Docks, had replaced the local militia. The following year, the security net tightened even further. Army officials restricted the number of entrances and exits to all docks, posted soldiers, and inspected membership cards or other identification of all longshoremen, screwmen, and other workers.[34]

With mobilization of public support for the war effort a top priority, government officials welcomed the support of the leaders of New Orleans' black community. From the outset, individual black ministers joined the Inter-Denominational Negro Ministerial Alliance in calling upon young black men to register. Prominent black Republicans were active as well: Walter Cohen headed the black military recruitment drive, and J. Madison Vance, the attorney who had long represented the city's black unions, promoted registration and led a campaign in black New Orleans to raise money for the Red Cross. Black longshore union president Albert Workman, who was also active in registering black workers, eventually came to head the Negro division of the U.S. Employment Service's local office. When the government announced that among the first companies in the American Expeditionary Force to arrive in France would be a battalion of one hundred black longshoremen from New Orleans, black labor leaders sponsored public meetings to encourage black enlistment in stevedoring regiments. In early June, hundreds of blacks gathered to send off the first black longshore volunteers; later, in October, 228 black draftees assigned to stevedoring regiments, led by a contingent of black Spanish-American war veterans, marched from their union hall to the New Orleans Terminal Station. There, an estimated 20,000 city blacks gathered to see them off.[35] Moreover, the black Central Labor Union and ILA locals strongly condemned the alleged "efforts and unscrupulous and uncalled-for activities" by the IWW along the New Orleans riverfront, in particular at Stuyvesant Docks, the Southern Pacific wharves and the American Sugar Refining Company.[36]

But if black leaders backed the war effort, they also sought to use the national emergency to improve the situation of their community. In late April 1917, members of the Odd Fellow's Israel Lodge No. 1971 not only pledged allegiance to the United States and denounced reports of pro-German sentiment among blacks, but also requested that army and navy enlistments "be open without regard to race, color or restriction to every male citizen eligible." Entering the debate on the causes of the exodus of rural blacks to the North, the African Methodist Ministers' Alliance called on Louisiana sugar planters to abolish scrip payment, increase wages, and improve conditions for black field hands. The National Association of Colored Teachers in Colored Schools insisted on better educational facilities, and local black leaders petitioned the city's school board to expand facilities in order to relieve overcrowding and to increase the number of industrial and teacher-training classes.[37]

The labor movement represented a large constituency whose ideological and financial support the federal government also solicited strongly. Meeting in April 1917, the Louisiana State Federation of Labor declared that "unionism is the foundation of patriotism" and supported Wilson's preparations for war. The white Building Trades Council and the black Central Labor Union issued a joint statement affirming organized labor's full cooperation in carrying out war work in New Orleans. Yet labor's support was not unqualified. Long-time AFL organizer James Leonard pledged that labor would "pull off its coat and work for nothing if that is necessary to whip the Kaiser," but it would "not work for nothing or for less than it should receive for employers who are profiteering." For its part, the state labor federation went on record as opposing universal military conscription. Union leaders recognized that employers were far better represented than labor groups on councils and industrial boards overseeing the home front war effort. Though there was little they could do to change that situation, they concentrated on strengthening their existing unions, organizing new ones, and calling attention to abuses they uncovered. The white Central Trades and Labor Council, for example, petitioned the city's commission council to investigate bread prices in July 1917, and in the fall of 1918, strongly opposed a proposal by the superintendent of police to monitor city workers by issuing mandatory work identification punchcards, on the grounds that it would place the "loyal working man on the same basis with the gambler and loafer."[38]

But it was less the gambler and loafer than the larger question of the labor supply that worried city employers. New Orleans, like the rest of the South, confronted serious labor shortages during the war years, as private employers competed with each other, with the army, and with the municipal and federal governments for what seemed to be a perpetually inadequate supply of both skilled and unskilled labor. "The supply of labor was reduced even before the war began," the *Times-Picayune* reflected in May 1918, by the "activities of the railroads in securing men for their rough work, track-laying and repairs." The war severely exacerbated the shortages. Freight often accumulated on city wharves, and the city's park commission and public works department found their efforts seriously hampered. To alleviate the problem, Louisiana planters debated importing Mexican laborers, and a New Orleans shipyard company utilized Puerto Rican workers; women found employment in previously restricted occupations, such as streetcar conducting and janitorial service; and urban employers resorted to newspaper advertisements to attract a workforce. The crisis even reached the ranks of domestic labor. One white housewife complained to the Women's Division of the U.S. Employment Bureau in October 1918 that "[r]ecently, the shortage almost has reached the percentages of a strike," as large numbers of black women voluntarily left white households for factory work that paid higher wages for shorter hours.[39] But following a two-day survey of the city, a senior examiner for the U.S. Employment Service concluded that, overall, New Orleans faced no real shortage of labor. "The reason so many men have left New Orleans to work at other places," concluded Frank Goodman, "is because they are paid better wages." It was the absence of *cheap* labor that employers bemoaned.[40]

And cheap labor was black labor. The cutoff of European migration to the United States, the wartime draft, and the expansion of northern industries pulled like magnets, drawing southern blacks (and whites) out of their agriculturally depressed region. Although no precise figures are available, nearly half a million black men and women migrated to the North during the decade of the 1910s in search of higher wages, better treatment, and greater social and educational opportunities for themselves and their children. Louisiana's labor commissioner estimated conservatively that 12,000 men, mostly black, emigrated during the first eight months of 1917 alone.[41]

The "Negro exodus," as the black migration to the North was called, was a source of great worry to southern planters, industrialists, politicians, and publicists. White southerners often invoked a conspiracy theory to explain the exodus: an organized effort on the part of northern industries was drawing off rural and urban southern black labor. Northern agents, some believed, were active in the Crescent City, employing black sub-agents who "have gone among their people here and made remarkable promises of wages for jobs in Chicago and Kansas City packing houses." But migration involved far more than the agents' promises. White southerners slowly realized that the migration had its own momentum—blacks who had gone north "are instrumental in persuading many who are still here to come on." The African Methodist Ministers' Alliance and the National Association of Colored Teachers in Colored Schools offered their own explanation: low wages, poor housing and schools, and a "lack of consideration" motivated the migrants.[42]

Whatever the specific mechanisms through which Crescent City blacks learned of northern employment opportunities and whatever their motives, the effects of the migration keenly affected the waterfront. As early as 1916, the *Item* noted that along the docks, "where negro labor is much depended on, workmen are becoming more scarce every day. To such an extent has the labor supply been drained by the North that steamship men find it a considerable problem to facilitate the loading and unloading of cargoes." The problem only seemed to worsen with time. In early 1918, managers on the Southern Pacific docks found it "almost impossible to get workers . . . [E]very means to get dock workers, except the drafting of them, has been tried."[43]

But the problem was not rooted solely in the fact of black migration; nor was the shortage restricted to black workers. White longshoremen's union president Harry Keegan testified before a government agency in October 1918 that the irregularity of work and the inadequacy of pay were responsible for driving black and white dock workers away from the levee to other, steadier, better-paying jobs. The labor situation on the [river] front has got to be very scarce, because there is no inducement to keep the men out there at work. Some weeks . . . we are pitching from six o'clock in the morning, or seven o'clock . . . until ten and eleven o'clock at night. If the inducement is not offered to those men to stay there and bring back the men who have left, then I say the labor condition is going to be 100 percent worse.

Keegan estimated that between 250 and 300 white longshoremen were working on the Industrial Canal for daily wages of $5 and $6, while other men found jobs in a nitrate plant in Nashville, at Muscle Shoals in Alabama, and in Lake Charles, building cantonments.[44]

Keegan had identified two central problems that the federal government had to address—irregularity of work and inadequate pay—but there were additional factors at work. Labor shortages, and unrest, and the need for coordinated mobilization of economic resources and uninterrupted production prompted new thinking and action at the national level regarding labor relations. To eliminate strikes, ensure continuous production, and reduce labor turnover and mobility, the government established a wide array of official agencies to monitor, investigate, and adjudicate conflicts between employers and workers. To secure the cooperation of the labor movement, the government relied upon wage increases, the recognition of unions, and the principle of collective bargaining. Moreover, it accorded to AFL leaders the status of junior partner in tripartite bodies that included representatives of government, business, and now labor.[45]

Longshore labor became an early focus of wartime federal concern. "There are few branches of industry where continuity of operation is more essential than in marine and longshore work," a federal agency noted. "A few hours' delay involves very heavy loss to the ship owners and unrest and discontent on the part of the employees, which eventually results in large increases in operating costs. The industry is particularly open to retaliation and labor reprisals." Given the importance of prompt and uninterrupted loading and unloading of vessels to the military effort, one academic observed in 1919, all "our efforts in prosecuting the war would have been futile had our machinery broken down on this point."[46] That machinery—consisting primarily of the National Adjustment Commission, formed by representatives of the U.S. Shipping Board, the Departments of War and Labor, the ILA, the AFL, and principal Atlantic and Gulf coast shipping operators—assumed responsibility for longshore labor policy in August 1917. The Commission oversaw the creation of some twenty-four local bodies, each responsible for hearing testimony and issuing rulings in specific cases. Accepting the union wage scales, hours, and conditions that prevailed on August 1, 1917 as basic standards, the Commission hoped its local boards could bring employers and workers together on a voluntary basis and mediate all differences between them. Although the Commission lacked enforcement powers, it required participating parties to accept its awards as final and binding.[47]

The federal government's goal of continuous production encountered numerous obstacles, not the least of which was the simple fact that the second decade of the twentieth century witnessed waves of strikes and working-class militancy of unprecedented scale. The causes were many: workers' need and desire for higher wages, shorter hours, and union recognition; their resistance to the reorganization of the labor process by employers committed to some variant of scientific management; and their commitment to maintaining or extending control over the labor process through work rules. American workers during the 1910s experimented with new organizational forms and resorted to strikes to achieve their aims, often in defiance of their own union leaders. The United States' entry into the war did little to calm industrial unrest. The number of strikes and lockouts in 1917 surpassed even that of 1916—itself a record year—and the numbers remained high well into the early 1920s.[48] New Orleans was hardly immune from the national epidemic of strikes. If the framework of labor and management cooperation established in 1908 kept New Orleans dock workers quiescent during the wave of labor militancy for the first part of the decade, wartime conditions brought them back into the national pattern. Soaring inflation and employer efforts to intensify the labor process to compensate for labor shortages quickly undermined the earlier system of labor relations and inaugurated a new, more strained era in industrial relations on the docks.

IV

For the duration of the war, the federal government's formula for handling conflict on the docks—voluntary arbitration by local boards of the National Adjustment Commission (NAC)—served as a foundation for labor relations. In general, it accomplished its goal of substituting the peaceful resolution of differences for strikes. One U.S. Shipping Board official in New Orleans boasted in October 1919 that because of the Board's efforts, "the past two years the affairs of the longshore industry have been conducted in an orderly manner and agreements made have been lived up to by all concerned." Labor market conditions clearly played a crucial role as well. Louisiana's Bureau of Labor Statistics attributed the relative calm following the 1917 unrest to the fact that "the demand for labor of all classes and vocations far exceeded the supply, and . . . the employing interests, whether or not they were inclined to do so, of necessity were compelled to be fair and respect the laws."[49] But few workers on the New Orleans waterfront would have characterized their employers' behavior as either fair or respectful. In the context of wartime inflation, labor shortages, and the introduction of new kinds of freight and scheduling demands, neither unions nor employers remained satisfied for long with the NAC's rulings. While the new system of labor relations successfully substituted arbitration for overt conflict, it hardly eliminated the tensions between employers and employees that simmered just beneath the surface.

The era of collaborative labor relations came to an abrupt end in the fall of 1917. On September 1, the 108 men in the black teamsters and loaders union fired the first salvo by striking for a wage increase from 28 to 50 cents an hour.

Ten days later, they returned to work with a compromise hourly wage of 36 cents.[50] Longshoremen struck next. Although shipping agents and stevedores readily acceded to their demand for a 25 percent wage increase (raising deep-sea longshore wages to 50 cents an hour, 75 cents for overtime, and $1 an hour for Sunday work), they objected to the "custom" of the five-year contract, which had only recently been the cornerstone of the collaborative system. Now, in an uncertain age, the Steamship Association preferred the risk of renewed strife over long-term planning.[51] A third group of workers—600 black freight handlers, called car loaders and unloaders—also walked off the job for higher wages. The actions of the longshoremen and car freight handlers effectively halted all riverfront work outside the railroad docks.[52]

These early wartime conflicts provided the first opportunity for the NAC's local adjustment board to establish itself as a key institutional player in labor relations. Initially viewing the board as a potential means for disciplining unruly workers, some agents hoped that decisive government action would force longshoremen to arbitrate. The solution to the whole problem, one prominent agent suggested, would be an "appeal . . . to the United States government to put two thousand men in uniform on the docks to relieve the situation" if strikers refused to return to work. While not a representative opinion, the specter of state intervention was the main topic of discussion at an overflow meeting of the black and white unions; it probably convinced a bare majority that arbitration and minor concessions were the lesser of two evils. As one union leader put it: "The steamship agents have forced us to lay the matter before the local adjustment commission. If we refuse to let the local board adjust matters, the United States government will start loading the ships that are now tied up." Preferring not to test their own strength, longshoremen returned to work under a three-year contract granting them a dollar a day wage increase, and agreed to submit outstanding differences to the local adjustment board.[53] The arbitration machinery also provided the basis for ending the car loaders' and unloaders' strike. Organized in 1913 as Local 854 of the ILA, these black men worked for private contractors handling goods from the city's Public Belt Railroad and occupied a rung on the waterfront's occupational ladder far below that of the black and white screwmen and longshoremen. From the start, this weaker union welcomed government intervention and promised to return to work in exchange for higher wages pending arbitration. Employing car contractors, with little experience in handling unions, initially rejected this plan. But when they finally relented, they were not disappointed by the local adjustment board's ruling, which granted the handlers a 25 percent increase—an amount substantially lower than that of the longshoremen.[54]

In its deliberations and decisions, the local board accepted entirely the existing occupational hierarchy, taking careful account of the historical development of the city's economy and docks, as well as the balance of power between different groups of workers and their employers. This meant that longshoremen and screwmen continued to exercise considerable power and earn high wages. At the end of September 1917, the board averted a strike by cotton screwmen, raising their wages to $31 per gang per day. The following

year, it increased screwmen's wages to $36 a day. But workers at the lower end
of the port's occupational hierarchy—car loaders and unloaders, coastwise
longshoremen working for railroad and steamship companies, and the various
freight handlers' organizations—continued to labor at lower wages and under
harsher working conditions. Even in wartime, wage differentials between
different groups of dock workers remained wide. Board members accepted
traditional employer explanations for the disparity in the status of longshore
occupations. To take one example, because of the greater regularity of their
work and the lower skill levels required in handling freight, coastwise longshoremen received lower wages than their deep-sea counterparts. By late
1918, deep-sea longshoremen (that is, regular longshoremen on transatlantic
steamships) were working an eight-hour day with a Saturday half holiday, at a
basic rate of 65 cents an hour. In contrast, over the course of 1918 the board
raised the wages of Southern Pacific coastwise longshoremen from 48 cents to
60 cents an hour (a wage that held steady even after deep-sea wages increased
to 85 cents in 1919) and of Illinois Central freight handlers from 28 cents to 37
cents an hour. Farther down the occupational hierarchy, car loaders and unloaders and Southern Pacific railroad shed yard freight handlers received
roughly 40 cents an hour in early 1920.[55] Labor shortages and the need for
sustained production enabled those at the bottom to make modest wage gains.
But the board's decisions to grant increases that matched the rise in the cost of
living hardly satisfied the aspirations of workers at the lower end of the port's
job ladder.

Publicly and privately, the New Orleans adjustment board felt conflicting
pressures from concerned parties. In late September 1917, black car loaders'
union president Alexander Paul successfully challenged the board's composition, claiming that the disproportionate representation of employer interests
undermined its authority. Paul stood on firm ground, for guidelines established
in Washington accorded labor and capital equal representation. Paul and
Thomas Harrison, the Dock and Cotton Council's man on the local commission, forced the reconstitution of the board through the resignation of the head
of the Steamship Association, S.T. DeMilt. At the same time, employers found
a strong friend in the local board's chairman, Captain P.T. Murphy of the Army
Quartermaster Corps, who shared the agents' and stevedores' frustrations. In
response to a series of short walkouts by Southern Pacific freight handlers in the
summer of 1918, Murphy unsuccessfully sought authority to take legal action to
restrain the leaders of the two freight handlers' unions from interfering with
government work. Employers lost this ally, however, when a Washington directive barred military personnel from serving on such bodies, thus relieving Murphy of his position after only eleven months on the job. Murphy's temporary
replacement, Mayor Martin Behrman, alarmed employers and pleased workers. One of Behrman's first moves was to order the Southern Pacific line to
recognize its freight handlers' unions, a move that the company's managers
immediately appealed. In private correspondence, they made it clear that they
found Behrman wholly unacceptable.[56]

Waterfront employers also learned quickly that not only did wartime condi-

tions wreak havoc with the labor supply, but in many cases managerial authority weakened and labor discipline suffered as well. As usual, workers at the
upper end of the occupational hierarchy were in a solid position to help
themselves. No sooner had the three-year longshore contract been signed in
September 1917 than stevedores noticed a shift in their workers' behavior.
Employers' chief complaint centered on the amount of work performed. Ignoring "custom," longshoremen reduced from eight to seven the number of
140-pound sacks of flour and 128-pound sacks of oats that they carried per
two-man truck; they similarly reduced the number of railroad iron bars handled per sling load. Longshoremen were "making the loads of practically all
commodities smaller, the men in the holds actually refusing to land loads that
was [sic] a regular thing during the life of their previous contracts with us,"
testified John Heyn in October 1918. Moreover, union longshoremen constantly demanded more men per job. Complained stevedore Kearney, union
workers "come along and demand more men . . . on the truck; they will
demand more men in the hold, and if you don't put them on, they won't
work."[57]

Stevedores and agents found longshore labor increasingly unreliable and
uncooperative. Employers chafed at the strategies longshoremen devised to
maximize their wages and cope with the irregularity of work. "[T]here has
been so much night work here," agent W. H. Hendren wrote to the head of
the National Adjustment Commission in October 1918, "that the laborers in
many instances have declined to work in the day time, laying back for night
work for the reason that they get a higher rate of pay for such work than the
day work."[58] Verbal agreements between gangs and stevedores to perform
certain tasks no longer held much weight. In one case, longshoremen quit for
the day late one Saturday afternoon with only four cars of flour left on the
wharf. Although they were receiving time-and-a-half overtime, they preferred
the guaranteed time at double pay the following Sunday morning. Other
gangs of workers might not finish a job, even with overtime pay, if by quitting
they could create more work from themselves later.[59] It appeared to employers that the union workers were taking advantage of the labor shortage to
change their working rules and conditions. Union workers had "grossly and
persistently violated the provisions of the contract and used it more as a cloak
for slacking than as a guide for duty," argued agent N. O. Pedrick. The "big
feature with labor out there," concluded W.S. Smith, "is to get the individual
members to feel that they are morally bound by this contract."[60]

Employers found no satisfactory answer to that problem. Even the standard arbitration process almost ceased to function. The controversy over the
loading of flour sacks in late 1917 and early 1918 is a case in point. Stevedores
viewed the reduction in the sack load as a violation of "custom" (and hence
the contract), and demanded that the local adjustment commission consider
the matter. Longshore leaders Keegan and Workman waited months before
responding to the employers' charges, and then insisted that employers utilize
the tried and true arbitration clause for overwork in the contract. On numerous occasions, longshoremen simply ignored requests for arbitration.[61] In

1917, however, longshoremen did make a concession that made the arbitration committee, in theory at least, more equitable. Under the earlier system established even before the 1908 settlement, a committee was composed of three longshoremen and two stevedores; during the era of labor peace, union leadership was held responsible for policing its rank and file. In 1917, employers demanded equal representation. Under the new terms, two longshoremen, one black and one white, and two stevedores made up the committee, and the four men would chose an umpire. But in practice, the new system failed to function. Stevedore Frank Thriffley summed up the manager's dilemma: "If they refuse to carry what you tell them to carry, or to take a load that you send down to them, they cease work, because I say: 'You must take it,' and they say they won't. Therefore, they will cease work." At that point, an arbitration committee was formed, but the umpire issue was the stumbling block, with both sides vetoing the choice of the other. "Therefore," Thriffley noted, "rather than to call an arbitration, I have continued to work, at a loss to myself."[62] This lesson was not lost on stevedores, who preferred to grant the workers' demands to losing time while waiting for arbitration. Captain P. T. Murphy, the local board's first chair, was shocked to learn just how arbitration in New Orleans actually worked, and he tried unsuccessfully to change the contract terms to maintain operations while a dispute was arbitrated.[63] He then requested legal power from the National Adjustment Commission to compel adherence to the board's decisions. "If we can not be so equipped there is little or no use having the Local Adjustment Commission, its meetings and deliberations being a waste of good time."[64] Surely, that was a point that crossed the minds of at least some employers.

What appeared to employers as a managerial nightmare was nothing more than labor's pragmatic response to the changing material conditions. The American entry into the First World War had a significant impact on the world of waterfront work. Wartime commerce accelerated the diversification of the import-export trade begun earlier in the decade. The U. S. Shipping Board designated New Orleans as one of five major centers for the transcontinental shipment of foodstuffs, especially grain, for export, and a great deal of army equipment, much of it large and bulky, passed over the city's wharves. These new types of cargo, particularly military goods, were generally heavier and more dangerous to load, argued black longshore president Albert Workman. The work demanded of the men, Keegan noted, was at times "a physical impossibility." Even before the U.S. entry into the war, the introduction of machinery on the docks had intensified the labor process. Complained one longshore union official, the "machines that are being used on the docks take off or put on cargo faster than the present number of men can handle it."[65] In fact, the stevedores' refusal to employ more men had contributed to the 1917 walkout. But what their 1917 contract failed to address the longshoremen accomplished by their own efforts. The shortage of labor enabled them to demand more men per job and to impose some new limits on the amount of work performed, affording them increased protection against changing conditions.

Despite their new bargaining power, longshore unions only partially addressed the problems of wartime commerce. While agreeing to submit grievances to arbitration, they made it clear that they were "not willing to sacrifice that the employer may build up immense fortunes."[66] Their primary complaint during and immediately after the war was the relationship of wages to inflation. In 1918 and 1919, longshoremen petitioned and received several wage hikes from the local adjustment commission. Local commissioners understood well that wage increases were absolutely necessary to retain a longshore labor force in a period of labor shortage. Without monetary "inducements," there was little to prevent experienced dock workers from seeking employment in shipbuilding, construction, or other fields that offered more money. But higher wages for deep-sea longshore work were related to a second, persistent problem—the irregularity of dock labor. The "genius has not been created as yet who can control the movement of ships in such a way where the longshoreman or the man who follows the water front for a living can be employed regularly every day," ILA vice president Anthony Chlopek complained before a New Orleans adjustment meeting. Only during the cotton shipping season did longshoremen find steady jobs. Wartime developments did little to change this condition.

Longshoremen devised their own strategies to cope with the problem. Most important, they maintained their own ethical codes about the performance and division of work. There "is a feeling that has got into most of the men," one stevedore complained, that any man who "attempts to do more than the man next to him is considered a disloyal man, a disloyal union man."[67] If unappreciated by managers, such behavior protected men from overwork and stretched out existing work. On the rare occasions when labor supply exceeded the demand, union men reduced their overtime and Sunday work and divided the work as evenly as possible. More often, they preferred higher paying night and Sunday work to day work, staying away from the docks until evenings and Sundays. While frustrating to employers, longshoremen were simply exercising their market power in a way that was logical and beneficial to themselves.[68]

Agents and stevedores were of two schools of thought on the question of labor's efficiency, quality, and character. The majority blamed workers outright. Argued Crawford Ellis: "A great many men do not want to work a whole week. They will work three or four days and make some money and lay off and come back again." Kearney, stevedore for the Harrison line for a quarter-century, concurred: "There are plenty of men. You can find lots of men knocking around, possibly at the corners or around the wharf talking, and you will ask them to work and they will say: 'Well, we don't care about working today.' It is a very, very hard thing to get at. . . ." "I cannot explain it and don't understand it, why the men will not work steady, why they will walk out there in the morning and refuse to work. . . . Well, the majority of them out there in that particular case were colored men." The availability of alternative employment only aggravated matters. Stevedores and agents were well aware that some longshoremen labored in other lines of work, such as shipbuilding and carpentry; other workers revealed a kind of footloose quality, moving up and down the river in search of easier work and better conditions. And wartime wages were part of the problem. With every increase in the wage scale, complained Pedrick in 1919, there "has been, if anything, a falling off of the efficiency of the work." A smaller number of employers believed that the problems of discipline were directly related to the inexperience of many of their new workers. Neither the black nor white longshore locals could provide a sufficient number of experienced workers. "The efficiency of the men has been materially reduced, due to the fact . . . that we had to bring in a lot of strange men to take the place of the regular longshoremen from time to time," explained W. S. Smith. Upon closer examination, stevedore Frank Thriffley recognized that many of his troublesome men were in fact green hands. But he too blamed the union. Regular union longshoremen did not instruct the inexperienced union men, who "do not understand the fundamental ideas of unionism, they think that once they get into a union they don't have to do anything more."[69]

Whatever the cause of inefficiency and hostility, all employers of waterfront labor understood that the war had ushered in a new era in labor relations. Their wartime experiences taught them to place a priority upon the very issues of managerial authority and control that they had ignored during the previous decade. Employers entertained a number of different proposals. Local Adjustment Commission chairman Captain P. T. Murphy recommended to his superiors that the only remedy would be to station a stevedore regiment in the suburbs of New Orleans. "With such a unit," he argued, "there would be no question as to what the Longshoremen would do, and this mulish obstinacy and procrastination on their part would instantly disappear." Few went that far; most employers instead sought to enlist the local adjustment board to their cause. When the local commission convened in October 1918 to address labor's demand for higher wages, Steamship Association officials petitioned to link wages and work rules, arguing for a sweeping revision of existing contracts to restore managerial authority. It was imperative, they believed, that employers regain control over the direction of work, the power to hire and discharge workers, and to determine the number of men required per job. But with stability a central priority, the adjustment commission refused to tamper with the traditional work rules that agents and stevedores now found problematic. The wartime experiences of New Orleans waterfront employers called into question the very foundations upon which union labor rested.[70]

V

No one could anticipate the precise course the American economy would follow or the shape that labor relations would assume in the postwar period. The nation entered the reconversion era with few planning or control mechanisms intact. Both the Wilson administration's dislike for peacetime economic

planning and pressure from industrial capitalists prompted the rapid dismantling of wartime regulatory bodies. The anticipated economic crisis, however,
did not crystallize immediately. A boom in the construction industry, strong
overseas demand for American exports, and high levels of domestic consumer
spending sustained the economy throughout 1919 until early 1920, despite the
cancellation of massive war orders. At the same time, the federal retreat from
a planned economy involved the abandonment of many of the mechanisms of
labor-capital mediation that had afforded the AFL enhanced recognition and
some limited institutional power. The signing of the Armistice in late 1918
ended both the war in Europe and the tenuous and frequently unobserved
truce between labor and capital in the United States. Workers and businessmen alike recognized the importance of the coming struggle. Labor sought to
consolidate, preserve, and extend its wartime gains, and capital sought to roll
back the advance that the national emergency had made possible.[71]

A great deal was at stake. From 1918 through 1920, New Orleans witnessed the strongest wave of union organizing since the turn of the century,
embracing white and black, skilled and unskilled, and private and public
sector workers. The city's labor movement began a drive in late 1918 to
organize new unions and strengthen existing ones. Hotel waiters and bartend-
ers, blacksmiths, carriage and wagon makers, butchers, and banana carriers
were among the trades to form new unions. Given the importance of the
municipal and state sector, organizing public employees was a high priority for
the Central Trades and Labor Council. In November 1918, the Council inaugurated plans to organize policemen, firemen, and civil service employees.
Those efforts bore some fruit: for example, Dock Board employees in the
cotton warehouse unionized, as did clerks of the Public Belt Railroad and 300
city street workers.[72] Efforts of the local labor movement to organize from
above were matched by a groundswell of activity from below which extended
even to the ranks of domestic laborers. Led by Ella Peete, wife of freight
handlers' union president Sylvester Peete, black women organized a Domestic
Workers Union in July 1918.[73] The task of the labor movement during peacetime would be to consolidate and extend such organizational gains achieved
during the war.

At the war's end New Orleans labor leaders, awaiting a new era in industrial life, prepared for the worst. Three hundred delegates representing some
75 unions inaugurated a "preparedness movement" to combat the anticipated
"concerted action among employers to lower wages when business conditions
return to normal."[74] Labor's goals were two-fold: to resist any wage reductions
before inflation abated, and to organize the city, top to bottom. They had their
task cut out for them. For the next year, inflation maintained the high cost of
living, while numerous employers attempted to cut wages. For the first half of
1919, conflict raged intermittently on the large municipal, state, and federal
construction projects. Thousands of carpenters, piledrivers, structural engineers, and others battled private employers and government managers of the
army supply base and the Industrial Canal over the issues of non-union workers, job definitions, wage increases, and work rules; by late summer and

autumn, the conflict had shifted to the city's wharves and waterfront warehouses. But continued labor shortages contributed to many union victories in
this first year after the Armistice. Even the powerful Dock Board was brought
to heel by union piledrivers. Although it withheld recognition of the union, the
Board had little choice but to capitulate to wage and work rule demands that
June. By early 1920, however, public and private employers were in a better
position to fight back, as economic expansion came to a halt. Well-established
and stronger white unions in the building and metal trades weathered their
employers' assaults, but they failed to secure their desired wage increases or
the closed shop. Newer, weaker unions fared worse. To take but one example,
the huge American Sugar Refinery at Chalmette easily replaced black workers
on strike for union recognition and higher pay in early 1920.[75] The success or
failure of union demands corresponded largely to the strength and age of the
organization, and by the early 1920s, little remained of the associations formed
during the wartime organizing wave.

The postwar labor crisis hit the New Orleans waterfront with full force in
the fall of 1919. For the next eight months, employers and workers struggled
to define the terms of the new era of industrial relations. A strike wave began
in September when newly organized Dock Board employees at the cotton
warehouse struck for higher wages. The Dock Board relented only after
sympathetic action by longshoremen and screwmen halted its operations, but
shortly thereafter it began firing its union members.[76] Far larger in scope was
the strike of 1500 banana workers—inspectors, checkers, packers, classers,
and messengers—who demanded higher wages, standardization of hours, improved working conditions, union recognition, and the reinstatement of some
twenty employees discharged for union activity. Little had changed on the
city's fruit wharves over the decade, and the powerful United Fruit Company
led the anti-union counterattack by refusing to negotiate, employing strikebreakers, securing a court injunction against alleged union intimidation of
scabs, and installing new power-driven conveyors on their wharves to reduce
the need for human labor. Only one company eventually came to terms with
the banana workers' union; operations at the other three continued with little
problem, and the city's fruit market remained largely undisturbed.[77]

Neither of these strikes—nor a number of short-lived walkouts by public
grain elevator workers, car loaders, Public Belt employees, and tugboat
men—halted the flow of commerce as successfully as the longshoremen's
strike did. In October 1919, the black and white longshore unions demanded
that their hourly rate be raised from 65 cents to $1. In the first postwar
contest, waterfront employers assumed that the fight would follow the rules
established during the war: both sides would present their case to the adjustment commission, whose judgment would be binding on both parties. A Joint
Shipping Industrial Conference, held in Washington in June 1919, had extended the life of the arbitration agreements in the longshore and maritime
industry, an act approved by the ILA and by major Atlantic and Gulf coast
shipping interests.[78] Whatever problems New Orleans employers had with the

wartime system of labor relations, they appeared unwilling to chart unknown waters in the reconversion period by abandoning it.

Longshoremen, however, had other ideas. Understanding that arbitration does not take place in a social or political vacuum, they had little trouble reading the writing on the wall. A federal wartime priority had been uninterrupted production; to achieve that goal, local adjustment boards often accommodated the demands of powerful waterfront unions. By the fall of 1919, the government's priorities had shifted to lowering the cost of living. "Any substantial increase of wages in leading lines of industry at this time," President Wilson made clear, "would utterly crush the general campaign which the government is waging, with energy, vigor, and substantial hope of success, to reduce the high cost of living." Under these circumstances, the Administration called on "every thoughtful American" to accept a truce between labor and capital by postponing the wage issue. National Adjustment Commission chairman William Z. Ripely, a Harvard economist, took his cue from the Wilson Administration. In its first major postwar ruling in 1919, the National Adjustment Commission demonstrated its determination to hold the line on wages when it awarded New York longshoremen a mere five-cent an hour increase—an action which prompted massive walkouts in that port.[79] The New Orleans Steamship Association applauded the federal efforts to freeze wages, seized upon Wilson's proclamation to justify their refusal to grant wage increases to longshoremen and screwmen, and turned to the National Adjustment Commission as an important ally in keeping their unions in line.

Whatever faith New Orleans longshore unions might have placed in the adjustment system vanished with the Commission's New York wage ruling. Repudiating arbitration altogether, longshore workers instead fell back upon their raw power. Their five-week strike, executed with iron-clad discipline, tied up hundreds of ships, diverted hundreds of others, and left agents and stevedores stunned and angry. On several occasions, black and white workers attending joint meetings overwhelmingly rejected proposals for submitting to the Commission, and shouted down compromise proposals, even those made by officers of their International. Dock men argued that their officials were "not in touch with local conditions and do not know what the men want."[80] For the duration of the strike, the black and white longshoremen maintained a solid front, voting repeatedly against arbitration.[81]

Despite the massive tie-up, government officials, steamship agents, and stevedores preferred a wait-and-see attitude over industrial warfare. On several occasions, though, they threatened forceful intervention. Calling on New Orleans workers to end their "unauthorized strike," the U.S. Shipping Board announced in mid-October that the government was contemplating the use of U.S. soldiers to handle the port's cargo; less than a week later, the *States* reported that agents and stevedores were ready to employ strikebreakers guarded by U.S. soldiers if the strikers rejected a ruling by the National Adjustment Commission, which was scheduled to meet in New Orleans. When the Gulf Navigation Company settled with the strikers on a compromise basis in early November, the Shipping Board intervened decisively, ordering it to discharge some 900 longshoremen who had begun work under the separate peace.[82] But the Steamship Association had no desire to risk the full wrath of the Dock and Cotton Council by employing strikebreakers. For the moment, its goal remained more modest: to re-establish some mechanism to eliminate strikes and ensure steady work in a period of expanding commerce. For all the headaches it had caused the employers in the past, the adjustment machinery was the best, and only, tool they had at hand.

And it eventually, if indirectly, secured the strike's termination. The National Adjustment Commission had convened in New Orleans to hear the proposals of other Gulf coast longshoremen—from Galveston, Pensacola, Mobile, Gulfport, Port Arthur, and Beaumont—who demanded the same wage scale sought by the New Orleans unions. Employers and commissioners alike hoped that a favorable settlement for these men would induce the New Orleans strikers to return to work. Attempting to prevent a Gulf-wide strike, the Commission established a uniform pay scale for the region, giving longshore workers an increase to 80 cents an hour for day work, $1.20 for night work, and $2 an hour for Sundays. Although it fell far short of their initial demands, Gulf coast longshore leaders accepted the offer. So too did their New Orleans counterparts. Pressured by the Dock and Cotton Council to end the five-week strike that had paralyzed the port and thrown most Council union men out of work, longshoremen voted by secret ballot to accept the compromise and return to work.[83]

The 1919 settlement was cause for only a temporary celebration in employers' ranks, however. Waterfront labor conditions remained highly unsettled. In late December conflict threatened to erupt once again, when 400 longshoremen initiated an unauthorized strike on two steamships, refusing to unload the cargoes of kainite at the new wage rate, 80 cents an hour, established in their recently signed contract. The men refused orders by union officials to resume work, arguing that unloading such extremely unpleasant cargo as fertilizer should be remunerated at the wage of 90 cents per hour. When a joint session of the two unions decided to resume work pending arbitration, members of the New Orleans Steamship Association were hardly impressed. If shipping in New Orleans was to continue, they believed, the Association "must deal with a dependable, well-organized body with whom contracts will be as binding as they are with the employers."[84]

Shipping operators proposed a number of answers to the question of just how to obtain that dependable body. They turned first to the court system. Alleging breach of contract for the late December walkout, the Holland-American Steamship Company sued the unions for over $18,000 in damages. That spring, Judge Rufus Foster ruled in the company's favor, agreeing that contracts could not be treated as mere scraps of paper. But the penalty awarded fell far short of the punitive damages initially sought, as Foster criticized the company for making "no effort to employ non-union men, or to minimize the damage." Employers rejected Foster's reasoning. George Terriberry, Holland-American's attorney, argued that the unions' hold on the labor situation was so great that the strikebreaker option was unrealistic. Nonetheless, the lesson was not lost on New Orleans' men of commerce, who filled the jury boxes as spectators to the proceedings. Contracts may have been legally binding, but by themselves, they could not govern successfully the behavior of the stronger waterfront union men on a day-to-day basis.[85]

Some agents attempted to resurrect a system based on cooperation, uninterrupted work, and compulsory arbitration. In mid-January 1920, representatives of every commercial exchange met with local, state, and federal officials and longshore union leaders to hammer out new rules of industrial relations. Recognizing that inflation undermined stability on the docks, businessmen established a committee pledged to expediting the involvement of the National Adjustment Commission; in exchange, longshore presidents Harry Keegan and Albert Workman agreed in principle to submit all differences regarding wages, working conditions, and other grievances to that new committee.[86] Speaking in a unified voice, this joint exchange committee assumed an increasingly prominent role in the settlement of labor crises. But while businessmen publicly remained committed to the peaceful resolution of difficulties, their private thinking on the labor question was hardening. The experiences of the winter and spring of 1920 would convince many members that the stick was mightier than the carrot.

The question of wages kept the waterfront highly unstable in the opening months of the new decade. The National Adjustment Commission had failed to act on the demands of the port's less powerful unions at its November 1919 hearings. Frustrated by the long delays and apparent lack of federal interest in their condition, approximately 1000 car loaders and unloaders and railroad terminal freight handlers struck in early January 1920 to force the Commission to address their demand for a 50 percent wage increase. Their target was U.S. Railroad Administration director General Hines, who at this point claimed jurisdiction over all railroad employees. The Association of Commerce members pleaded with Sylvester Peete, black president of the Terminal Freight Handlers, and with other union leaders, to resume work pending a government ruling, pledging their full effort to secure that ruling. One week after the walkout began, the black union men relented and returned to work. In mid-January, however, another group of workers struck. Southern Pacific steamship line's 900 to 1000 coastwise longshoremen (two-thirds of whom were black) walked out because of the government's failure to rule upon their demands for a wage increase from 60 to 75 cents an hour. Agreeing that the strikers had been short-changed by the government delay, the joint exchange committee promised to use its influence to speed up the adjustment process. Its intervention succeeded. The coastwise strikers returned to work days later on the understanding that the Railway Wage Commission would soon adjudicate their dispute.[87]

New Orleans labor, one observer noted, seemed "to be set on more of a hair trigger than elsewhere." The Railway Administration's failure to take any action on the New Orleans wage issue, and employers' refusal to increase wages outside of the framework of the formal adjustment system, soon ignited yet another round of waterfront strikes. The unions rightly feared that the Railroad Administration's strategy was to delay granting any further increases before the government ended its wartime controls and restored the national rail systems to private hands. In early February, 2000 black car loaders and unloaders and railroad terminal freight handlers joined 800 white and black Southern Pacific coastwise longshoremen, halting work on the Southern Pacific and Illinois Central wharves and stopping the flow of rail car freight to the public docks. This time, the businessmen's joint exchange committee adopted a two-track policy. It blamed the strike upon the Railroad Administration's delay and again called for a prompt wage ruling; at the same time, it condemned the strikers for imperiling the port's prosperity and future development. When strikers ignored the businessmen's entreaties, employers adopted a harder line. The joint exchange committee requested formal permission from the Railway Administration to employ outside, non-union workers, and Morgan's Louisiana and Texas Railway and Steamship Line, the Illinois Central, the Southern Pacific, and the Southern Railway put strikebreakers to work. Whatever the cause of the strike or the justness of the workers' complaints, employers argued, they would no longer permit labor disputes to threaten the port's prosperity.[88]

At the same time, the joint exchange committee solicited the assistance of New Orleans' labor leadership in ending the walkouts. Earlier, Cotton Council president James Byrnes and ILA vice president Thomas Harrison had made clear their opposition to the strikers. Harrison's support of the exchange committee's proposal was not unexpected; long an advocate of labor-capital cooperation, he had counseled the longshoremen to terminate their fall 1919 strike, and his own union, the white cotton screwmen, had accepted the authority of the National Adjustment Commission. Harrison now offered the same advice to the new strikers: "The exchanges have shown they are working for you," Harrison told the black workers. "Do not antagonize the exchanges, do not antagonize the rest of the city." The Central Trades and Labor Council also took a strong interest in the deepening crisis, with its delegates unanimously pledging their cooperation in ending the waterfront strikes. The Metal Trades Council and Building Trades Council passed critical resolutions vowing to support the strikers if they were right but to abandon them if it could be shown they had violated earlier agreements to continue working. Clearly, what appeared to be almost continuous strikes on the docks worried not only shipping interests but other workers as well. Pressed by labor leaders and employers to resume working, coastwise longshoremen, freight handlers, and car loaders and unloaders called off their strike.[89]

But matters only grew worse. Contractors and railway officials complained bitterly that most car loaders and unloaders refused to show up for work, that only the handful of employers who agreed to the higher wage scale seemed to get the men they wanted. By late March, contractors at the public docks thoroughly capitulated to the black union and agreed to the 25 percent increases. A far more threatening problem centered on the Illinois Central's

Stuyvesant Docks. Returning strikers found that half of their places remained filled by strikebreakers and that the company would only reinstate the men as individuals, not as a union. This explicitly violated the terms of the agreement worked out by the businessmen's joint exchange committee. If Dock and Cotton Council members had been less than enthusiastic about the initial strikes, they now expressed outrage at the railroad's undisguised assault on union labor. The Council supported "these men and we don't care how the government feels about the matter or the way you gentlemen feel about it," a spokesperson declared. "It is then our turn to show that we want justice, and we will get it in the only way we have." Harrison abandoned his earlier caution. The men could not very well return to work at Stuyvesant Docks "unless they were recognized and recognition meant that they must all be taken back." Unless the Illinois Central dismissed its scabs and reinstated union members, the entire Dock and Cotton Council—now numbering twenty-seven locals with as many as 20,000 men—would call a general sympathy strike. Not since 1907 had the Council invoked its ultimate power in such strong terms.[90]

For a week the threat of a general strike hung over the port. The joint exchange committee prepared for battle even as it sought some middle ground. When workers rejected the Illinois Central's offer to take back 150 men—at 25 a day for 6 days—and give union men preference for jobs as they became vacant, the city's commercial elite, exasperated, readied itself for the coming "test of power." Claiming the support of every businessman in the city, Association of Commerce president Arthur Parker told union leaders: "The business men of New Orleans will fight you to a finish. They will close up their offices and warehouses and factories . . . entirely for two months. Either we are running our business or you are, and we are going to find out without any more pussy-footing." Warren Kearney invoked the specter of the open shop: "remember what happened in Seattle," referring to the recent destruction of that city's unions. Unless waterfront labor untangled its own mess and adopted a "stable working policy," the city's employers threatened industrial warfare. Employers soon would have an example closer to home than Seattle to draw upon. A black coastwise longshoremen's strike on the Mallory Line in Galveston was crushed when the Texas governor declared martial law and sent in the Texas Rangers to occupy the docks, to the deep satisfaction of the newly formed Galveston Open Shop Association. In that strike's aftermath, a special session of the Texas legislature passed an Open Port Law which barred workers from interfering in any way with the free passage of commerce in that state.[91] But as so often was the case in New Orleans, principle receded before practical compromise. Mayor Martin Behrman counseled union leaders against striking while he opened talks with railroad officials. At Behrman's behest, Illinois Central officials agreed to reemploy 200, not 150, former strikers while giving preference to union men in the future. The threat of a general strike evaporated, but it was clear that on the question of union power, neither labor nor capital had won a clear-cut victory.[92]

If this settlement ended the "era of chaos" that had characterized water-front labor relations since September, it did not end the era of instability. Employers dug in their heels and refused to accede to any new labor demands. Late April and early May witnessed an unsuccessful week-long strike by a newly formed Deep Sea Marine Clerks and Checkers Union, affiliated with the ILA and the Dock and Cotton Council, which struck for higher wages, union recognition, and the closed shop. Extending to the clerks the solidarity each Council union was entitled to, union deep-sea longshoremen and car loaders and unloaders refused to handle any freight touched by non-union clerks; when union officials called on these men to remain at work, individual longshoremen and screwmen walked off the job. The following fall, the Dock and Cotton Council pressed for recognition of the clerks' union as part of a general settlement of labor issues, but failed to impress either the U.S. Shipping Board or steamship agents.[93] As well, coastwise longshoremen ignored the advice of their mayor and fellow unionists. Men who struck Morgan's Louisiana and Texas Steamship Line at the Southern Pacific docks in mid-March remained out when other freight handlers and car loaders returned to work. This coastwise strike extended along the entire Gulf and Atlantic coasts, grinding Southern Pacific's freight traffic to a virtual halt. But after several months, the company had assembled a non-union labor force in New Orleans. With the joint exchange committee's endorsement—the "time for pussyfooting with outlaw strikes is over," Association of Commerce president Arthur Parker declared—Southern Pacific secured an injunction from Federal Judge Rufus Foster, enjoining eleven strikers, including the black and white presidents of the coastwise unions, from interfering with strikebreakers in any way. Long one of New Orleans' most anti-union steamship companies, Southern Pacific resumed operations on an open-shop basis.[94] A third case proved equally futile for workers. In the fall of 1920, a ten-day strike for a 25 percent wage hike by teamsters hardly made a dent in the flow of commerce, and these black workers returned at the old scale of wages.[95]

Yet the increased hostility toward waterfront labor unions had not led the port's employers to the point of breaking off all relations. While seriously shaken, the broader framework for resolving conflict peacefully remained intact through the year's end. Despite the formal dissolution of the National Adjustment Commission in the summer of 1920, steamship agents and the U.S. Shipping Board agreed to negotiate with Gulf coast longshoremen and screwmen that autumn. Upon the expiration of their three-year contract, signed during the fall of 1917, a one-year contract specifying minor wage increases secured a tentative peace on the docks.[96] But the key word was "tentative." The new contract constituted less a settlement than a truce in what appeared to be guerrilla warfare on the docks. On several important fronts, the balance of power between the Cotton Council and the Steamship Association was shifting, from the former to the latter. As the conditions that had promoted union power for some two decades began to deteriorate, the Steamship Association prepared to do battle once and for all for dominance of the city's waterfront.

V

The political balance of power that had proved favorable to New Orleans dock workers before and during the war began to shift in the postwar era. While some federal agencies continued to promote uninterrupted production through arbitration, adjustment bodies assumed a harder stance against workers' wage and work rule demands. The most decisive changes came at the level of local and state politics, as the political environment that had long favored the Dock and Cotton Council rapidly deteriorated. For four terms—from 1904 to 1920—the Choctaw Club's leader, Mayor Martin Behrman, had dominated city politics. Patronage remained the essential glue of Behrman's electoral coalition, even though civil service reform had reduced the patronage power of political parties. Behrman's extensive civic development projects generated numerous contracts for local companies as well as jobs for unemployed workers. Important segments of organized labor too found reason to ally with the mayor. While the relationship of the twentieth-century Democratic machine and the labor movement was never unproblematic, waterfront workers frequently found assistance from the city government in exchange for their electoral support.[97]

Behrman and his Democratic party regulars found consistent backing from white Dock and Cotton Council leaders. A short-lived Democratic Labor League, composed of several union presidents, threw its support behind Behrman's first mayoral campaign in 1904, and Thomas Harrison, Harry Keegan, and other white dock leaders campaigned hard for the election of machine candidates. White labor leaders were rewarded for their efforts. By 1905, James Hughes of the longshoremen, a former court recorder, had become Behrman's chief clerk; Harrison and Central Trades and Labor Council president Robert E. Lee both served as state Commissioner of Labor Statistics (a position created by the state machine explicitly for labor); and other union heads found employment in the Department of Public Works as the general superintendent, in Civil District Court as deputy clerk, and in a range of other low-level political positions. The machine continued to reward loyal labor leaders with minor positions in the municipal government throughout Behrman's rule. In 1920, for example, a number of union chiefs could be found in such positions as inspector of the city Board of Health, building inspector, comptroller, constable of a city court and assistant to that constable, and as an elevator tender at City Hall. These posts hardly represented the commanding heights of political influence.[98] Nonetheless, numerous white union leaders could be found in the mayor's camp every election day. Even the machine's temporary fall from power in 1920 did little to disturb its alliance with some union leaders. The Central Trades and Labor Council "is not a representative body of labor," an IWW journalist complained in 1923, "it is political and is dominated by politicians, who though carrying cards in their respective crafts, have not done an honest day's work in the past ten years or over."[99]

That some workers derived benefit from the Behrman machine is not in doubt. The Dock and Cotton Council relied upon the mayor's support during its major confrontations with steamship agents and stevedores; in 1907, 1917, 1919, and 1920, as well as in other years, Behrman refused to break the Council's waterfront strikes, preferring to play the role of mediator. Moreover, the New Orleans machine succeeded in frustrating the political plans of elite, anti-labor reformers. When members of the city's commercial elite elected their anti-Ring candidate, Luther Hall, to the governorship in 1912, they found Themselves checked by New Orleans politicians, unable to implement parts of their program. They were further dismayed at their candidate's cooperation with the New Orleans machine, which exerted strong influence over Louisiana's affairs. Similarly, while the Good Government League finally imposed a commission form of government on the Crescent City in 1912, Behrman and the Choctaw Club thwarted the reformers' goal of non-partisan government. Nominating four leading businessmen as Ring candidates for the new Commission Council, the flexible Choctaw Club proved it could beat the reformers on their own terms.[100]

Whatever political advantage labor possessed vanished by 1920, an election year that completely disrupted old patterns. At the state level, reformers elected to the governorship John M. Parker, a man whose hostility to both the New Orleans machine and the labor movement had remained constant for over three decades. Parker's campaign received the strong endorsement of the city's press and the backing of Governor Rufus Pleasant, who, it was reported, undermined the machine by playing at its old game—denying patronage on the order of several thousand jobs to Behrman's forces and swelling the Dock Board payroll with anti-machine voters (historian T. Harry Williams noted Behrman's complaint that "the wharves were so full of workers that cargoes could not be loaded or unloaded").[101]

A Progressive-era reformer, John Parker called on "thinking men on both sides" to "devote their best brains and energies" to the solution of the labor problem. He wasted no time in exercising his power, even months before his inauguration. At an annual banquet of a bankers' association in March 1920, Parker presented with much fanfare a plan to end all industrial strife in Louisiana. Both capital and labor would be bound to provide 90 days' notice (later changed to 60 days) before commencing a strike or lockout; during that time, a committee of three "high-grade, disinterested men" could serve as an all-purpose arbitration committee, with the power to summon witnesses, examine books and records, and hold public hearings, at the request of either party. "We are living in a new world," Parker declared, and there "is no more important question today before the American people than an earnest patriotic effort to bring capital and labor together and prevent those terrible conditions which are now existing in the nations of the old world." It did not take long for the Association of Commerce to endorse the program or for the white Central Trades and Labor Council and the interracial Dock and Cotton Council to reject it. Despite his fierce rhetoric, Parker failed to impose a new legal framework on his state's industrial relations. The arbitration plan—and additional legislation barring strikes by public utility workers—died at the State House.[102]

The New Orleans labor movement saw strong anti-union biases in the new governor's call for a "square deal." Before his inauguration, Parker announced that he would bar labor leaders from holding political jobs and union cards at the same time—a move directed against the Behrman machine. During the winter, the governor-elect repeatedly condemned the waterfront unions' continuous walkouts as damaging to the port's development. More ominous was his approach to resolving labor conflicts once in office. Parker assumed his post at a critical moment. In what was widely viewed as a major confrontation between the forces of labor and capital, thousands of boilermakers, machinists, and helpers in the newly formed Federated Metal Trades Crafts struck twenty-six large repair plants, foundries, and metal works in mid-May; on June 1, they were joined by thousands of union carpenters whose strike halted what was perhaps New Orleans' largest building boom. Both groups sought to increase wages to $1 an hour, but more important, they demanded the closed shop.[103]

Parker targeted only those striking against the Industrial Canal and Dock Board projects. But his intervention "fell like a bombshell" on the entire New Orleans labor movement, tipping the scales of power decisively and ushering in "a new epoch" in handling labor controversies on public works. The Dock Board and other public governmental bodies had long opposed trade unions, dismissed union workers, and broken strikes. Never before in peacetime, however, had labor's challenge been so large and the institutional response so determined. Parker condemned the "tyranny of labor in the hands of men who do not hesitate to rule and destroy without question," and quickly laid down his law. Unless the 300 strikers who worked for the Dock Board and Industrial Canal resumed work immediately as individuals, "so help me God . . . I shall do my duty absolutely and fearlessly." That duty involved replacing every last striker and protecting every strikebreaker. The carpenters backed down and returned to work. As for his "square deal" for labor, Parker assured the strikers that he personally would investigate their grievances. After holding a series of public meetings with public workers and Dock Board officials, the new governor issued a wage decision that granted to Industrial Canal workers a graduated scale that fell short of the strikers' demand. Reluctantly, the carpenters accepted the governor's award while continuing their fight against private contractors.[104] Parker's actions toward the municipal employees differed little from those of his predecessors; his innovations lay only in his personal involvement, in the publicity with which he accomplished his task, and in his making an ideological virtue—union busting—out of what earlier officials saw as merely a necessity.

The municipal elections of September 1920 enabled Parker and his anti-Ring allies to consolidate, at least temporarily, their political hold. The Orleans Democratic Association (ODA) offered the latest and most successful challenge to "Ring rule" since the Choctaw Club's formation in 1897. Personality conflicts among ward leaders, and divisions between local and state officials, complicated postwar political alliances and seriously weakened Behrman's hold over the city. The destruction of the New Orleans machine had long been a goal of John Parker, and the governor now threw all the state resources he could—including considerable Dock Board patronage—behind the ODA's nominee, Andrew J. McShane, an undistinguished hides merchant. Denouncing the Choctaw Club as a political dictatorship and association of political job holders, the ODA and the governor had a long history of elite opposition to the machine to draw upon in their fight against "bossism and Behrmanism."[105]

In addition to the charges of moral laxity, protecting prostitution, inefficiency, and corruption—what political scientist George Reynolds earlier called "every known evil of city government"—the ODA made a direct appeal to the labor movement. It was a move that paid off. While many union officials remained loyal to the mayor, labor's ranks provided not an insignificant number of ODA backers. Dissatisfaction with the Behrman regime was strong. In February 1920, one unnamed union official told the *Item,* "It has taken the Ring about 20 years to find out that there is such a thing as the working man's vote. . . . It won't go down." Behrman was "labor's bitterest enemy," a leader of the boilermakers' union declared in August. As "unfriendly as he has been to organized labor, he now seeks our support, blowing hot and cold, and caring for the laboring man only when he needs his vote, and never at any other time."[106]

It was not difficult to hold up the Choctaw Club's labor record to public scrutiny and find it wanting. Some union men scored the composition of the machine's leaders. During the past eight years of commission government, McCable of the boilermakers complained, Behrman had "surrounded himself with architects, wholesale merchants, insurance men and bankers. Not a labor man ever sat in the commission council." Worse was the mayor's behavior toward unions. Under Behrman's administration, municipal organizations like the Dock Board, the Public Utilities Commission, and the Public Works Department had refused to recognize or engage in collective bargaining with public employees' unions, had broken strikes of public workers, and had offered increased police protection to private employers confronted by strikes.[107] In an anti-machine editorial, the *Item* denounced the arrogant, "blundersome, Hypocrisy of the Machine and the Mayor in undertaking to INFLAME CLASS-FEELING and RIDE Labor as a beast-of-burden back into office" and condemned Behrman "for trying to prejudice Mr. McShane by DENOUNCING in Mr. PARKER a course of action that Mr. BEHRMAN himself has FOLLOWED more ruthlessly, with a train of blasted laborite hopes behind him as mute witnesses of his present fraud upon the voters."[108]

Many of the ODA charges of anti-labor behavior by the machine rang true. There is no doubt that the twentieth-century Choctaw Club was far less attentive to the needs of organized labor than its nineteenth-century predecessor. The Behrman administration had courted the city's business and commercial leaders far more assiduously than either Guillote or Fitzpatrick ever had. Its port modernization program had stolen much of the reformers' thunder on economic issues, and its policy against recognizing unions on public works had won the praise of many businessmen. For most of the 1910s, then, workers in the large and growing municipal sector confronted powerful administrative boards whose behavior was indistinguishable from that of the most anti-union private employer. As the city's chief executive during these years, Behrman now found that his neglect and hostility toward municipal unions made him a lightning rod for anti-Ring labor sentiment.

Yet with a social base that cut across class lines, the machine was a political creature that recognized the necessity of carefully and pragmatically balancing competing interests. The Choctaw Club "had its progressive and conservative sides," historian Matthew Schott has noted, but it functioned "not as an initiator of political action but as a mediator among pressure groups." Indeed, Behrman respected power, especially when it impinged upon his rule. When union strength rendered the forceful resolution of labor conflicts impossible or too costly, he would often employ his skills as negotiator and broker. On countless occasions, the mayor crafted compromise measures at least minimally acceptable to the representatives of labor and capital. Most recently, for example, he had averted a threatened general strike by the Dock and Cotton Council in the spring of 1920 by pressuring both labor leaders and the managers of the Illinois Central. During the war, Behrman sometimes placated labor (such as when he ordered the Southern Pacific to recognize the union of its freight handlers), leading shipping operators to protest to federal officials that the mayor was an unsuitable mediator.

The powerful Dock and Cotton Council had reaped perhaps the largest rewards for its support of the Behrman machine during its sixteen-year rule. Behrman's overriding goal, beginning in 1907, had been to minimize the disruption that waterfront strikes caused to the city's commerce. Toward that end, the mayor would sit down with waterfront labor leaders, black and white, and run interference for striking riverfront unions. While his police force gave protection to employers during strikes and harassed strikers in smaller or weaker unions, Behrman left Cotton Council members to pursue their own strategies vis-à-vis shipping agents and stevedores. In moments of crisis, it seemed safer to grant dock union demands than to permit an all-out fight that would immobilize the city's commercial life. Not surprisingly, the Dock and Cotton Council formally endorsed the municipal machine ticket in August 1920.[109]

But in a competitive political marketplace, the machine could not easily rally its working-class supporters with such a mixed record. It did try, though. Following Parker's victory, the machine made overtures to the white labor movement, offering patronage and "some of the juiciest fruit on the political plum tree" in exchange for support. In September, Dock and Cotton Council president James Byrnes and ILA vice president and screwmen's leader Thomas Harrison appeared on the machine ticket as labor candidates for clerk of the Criminal Court and Commission Council respectively. But such efforts were too little, too late. The white labor movement approached the September municipal elections a divided force.[110]

Following a national program initiated by the AFL, the New Orleans Non-Partisan Political Campaign Committee, formed in the spring of 1920, brought together hundreds of delegates representing some 150 unions who pledged to investigate the labor record of all candidates. The participants' one point of agreement was the endorsement of all union members running for political office. In August, it formally backed the candidacies of machine-supporters Thomas Harrison for Commission Council, Pat McGill for constable, and James Brynes for clerk of the Criminal Court, as well as ODA candidates Richard Dowling for judge of the First City Criminal Court and Wilbert Black for Commission Council. The Non-Partisan Committee adopted no position on the mayoral race, however, and efforts to promote Martin Behrman generated intense opposition within the Committee's ranks. Speakers denounced the Dock and Cotton Council's endorsement of the entire machine ticket and individual unions "repudiated" what they saw as efforts to "deliver the trade union vote" and plunge their organizations into partisan politics. By September, labor's divided mind on the mayor's race was standard press fare. At least three unions withdrew from the annual Labor Day parade in early September over the selection of labor leader John Banville—a "zealous ringster" and machine comptroller of the city—as grand marshall. When Banville introduced Mayor Behrman—who had spoken every year to Labor Day crowds— a five-minute shouting match between McShane supporters and Behrman backers ensued, illustrating just how much trouble the machine was in.[111]

The September primary witnessed the Choctaw Club's first setback in two decades. In the closest vote ever, Behrman lost to McShane by less than 1500 votes. Many on the machine ticket went down with Behrman. Waterfront leader Brynes lost his race for Criminal Court clerk, and Thomas Harrison, who survived the primary, faced defeat in the general election weeks later. Ecstatic ODA leaders attributed their victory in part to labor's defection from the machine, the vigilant policing of polling places, and, in the *Item*'s words, the clearing out of registration rolls in the "rotten boroughs along the water front, out of which the Machine usually piles up its enormous majorities." The "anti-Behrman citizenry," the anti-Ring *Item* gloated on election night, struck the "final blow against boss mismanagement."[112]

The election of an anti-labor governor and mayor could not have come at a more auspicious time for city employers. Ideologically and programmatically, New Orleans businessmen prepared for an assault upon the labor movement. The "spirit of unrest"—the widespread unionization of new sectors of the labor force and the wave of strikes from late 1918 to early 1920— had seriously alarmed employers. The waterfront experience stood out as a negative example. "The fact that the supreme test must come some day between the labor and capital of New Orleans," the *States* reported, was foretold earlier that year at hearings between the joint exchange committee and the dock workers. At that time, employers believed "it was made evident by the union men that they would never be satisfied until they dominated the shops where they are employed and say who should and who should not work." By May 1920, adherents of the open shop drive, or the American Plan, drew the line—now successfully—against labor's demand for the closed shop. Endorsed by the Association of Commerce, the Southern Pacific Steamship line, the Metal Trades Association, and the General Contractors' Association of the city, open shop proponents urged all New Orleans industries to join their movement.[113]

Shifting political currents alone did not determine the fate of New Orleans workers on and off the docks. Postwar economic developments further set the stage for the labor movement's decline. For a year after the Armistice, steadily rising wage rates had corresponded to the rising cost of living. But after early 1920, federal policy and the end of the economic boom brought prices down quickly (the fastest decline in American history took place between August 1920 and February 1921). Beginning in 1920, pressures to lower wages intensified. In a number of occupations, the Louisiana labor commissioner later reported, employers reduced wages to "virtually pre-war levels." While the most rapid and severe drops hit non-unionized common laborers and semi-skilled workers, even skilled union workers suffered sharp reductions. As unemployment rose substantially in 1921, emboldened employers cut union wages by 10 to 20 percent. The Metal Trades Association and the General Contractors' Association reduced wages during the spring and summer of 1921, despite massive strikes. In September, the Dock Board ordered an 18 percent wage reduction for all of its Public Cotton Warehouse employees. When some 1500 black and white warehousemen, teamsters, loaders, samplers, markers, and inspectors struck in protest, the Board hired strikebreakers. Only a sympathetic strike by longshoremen and screwmen, who refused to accept bales handled by non-union men, led to a compromise settlement in which the Dock Board—utterly opposed to negotiating with unions—took back most of the strikers while retaining some of their scabs.[114]

New Orleans employers had not only finally found a friend in the governor's office, but steamship operators gained their own ally in the U.S. Shipping Board as well. Although the policies of the Board and its National Adjustment Commission had often favored longshore laborers during the war, the Board's postwar agenda more often coincided with that of employers. Its dual goals were to promote a reduction in the cost of living and to ensure profitable production of the ships in its fleet. As early as 1919, the Board's modest wage awards had aroused the suspicion of New Orleans longshoremen. In 1920 the Board held wages constant, and the following year it proposed substantial cuts in longshore wages. Its behavior toward the International Seamen's Union indicated the new approach. In May 1921, it reduced wages, eliminated union preference in hiring, and barred union representatives access to their members on board ships. When seamen struck in protest, Admiral William Benson, the Board's chief, placed U.S. troops on government-run merchant ships and declared that Shipping Board vessels would be run open shop.[115] In the ports along the Gulf and the Atlantic and Pacific oceans, the government assault delivered the seamen a virtually fatal blow. That autumn, Board officials took another strong anti-union step. Infuriated by a sympathy strike of New Orleans longshoremen and screwmen on behalf of striking public cotton warehousemen—which was sanctioned by their contract—that halted the loading of Shipping Board vessels, the government body instructed its district directors to "use whatever labor available" when unions violated their agreements.[116]

VI

It was in this context—of economic downturn, a broad anti-labor employer offensive, the election of pro-business state and municipal officials, and the deepening hostility of the U.S. Shipping Board and the local Dock Board—that the battle over waterfront wages and work rules erupted in the fall of 1921. There was no question that wages would be reduced; the ILA had even counseled its locals to accept reductions to avoid a larger onslaught. In early October, black and white longshoremen voluntarily agreed to cut their hourly wage from 80 to 65 cents, pending further negotiations. Following a three-day longshore walkout in early November, employers retracted demands for a ten-hour day and other rule changes in exchange for the 65-cent wage rate.[117] More difficult were the negotiations between operators and screwmen. Under the contract that expired on September 30, 1921, gangs of five screwmen received $36 per eight-hour day for hand stowing 187 bales of cotton. For two months that fall, steamship operators and their workers negotiated the terms of a new agreement. The Steamship Association demanded that screwmen increase the 187-bale limit (established by the Port Investigation Commission in 1908 and reaffirmed during the war by the National Adjustment Commission) to 211 bales, and that they assume the responsibility and cost for rolling cotton bales to the ship's side—a task previously performed by a longshoreman at the employer's expense—as part of their daily work. The unions' counter-offer to increase their limit to 200 bales only heightened the operators' impatience. Employers upped the ante by demanding parity with Galveston, where workers received 20 cents for every bale of cotton stowed and 40 cents for any bale screwed, with the men rolling their own cotton to the ship's side. On November 20, with negotiations deadlocked, the black and white screwmen voted to strike.[118]

For the first time since the 1907 general strike, steamship operators turned their full attention to the port's highest paid workers. Their objections to the screwmen had changed little in almost a decade and a half. Once again they insisted that the two unions were irrelevant in the modern age. To a much greater extent than in 1907, technological and commercial developments had rendered the screwmen obsolete. The advent of larger ships, high density compresses, and lower freight rates had led steamship operators to abandon almost entirely the screwing of cotton in ships' holds. But notwithstanding the radical change in the work process, the screwmen maintained both their higher rate of pay and dominant position within the Dock and Cotton Council.[119] Worse still was their disruptive power. The port of New Orleans, read a statement from nine leading commercial organizations, "is now and has repeatedly been menaced by sympathetic strikes among its dock and shipping labor, often many men who have no cause or complaint against their employers refusing to handle the commerce of the port because other workmen, few in number, decline to accept the wage and working conditions obtaining among the highest paid workers at competitive ports." So menacing and intolerable had this recurring situation become, they argued, that they called for the formation of "a permanent organization to guard against labor revolts" that were continually "menacing our commerce."[120] With the strong and unqualified approval of the city's business leaders, the Steamship Association vowed to resolve these issues with all the tools at its disposal.

While coastwise steamship lines, railroad dock managers, the Dock Board, and smaller middlemen had resorted at times to "outside labor" during labor crises, not since 1907 had the steamship agents utilized strikebreakers to replace either longshoremen or screwmen. In 1921, the Steamship Association wasted no time in assembling an "independent" labor force. Its move, which struck directly at the Dock and Cotton Council's power, was facilitated by postwar unemployment. Noted Shipping Board representative Jenkins: "[Y]ou can blow a horn on the streets here and get a thousand idle men in a minute who will be glad to get work." In the depressed New Orleans economy, hundreds of local non-union laborers immediately responded to advertisements in the city's papers for work on the docks. Many of the new recruits—three-quarters of whom were estimated by the *Item* to be black—were ex-servicemen.[121]

To minimize contact with the unions, the Steamship Association chartered four tugboats to carry the men from the Poydras Street freight shed where they were instructed to assemble each morning, to the ships docked along the levee. The Dock Board, never a friend of labor, granted its harbor and river front patrol of 100 men the full power of arrest and concurrent jurisdiction with the city police and further instructed its force to pick up all waterfront loiterers. Police officials stationed not only extra squads of men armed with sawed-off shotguns but motorcycle police in the area as well. In addition, the Steamship Association employed large numbers of private armed guards to patrol the riverfront. With these elaborate precautions in place, 950 strikebreakers began work on 16 ships on November 24.[122]

The union response was immediate and determined. Delegates to the Dock and Cotton Council—now composed of some 24 waterfront unions, representing between 12,000 and 15,000 men—voted unanimously to order a general strike.[123] On November 24, black and white yardmen and longshoremen, black teamsters and chauffeurs, warehousemen, cotton handlers, stave classers, car unloaders, coal wheelers, and coal trimmers quit work in sympathy with the screwmen; most remained out until the strike ended on December 3. On the first day, hundreds of union pickets stood silently near the docks, watching almost 1000 strikebreakers behind a cordon of police register for employment and board the tugs that carried them to their new jobs. Yet the general strike failed to deter the steamship operators. Exuding confidence, they declared satisfaction with their new labor force and rebuffed all recommendations to arbitrate.[124]

If the Steamship Association had abandoned restraint and ignored all channels of mediation, other voices counseled a somewhat more cautious path. Initially, J. C. Jenkins, the U.S. Shipping Board's director of industrial relations, endorsed the Steamship Association's wage and rule demands and approved the hiring of "any sort of labor obtainable," that is, strikebreakers. "It's a fight to the finish between the steamship operators of New Orleans and the port workers, with the government behind the employers," Jenkins noted early during the strike. George Santa Cruz, the Shipping Board's New Orleans director, concurred, promising "Whatever the Steamship Association committee wants us to do, we shall do." But there was a limit to the Board's support in the struggle against labor. While the Shipping Board strongly backed lower wages, more flexible work rules, increased managerial authority, and uninterrupted production, it nonetheless had important differences with the operators. In a letter to T. V. O'Connor, the former ILA president who now served as a Shipping Board commissioner, Jenkins rejected the Steamship Association's request that the Board assume responsibility for crushing the screwmen. He argued that there was "no justification" for the government "becoming a union destroying agency," that opinion on the issue of the screwmen's destruction remained "greatly divided" outside the ranks of the foreign steamship interests, and that it was doubtful that the screwmen's elimination could be accomplished without a long and bitter fight in which the Shipping Board would bear the brunt. Jenkins believed that the screwmen could be induced to adopt more liberal arrangements comparable to those followed in Galveston. With a negotiated defeat in mind, Jenkins attempted to bring the agents and strikers together.[125]

No one doubted that this dock strike took place in an atmosphere that was very different from the pro-labor one two years earlier. Never before had the Dock and Cotton Council confronted a group of employers so united and so solidly supported by state and federal organizations in such an inhospitable economic environment. Following a November 28 waterfront riot in which strikers and their sympathizers battled almost 250 policemen and non-union workers, the unions voted to reopen negotiations with the Steamship Association. In conceding to a settlement based on the Galveston model, the screwmen accepted many of the Association's terms, ending the strike. Under the new contract, they would receive 18 cents a bale for hand-stowed cotton and would stow 225 bales per day per gang; in exchange, steamship operators agreed to continue to pay longshoremen for rolling cotton to the side of a ship and to dismiss their non-union workers. Less than two years later, an ILA report attempted to put the strike into perspective. The 1921 settlement, which had left the screwmen's unions intact, was "considered by these two locals as a remarkable victory in view of the determined attempt to disrupt the two organizations."[126] It was hardly that. Rather, it represented the first major defeat for these powerful members of the Cotton Council. Despite a remarkable display of internal unity and labor support across craft and racial lines, the Dock and Cotton Council had entered a new era in which solidarity simply was not enough to ensure union power.

The 1921 contract merely had postponed the day of reckoning. When their contract expired in August 1923, longshoremen demanded a wage increase to 80 cents an hour, a reduction of the work week from 48 to 44 hours, and the addition of an extra cotton roller for every two gangs. Cotton screwmen similarly proposed new work rules, calling for an additional roller and an increase of three cents a bale for hand-loaded cotton. Steamship Association members were willing to negotiate on the question of wages, but they insisted that no working rules be changed. On September 13, some 1200 longshore workers walked off the levee; two days later they were joined by approximately 1000 screwmen. Although a more cautious Dock and Cotton Council declared no general strike this time, individual unions of car loaders and unloaders and others refrained from delivering freight to scabs. In late September, the ILA stepped into the fray, ordering its other southern locals to handle no freight loaded by the New Orleans strikebreakers; many southern longshore workers also struck for their own wage demands. Finally, the New Orleans branch of the IWW's Marine Transport Workers Union struck in sympathy and for the release of all "class war prisoners," calling out an unknown but probably small number of men from tugs and ships in port.[127]

Shipping operators now declared war. The issues remained the same as in 1921—what employers saw as costly and intolerable union work rules and the very existence of separate cotton screwing unions, which were "giving the port a reputation for irregularity and unreliability."[128] By 1923 steamship operators had the will, the power, and the allies to impose the reorganization of work that they had desired for decades. And they resolved to eliminate once and for all the "artificial, arbitrary and unreasonable division of work" between longshoremen and screwmen and replace it with but one class of waterfront labor, that of general longshore and ship work. Following the pattern set during the strike of 1921, the Steamship Association put the crews of its ships to work loading and unloading cargo, and began hiring strikebreakers—many of whom reportedly came from rural sections of the state—in large numbers. By early October, over 1000 men were working; within a week that figure had nearly doubled. By the middle of the month, the Association claimed that it was through with the unions and that labor conditions had normalized. Despite a diversion of some trade, September set a record for the heaviest ocean traffic in dutiable goods in the port's history.[129]

As in 1921, the Association did not lack allies. Its efforts received support from the city's major newspapers, the commercial exchanges, the municipal and state governments, and the court system. Daily reports of alleged union violence against strikebreakers filled the pages of what the local IWW branch called New Orleans' "prostituted press" and what the ILA denounced as "the defender of corporate greed, [which] skillfully avoided publishing the true facts."[130] Hundreds of city and Dock Board police, armed with riot guns and under orders to disperse all groups of men in waterfront neighborhoods, protected the Association's strikebreakers. At the head of Canal Street, observed a Wobbly reporter, "two armored automobiles loaded with machine guns stood ready to mow down the workingman who asked for the privilege of existence and the right to educate and care for his loved ones." The Dock

Board again sided with the operators. With the approval of Governor John Parker, it used its own crews to handle cargo. More important, it barred all strikers from the public wharves and sheds. Lastly, the court system proved a most valuable ally. Federal court Judge Rufus Foster, who had long served employers' needs on labor matters, readily granted the Board an injunction prohibiting longshore workers, as individuals or as a union, from harassing strikebreakers or even from congregating near the public waterfront. Moreover, when a civil district court issued a restraining order blocking the Board from interfering with the strikers' right to gather on public property, the State Supreme Court immediately upheld the Board by sustaining the original prohibition. The strikers had little recourse but to comply, since the local police did not hesitate to arrest dock workers who violated the injunction.[131]

While the 1923 strike delivered a crippling blow to the Cotton Council, the intervention of the U.S. Shipping Board kept the longshoremen's and screwmen's unions alive. In contrast to the 1921 experience, the Shipping Board now exerted little influence over the Steamship Association. Unwilling to lead the open shop drive, the Board parted company with the operators over their goals and intransigence. Moreover, the operators' experiment with non-union labor had not impressed government officials. An investigation conducted by E.H. Dunnigan, a commissioner with the U.S. Department of Labor's Conciliation Service, and Immigration Commissioner W.W. Tuttle found "not only an insufficient number of men to man the work, but . . . a class of labor that is not more than fifty per cent efficient." The strike also was proving costly to the Shipping Board, which operated four shipping lines in New Orleans, accounting for between 30 and 35 percent of the port's tonnage. The strike idled thirty-four of its ships. About a month after the strike began, the Shipping Board took the bold step of opening negotiations with the striking unions. The Board, it stated, "is not warranted in carrying on this fight involving tremendous expenditure of public funds solely on the issue of union or no union, the longshoremen having surrendered on the original contention upon which they went on strike." Given the unnecessary cost, the chaotic conditions, and the goals motivating Steamship Association operators, the Board proposed to stablize labor conditions on its own ships.[132]

The move was not entirely unexpected, as the Board had just settled a Norfolk, Virginia, longshore strike by raising wages from 65 to 75 cents. But it had refused to sign any contracts with workers in that port. Similarly, in mid-October, the Board's secret negotiations with New Orleans unions paid off, resulting in a gentlemen's agreement to raise wages to 80 cents an hour. The Shipping Board's "separate peace pact" provided employment for less than a third of the port's union workers. Moreover, while the screwmen, who voted unanimously to accept the Board's offer, received a one cent per bale increase in wages, they now were entirely responsible for trucking their own cotton to a ship's side. The price of survival now appeared to require waterfront unions to surrender written contracts and important work rules, key aspects of their power on the docks for over two decades.[133]

Unimpressed with these changes, the Steamship Association continued its all-out struggle against the unions. The Dock Board reallocated wharf space, concentrating Association ships and Shipping Board vessels at different points along the riverfront. The strike, handicapped by the court injunction and the undiminished hostility of the Steamship Association and the Dock Board, "continued until there was apparently no hope of winning," the state's labor commissioner later reported, "and many of the strikers returned to work or accepted employment in other capacities." Union workers found limited employment on vessels operated by the U.S. Shipping Board for the next decade. In 1933, that relationship ended when three independent operators who leased the Board's ships finally broke with the waterfront unions.[134]

Although the Shipping Board had saved them from extinction, the screwmen and longshoremen had suffered a devastating blow. The factors contributing to the Dock and Cotton Council's fall were clear. Postwar politics and economics generated intense pressures that undermined union power across the nation; New Orleans fit the national pattern. Against the backdrop of an uncertain and fluctuating economy, workers seized the opportunity in 1919 and after to extend to labor's ranks the democratic ideal the nation ostensibly had fought for abroad. But as political reaction set in in Louisiana, state and municipal politicians—as well as the federal courts—threw their weight behind anti-union employers and public institutions like the Dock Board, whose power had grown steadily over time. In the upheavals of the immediate postwar years, the system for regulating labor-management relations—which had served employers well after 1908 and less well during the war—quickly crumbled. Steamship operators ended their long dealings with union workers and joined their counterparts in numerous other industries in a new-found adherence to the ideology of the open shop. Dock workers' defeat was not restricted to New Orleans alone. Associations of agents and stevedores in Galveston, Houston, Texas City, Gulfport, Mobile, Pensacola, and Newport News and Norfolk delivered crippling blows to their ILA locals, resuming operations on an open shop basis.[135]

One additional factor requires investigation: the status of race relations between black and white workers on the city's waterfront. Had deteriorating relations between black and white dock men contributed to union weakness? The question is an important one, especially as the answer offered in the secondary literature on the subject is an unqualified "yes." The only substantive analysis of racial tensions within the Dock and Cotton Council was offered in 1931 by Sterling Spero and Abram Harris in their now classic *The Black Worker*. Noting the "old tradition of labor solidarity cutting across race lines" on the docks, Spero and Harris attributed much of the success to the formula of the equal division of work between blacks and whites. But that principle, they argued, "by no means operates with full and equal justice." Because black longshoremen outnumbered white longshoremen across the South, there were "a great many more black applicants for a job than white."[136] If the two races had divided work equitably on the waterfront earlier in the century, by the 1920s this was no longer the case. Other historians have adopted this line of reasoning. According to one recent account, the "trend toward Black majority in the labor force had . . . quickened during the 1910s," causing a "dramatic rise in the ratio of Black dockworkers to white," a process which the First World War accelerated.[137] According to this analysis, an equal division of work guaranteed to whites a disproportionate and unfair number of jobs, undermining the alliance.

Little evidence supports this claim. New Orleans' white population retained a three-to-one numerical dominance through the 1920s. The city's population grew more slowly than other New South centers like Atlanta and Nashville, with black and white rates of increase almost equal (13 and 14 percent respectively) between 1910 and 1920. By 1916, New Orleans experienced labor shortages, not labor surpluses, and it was the shortage of black labor on the docks that caught observers' eyes. This problem, which predated the country's entry into the First World War, only intensified during the war, as frustrated shipping agents scoured the docks for a sufficient number of workers.

It could be argued that white workers had access to a far wider range of jobs than blacks, and hence unskilled black workers—barred by tradition, union regulations, or employer preference—were entitled to a higher percentage of riverfront work. Three points challenge this proposition. First, there is no evidence suggesting that black unionists complained of any injustices caused by the equal division of work. They certainly had the opportunity. The bi-annual conventions of the International Longshoremen's Association provided disgruntled blacks with a forum for airing their grievances, and black union representatives and black international officers of the ILA often raised problems involving race relations in various ports. But black delegates from the New Orleans longshoremen's and screwmen's unions voiced no complaints on this issue during the period from 1910 through the mid-1920s. Second, blacks did in fact possess a majority of New Orleans waterfront jobs. They constituted half of the better-paid screwmen, longshoremen, and yardmen, and they dominated the lower-paid ranks of teamsters and loaders, car loaders and unloaders, railroad terminal freight handlers, coastwise longshoremen, coal wheelers, and coal shifters. These latter jobs occupied the lower end of the waterfront's occupational hierarchy, where wages and working conditions were worse than in the better paying longshore and screwmen's trades. But this hierarchy reflected an institutionalized racism which the Dock and Cotton Council neither caused nor challenged. In its over two decades of existence the Cotton Council proved neither ideologically inclined nor, probably, able to improve conditions at the lowest end of the occupational ladder. At any rate, black longshoremen and screwmen remained as silent as their white counterparts about the occupational structure of waterfront employment.

A close analysis of union membership figures raises a third objection to the argument that the equal division of work deprived blacks of jobs. Spero and Harris, drawing upon information provided by a waterfront superintendent during the late 1920s, estimated that 4400 black longshoremen and 1400 white longshoremen were members of their respective unions in 1923. One reading of these figures would suggest that the different membership levels illustrate the unfairness of the equal division of work. But both unions set their own admission standards; lower white membership could simply mean that the union chose to admit fewer members, not that there were fewer white longshoremen available. More important, the numbers themselves are misleading and wrong. While black longshore Local 231 carried some 2200 members on its books at the time of the ILA's 1923 convention, not all of those members actually worked as longshoremen. Charged in a factional dispute that the union had failed to pay its full per capita tax to the ILA, Albert Workman, delegate and president of the local, challenged the relevance of the total figure. At its peak in 1918 or 1919, he stated, the black longshoremen numbered no more than 2,006 men; by 1923 the number had fallen to 1900. And of that number, only 800 worked as longshoremen; the rest, in and out of the city, simply held other jobs. The Longshoremen's Protective Union Benevolent Association, chartered in 1874, had functioned for a half-century as a fraternal organization as well as a union. All members, dock workers and others, who kept current by paying their "medicine tax" and general dues were entitled to sickness and death benefits.[138]

The biracial union structure that had governed race relations on the docks had never eliminated racial inequality or tensions, and it did come under increasing strain during the depression of the early 1920s. U.S. Shipping Board records support Spero and Harris's contention that white longshoremen and screwmen called the 1923 strike over the objection of blacks (as well as the objection of white officers of the ILA). A labor spy—Investigator #36—reported to his unidentified superiors at the Maritime and Merchants Protection Company on September 13 that white union leader Harry Keegan and his men created a "general disturbance" at a joint black-white meeting to prevent blacks from voting against the walkout; "threats and utterances of violence came forth from the white longshoremen" who stated that "they expected the colored longshoremen to refrain from working while their white brothers were standing off." "None of the Vice-Presidents of the I.L.A. are able to prevail upon Keegan to have his men return to work," the district director of operations of the Shipping Board reported. "The colored longshoremen are all ready and willing to return to work at any time if Keegan and his bunch of hot heads will agree to it."[139] Clearly, the two unions sharply disagreed, and whites imposed the more militant option.

While the incident's racial dimension should not be obscured, neither should it dominate the story. Keegan's targets were not just black longshoremen but white officials of the ILA as well. His union had remained independent of the International for most of the twentieth century, occasionally flirting with affiliation but often remaining outside of the International's ranks. Not only did Keegan reject the advice of J. H. Fricke, the white chair of the ILA's Gulf District Branch, but he refused to permit Fricke to address his union's meetings. Moreover, whatever reluctance black longshoremen had expressed before the strike vanished once it began. Blacks maintained a solid front with whites for the duration of the battle; their president, Albert Workman, regularly represented both unions at meetings and in the press; and several other black unions—car loaders and unloaders and teamsters—joined a short-lived sympathy strike. The point is not that there were no racial tensions—there were. But if racial tensions were more pronounced in the economically depressed years of the early 1920s, they did not undermine the Dock and Cotton Council's strength. The challenge that destroyed the waterfront labor movement came from without, not within.

EPILOGUE

After exercising considerable control over the waterfront labor process for over four decades (with the exception of the late 1890s), the interracial federation of waterfront workers fell victim to an open shop drive that devastated much of the labor movement nationwide. The aftermath of the 1923 strike revealed in stark terms just what the loss of union power meant for the world of the waterfront. Employers, not surprisingly, viewed the era as a new golden age. Plagued with "almost yearly occurrences" of strikes and disorder prior to 1923, the Steamship Association believed that it had ended such disturbances once and for all. Consolidating its position without delay, the Association maintained an open shop, ruled the docks with an iron fist, and refused to deal with, much less recognize, unions. For the rest of the 1920s, E. J. McGuirk, the Steamship Association's labor committee chair, noted that "labor conditions on the river front have been infinitely better." The Association also relied heavily on a black labor agent, Alvin Harris, whom employers saw as "an unusually discreet and reliable colored man" and unions condemned as a "Professional Negro Scab Labor Agitator . . . who has openly admitted . . . that he is paid to suppress and hold black labor down in semi-slavery." Free from union interference, the Association implemented new procedures. It required its employees to be photographed, fingerprinted and numbered, and to "assume an obligation" not to belong to the ILA. In 1927, the Association quashed a challenge by a resurgent union movement by securing a permanent injunction from the U.S. District Court in New Orleans, protecting its employees from the "assaults and abuse of disgruntled and dissatisfied agitators." The injunction remained in effect in subsequent years.[1] As in the period following the 1894–95 race riots on the waterfront, the absence of union contracts, work rules, and control in the 1920s and early 1930s enabled employers to manage affairs with a free hand.

What of longshore laborers? How did the collapse of union power transform waterfront labor conditions? The "disastrous defeat for organized labor," Spero and Harris noted, broke its half-century hold on the port and sent union membership figures plummeting. But black journalist Robert Francis, writing in 1936, found a silver lining in the cloud of defeat. Although black longshoremen "sustained a loss from which they have never recovered," he concluded that they had made some noteworthy gains. In the immediate post-strike era, the labor force of the Steamship Association was overwhelming black, and a small number of blacks became gang foremen, a position long barred to them by the objections of white union men. If blacks had moved into foremen's ranks, however, their promotion proved temporary. In 1930, the chair of the Steamship Association's labor committee clearly stated that all foremen were white men.[2] And if, as Francis argued, blacks' moving into more and new positions represented a gain, the positions they moved into and the ways they moved into them gave them a great deal to complain about.

On all other counts, the new system of labor relations meant deteriorating conditions for waterfront workers. One of the first things to go were the remnants of union control or influence over hiring practices. During the war, for example, the two longshore unions exercised considerable responsibility for providing employers with gangs of men and for ensuring that work was shared equally among all members and between whites and blacks. One stevedore had testified that because the process was in the unions' hands, he had no personal knowledge of who actually worked for him. The open shop system changed that dramatically. In the glutted labor market of the late 1920s and 1930s, the Steamship Association implemented the notorious shape-up on the New Orleans docks for the first time in decades, placing individuals at the mercy of hiring foremen. In addition, wages fell considerably. While union men received 80 cents an hour under their gentlemen's agreement with the Shipping Board during the 1920s, non-union workers of the Steamship Association received 65 cents an hour; with the onset of the Great Depression in 1929, those wages fell even further, ranging from 40 to 65 cents an hour. The open shop policy of the Association had created intolerable "conditions of Semi-Slavery," a group of black and white union leaders declared, that had "cut the purchasing power of the River Front Labor so that they almost live from hand to mouth and in some cases are in actual want." Government investigators substantiated many of the unions' claims. A 1930 Bureau of Labor Statistics study reported that even employers admitted that longshore wages were "very low" and found that the average earnings in the port of New Orleans were "considerably lower than in any other large port in the United States." The shift of managerial control from union foremen to the Association's hand-picked men generated its own abuses. Workers, for example, complained of a system of "rebating," whereby longshoremen had to pay their foremen $2 or $3 a week if they expected to secure employment the following week. "In other cases," a government investigator discovered, "the colored men are merely paid a certain amount for the week, and the balance goes to the foreman, who is permitted to draw the pay on the brass checks turned over to him by the workers . . . [which] are often sold to saloon keepers and money lenders on the water front at very large discounts." Other complaints included forced loans at exorbitant interest rates; intimidation by private detectives; the arbitrary power of the Steamship Association's labor agent Harris to dismiss any man, black or white, who did not meet his approval; and discrimination against white workers and all union men.[3]

The formal alliance between the black and white longshoremen's unions survived the defeat of 1923, with the two organizations issuing joint statements and occasionally working together during the rest of the decade. But little remained of the interracial collaboration and power of the Cotton Men's

Executive Council or the Dock and Cotton Council. The open shop citadel erected by the Steamship Association proved too strong for the longshore unions to crack; the fragmented unions that did survive the 1920s exercised little control over either workplace conditions or wages. At the same time, their weakness fostered parochial outlooks.[4] The New Orleans Steamship Association had resolved decisively the generations-old struggle with the South's strongest labor movement, achieving at last untrammeled control over the labor process. A new generation of dock workers, inheriting the legacy of both the Dock and Cotton Council and the bitter years of the open shop, would challenge the Steamship Association with very limited success in the 1930s. But never again would their movement be characterized by the degree of interracial and inter-trade collaboration that had distinguished its predecessors. Born of economic necessity and strategic considerations in the decades before 1923, biracial unionism on the docks had never offered full equality to black workers. But the "industrial equality" it promoted, however limited in scope, served to bridge some gaps between blacks and whites by creating an arena for interracial contact, deliberation, and collaboration. The commercial elite's victory, in imposing the open shop and destroying the waterfront labor movement, buried one of the few significant exceptions to the rule of white supremacy in the Deep South.

NOTES

Preface

1. Herbert G. Gutman, "The Negro and the United Mine Workers of America: The Career and Letters of Richard L. Davis and Something of Their Meaning: 1890–1900," in *The Negro and the American Labor Movement,* ed. Julius Jacobson (New York: Anchor Books, 1968), 49–127. The important exceptions include: Paul B. Worthman, "Black Workers and Labor Unions in Birmingham, Alabama, 1897–1904," *Labor History* 10, No. 3 (Summer 1969), 375–407; Peter J. Rachleff, *Black Labor in Richmond 1865–1890* (1984; rpt. Urbana: University of Illinois Press, 1989); Dolores E. Janiewski, *Sisterhood Denied: Race, Gender, and Class in a New South Community* (Philadelphia: Temple University Press, 1985); Jacqueline Dowd Hall, James Leloudis, Robert Korstad, Mary Murphy, Lu Ann Jones, and Christopher B. Daly, *Like a Family: The Making of a Southern Cotton Mill World* (Chapel Hill: University of North Carolina Press, 1987); David Alan Corbin, *Life, Work, and Rebellion in the Coal Fields: The Southern West Virginia Miners, 1880–1922* (Urbana: University of Illinois Press, 1981); Ronald L. Lewis, *Black Coal Miners in America: Race, Class, and Community Conflict 1780–1980* (Lexington: University Press of Kentucky, 1987). On black workers in northern communties, see: Peter Gottlieb, *Making Their Own Way: Southern Blacks' Migration to Pittsburgh, 1916–1930* (Urbana: University of Illinois Press, 1987); Joe William Trotter, Jr., *Black Milwaukee: The Making of an Industrial Proletariat, 1915–45.* (Urbana: University of Illinois Press, 1985).

2. Stuart Hall, "Racism and Reaction," *Five Views of Multi-Racial Britain: Talks on Race Relations Broadcast by BBC TV* (London: Commission for Racial Equality and BBC Television Further Education, 1978), 26.

3. For an excellent discussion of the subject, see: Barbara J. Fields, "Ideology and Race in American History," in *Region, Race, and Reconstruction: Essays in Honor of C. Vann Woodward,* ed. J. Morgan Kousser and James M. McPherson (New York: Oxford University Press, 1982), 142–77.

Chapter 1

1. *New York Times,* December 21, 1863.

2. *New Orleans Tribune,* August 29, 1865.

3. Gerald M. Capers, *Occupied City: New Orleans Under the Federals* (Kentucky: University of Kentucky Press, 1965), 120; LaWanda Cox, *Lincoln and Black Freedom: A Study in Presidential Leadership* (Columbia: University of South Carolina Press, 1981), 46; Ted Tunnell, *Crucible of Reconstruction: War, Radicalism and Race in Louisiana 1862–1877* (Baton Rouge: Louisiana State University Press, 1984), 26.

4. New Orleans "went more gradually into the vortex of Secession than other Southern cities," Emily Hazen Reed wrote in her 1868 history of politics and war; "it contained more of the elements of Unionism than any other city." Reed, *Life of A.P. Dostie; or The Conflict in New Orleans* (New York: Wm. P. Tomlinson, 1868), 31. Also see James Parton, *General Butler in New Orleans. A History of the Administration of*

the Department of the Gulf in the Year 1862: With an Account of the Capture of New
Orleans, and A Sketch of the Previous Career of the General, Civil and Military (New
York: Mason Brothers, 1864), 253; Tunnell, *Crucible of Reconstruction*, 24, 26–27.

5. *Morehouse Advocate*, quoted in Frederick Law Olmsted, *The Cotton Kingdom*,
ed. by Arthur M. Schlesinger, Sr. (New York: Modern Library, 1984), 232; Slidell
quoted in Peyton McCrary, *Abraham Lincoln and Reconstruction: The Louisiana Experience* (Princeton: Princeton University Press, 1978), 56. New Orleans was center of
Unionist sentiment in Louisiana, and its residents voted heavily for the Constitutional
Union party's Bell and the northern Democrats' Stephen Douglas in the 1860 presidential election. Breckingridge ran third. See Capers, *Occupied City*, 20.

6. Earl F. Niehaus, *The Irish in New Orleans, 1800–1860* (Baton Rouge: Louisiana State University Press, 1965), 157; Sylvia J. Pinner, "A History of the Irish Channel 1840–1860" (unpublished B.A. thesis, Tulane University, 1954); Fredrick Nau, *The German People of New Orleans, 1850–1900* (Leiden: E.J. Brill, 1958), 30–34; Ira
Berlin and Herbert G. Gutman, "Natives and Immigrants, Free Men and Slaves:
Urban Workingmen in the Antebellum American South," *The American Historical
Review* 88, No. 5 (December 1983), 1190–91, 1195–200; Randall M. Miller, "The
Enemy Within: Some Effects of Foreign Immigrants on Antebellum Southern Cities,"
Southern Studies XXIV, No. 1 (Spring 1985), 34, 36, 45; Amos E. Simpson and
Vaughan Baker, "Michael Hahn: Steady Patriot," *Louisiana History* XII, No. 3 (Summer 1972), 229–52.

7. Leon Cyprian Soule, *The Know Nothing Party in New Orleans: A Reappraisal*
(Baton Rouge: The Louisiana Historical Association, 1961), 93–94, 104–5, 111–13;
Edwin L. Jewell, ed., *Jewell's Crescent City: The Commercial, Social and General
History of New Orleans* (New Orleans: 1873), 139; Robert C. Reinders, *End of an Era:
New Orleans, 1850–1860* (New Orleans: Pelican Publishing Company, 1964), 56–61;
Capers, *Occupied City*, 150–51.

8. *Address of Major Davis Alton, United States Army, Before the Working Men's
National Union League, Tuesday, June 23, 1863, at Lyceum Hall, New Orleans* (New
Orleans: "The Era" Book and Job Office, 1863); *Grand Mass Meeting of the Working
Men's National Union League, Held in New-Orleans, on the Night of July 11th, 1863, in
Commemoration of the Opening of the Mississippi River* (n.p.: 1863); Roger W. Shugg,
*Origins of Class Struggle in Louisiana: A Social History of White Farmers and Laborers
during Slavery and After, 1840–1875* (Baton Rouge: Louisiana State University Press,
1939), 198–99; Capers, *Occupied City*, 150–51.

9. The colonization plank in the League's preamble, however, was more an expression of racial ideology than a serious demand; it quickly disappeared from the
League's political agenda. But expressions of anti-black sentiment continued, as political candidates played heavily upon the racial fears of the white immigrant and native
working-class electorate. In the words of historian Roger Fischer, supporters of candidate Michael Hahn fed voters "a steady diet of dire prophecies that the preservation of
the white race in Louisiana depended upon his election" in the February 1864 campaign. See Roger A. Fischer, *The Segregation Struggle in Louisiana, 1862–1877* (Urbana: The University of Illinois Press, 1974), 24.

10. As Peter Rachleff has argued for Richmond, "Race overshadowed all other
features of the working-class experience. In the absence of black-white socializing, a
range of preconceptions, prejudices, and ideologically shaped viewpoints determined
attitudes . . . [The] connection between black labor and 'degradation' remained a
cornerstone of white working-class perceptions, intensified by the fears attending federal military occupation, the abolition of slavery as a system of social control, the

extension of suffrage to former slaves, and the persistence of unemployment." Rachleff, *Black Labor in Richmond,* 4–5. We have too few studies of nineteenth-century white working-class racism in the United States. Historians have explored in depth the overall American and northern forms of anti-black belief and have reconstructed the ideology of white craft workers (stressing the centrality of the labor theory of value, or producerist ethos, and artisanal republicanism). But the precise connections between the two, and the nature of the evolving contours of white craft and unskilled workers' beliefs, await detailed study. On the racial beliefs of white workers, see Alexander Saxton, *The Indispensable Enemy: Labor and the Anti-Chinese Movement in California* (Berkeley: University of California Press, 1971); Bernard Mandel, *Labor: Free and Slave. Workingmen and the Anti-Slavery Movement in the United States* (New York: Associated Authors, 1955). Edmund Morgan's *American Slavery, American Freedom: The Ordeal of Colonial Virginia* (New York: W.W. Norton & Co., 1975) remains unsurpassed as a model examination of the relationship between white freedom and black oppression.

11. *New York Times,* December 21, 1863.

12. According to historian Roger Shugg, the native and immigrant workers who had expressed a pro-Union sentiment and voted for Douglas in the 1860 election now overwhelmingly supported Hahn and his "moderate" Free State faction in the 1864 election. Not only racism but patronage played an important role in Hahn's election victory, as it would in New Orleans municipal politics in subsequent years. The support of General Banks, who encouraged Union soldiers to vote and liberally distributed jobs to white voters, proved crucial to Hahn's success. Unsuccessful radicals accused the general of buying Hahn's election. On election day in late February, 1864, Hahn received over twice as many votes as his two opponents combined. See Shugg, *Origins of Class Struggle,* 199–200; Fischer, *Segregation Struggle in Louisiana,* 24. On the 1864 election, see Eric Foner, *Reconstruction: America's Unfinished Revolution 1863–1877* (New York: Harper & Row, 1988), 46–49; C. Peter Ripley, *Slaves and Freedmen in Civil War Louisiana* (Baton Rouge: Louisiana State University Press, 1976), 169–70; Gilles Vandal, *The New Orleans Riot of 1866: Anatomy of a Tragedy* (Lafayette: University of Southwestern Louisiana, 1983), 27–28; James M. McPherson, *The Struggle for Equality: Abolitionists and the Negro in the Civil War and Reconstruction* (Princeton: Princeton University Press, 1964), 243–44; McCrary, *Abraham Lincoln and Reconstruction,* 212–36.

13. *Debates in the Convention for the Revision and Amendment of the Constitution of the State of Louisiana, 1864* (New Orleans: W.R. Fish, Printer to the Convention, 1864), 418–24, 430; Shugg, *Origins of Class Struggle,* 204; Vandal, *The New Orleans Riot,* 33–34; Fischer, *Segregation Struggle in Louisiana,* 24–25. The petitioners laid out the mechanics' arguments. Recalling for the delegates a past "when the capitalists could demand and exact . . . ten to twelve hours a day devoted to toil, (physical or mental, as the case required;)," these mechanics requested some act to relieve them "of the burden we have heretofore borne, that of working to suit the convenience of men who acquired wealth and position to the injury and oppression of us." A legal nine-hour day, they suggested, would permit them to devote a part of their time "to domestic scenes and the cultivation of the mind." Expressing considerable sympathy for the mechanics' position, a convention committee declared that the "homage that capital requires of labor is beginning to be insupportable and detestable." "Some efficient plan must soon be instituted to relieve the poor man from his manifold oppressive disadvantages; to give him a fair and equal chance to enjoy his existence; to

emancipate him from the mountainous interests and antagonisms that now oppress and keep in bondage to poverty." In this period of social upheaval and transition from a slave to free labor system, convention delegates revealed some understanding of the changes taking place in the country's political economy.

On the subject of black rights, the Free State representatives may have abolished slavery easily and diminished the power of the planter class, but their progressivism fell far short of sympathy for the cause of black rights and freedom. Although the delegates recognized the rights of blacks to education, equality before the law, and allowed blacks to serve in the militia, they did not enfranchise the black population. Instead, they empowered the legislature to extend the franchise to qualified black men at some point in the future. The civil rights advances that passed did so only because of the strong pressure exerted on the delegates by federal military officials and their allies.

14. Reed, *Life of A.P. Dostie*, 181; Fred Harvey Harrington, *Fighting Politician: Major General N.P. Banks* (Philadelphia: University of Pennsylvania Press, 1948), 166–67; W.E.B. DuBois, *Black Reconstruction in America 1860–1880* (1935; rpt. New York: Atheneum, 1985), 454–55.

15. *The Black Republican,* May 13, 1865; Reed, *Life of A.P. Dostie,* 181–82, 185. With control of municipal and state power in the hands of men hostile to labor and, many believed, to the federal Union itself, the city's white workers appealed directly to federal military authorities to take decisive action. On May 4, 1865, a labor committee presented a substantial petition to General Banks demanding the annulment of Mayor Kennedy's public works ordinances. The actions of the new regime fell "with great severity upon the laboring men, especially those who have large families, this reduction of their income about 30 per cent., without previous notice." Banks did not disappoint his working-class constituency; the following day, he ordered Kennedy to resign immediately. But the New Orleans conservatives found an ally in the new U.S. President, Andrew Johnson, who reinstated Kennedy and deprived Banks of his command. See: Clipping, and Labor Petition in the Papers of Nathaniel Banks, File: General Correspondence, May 1–6, 1865, Box 36, Manuscript Division, Library of Congress.

16. Harrington, *Fighting Politician,* 166–68; Reed, *Life of A.P. Dostie,* 181–88; W.M. Lowrey, "The Political Career of James Madison Wells," *Louisiana Historical Quarterly* 31, No. 4 (October 1948), 1031; Whitelaw Reid, *After the War: A Southern Tour, May 1, 1865, to May 1, 1866* (Cincinnati: Moore, Wilstach & Baldwin, 1866), 238; Fischer, *The Segregation Struggle in Louisiana,* 26; McCrary, *Abraham Lincoln and Reconstruction,* 335–38.

17. David Montgomery, *Beyond Equality: Labor and the Radical Republicans, 1862–1872* (1967; rpt. Urbana: University of Illinois Press, 1981), 234–35; Norman J. Ware, *The Labor Movement in the United States 1860–1895: A Study in Democracy* (1929; rpt. New York: Vintage Books, 1964), 4–6; John R. Commons, et al,, *History of Labor in the United States* (New York: The Macmillan Company, 1918), 87–96, 102–10; Steven J. Ross, *Workers on the Edge: Work, Leisure, and Politics in Industrializing Cincinnati, 1788–1890* (New York: Columbia University Press, 1985), 200–202.

18. *Daily Picayune,* December 14, 1865; *New Orleans Times,* December 14, 1865; *Daily True Delta,* December 14, 1865; *New Orleans Tribune,* December 15, 1865.

19. *Daily Picayune,* March 11, 1866. In the years after the war, Trevellick toured the United States speaking on the subject of the eight-hour day and exhorting workers

to turn toward electoral action. See: Obediah Hicks, *Life of Richard F. Trevellick, The Labor Orator, or the Harbinger of the Eight-Hour System* (Joliet, Illinois: J.E. Williams & Co., 1896); Montgomery, *Beyond Equality,* 171, 187–88, 192.

20. *Daily Picayune,* March 25, 1866; *Daily True Delta,* March 17, 1866.

21. New Orleans iron molders, in particular, fared poorly. In 1869, national union leader William Sylvis described them as the·"most miserable set of creatures as we have in the whole land." Jonathan Philip Grossman, *William Sylvis, Pioneer of American Labor: A Study of the Labor Movement During the Era of the Civil War* (1945, rpt. New York: Octagon Books/Farrar, Straus and Giroux, Inc., 1973), 187; Chicago *Workingmen's Advocate,* April 21, May 5, 1866. On the painters' cooperative, see *Workingmen's Advocate,* May 19, 1866.

22. *Daily True Delta,* March 14, 15, 17, 1866; December 14, 1865; *New Orleans Times,* December 14, 1865; *Daily Picayune,* December 14, 1865.

23. Eight-hour advocates fared poorly in their first political test, the municipal elections of March 1866, held one day after the huge eight-hour-day procession. A small independent party, composed of "hard-fisted workingmen and mechanics," nominated a mayoral candidate, George Purves. Conservative Unionists, backed by old, moderate Free State leaders, sought working-class votes by running two laborers and six skilled workers on their ticket. Both parties lost to the pro-Confederate National Democratic party, which successfully ran former Mayor John Monroe for his old post. Monroe moved quickly to secure his party's dominance by exercising his patronage prerogatives; he dismissed large numbers of police officers, including most remaining Unionists, and appointed his own ex-Confederates to over half of the 500 newly available positions. The relationship of this independent labor party to the eight-hour movement remains unclear. Press accounts which gave extensive coverage to the procession and the speeches made no mention of the party, much less any workers' endorsement of it. See Lowrey, "The Political Career of James Madison Wells," 1075; Joe Gray Taylor, *Louisiana Reconstructed 1863–1877* (Baton Rouge: Louisiana State University Press, 1974), 81; Vandal, *The New Orleans Riot,* 116–20, 204–6. According to the *Delta,* there were three tickets: the National Democratic party, the National Union party, and an independent party composed of mechanics. *Delta,* March 11, 1866. Vandal states that there were two parties: National Democratic party and the Conservative Union party. According to Willie Malvin Caskey, the Workman's Democratic ticket polled a highly respectable vote, despite its loss. Caskey, *Secession and Restoration of Louisiana* (Baton Rouge: Louisiana State University Press, 1938), 201–3.

24. The Young Men's Democratic Association, a body closely tied to the National Democrats, accused the Workingmen of factionalism and denounced the defections of "those not nominated" at their convention to the new party. For their part, Workingmen's party delegates divided over their relationship to the regular Democrats. At their nominating convention in April, they debated whether candidates for sheriff or district attorney who had attended the Democratic Convention should be eligible to run on labor's ticket. After tabling a motion to deny such eligibility, the Workingmen chose as their candidate for sheriff General Harry Hays, a Confederate general President Johnson had pardoned a week before the election and who headed the National Democratic ticket. Hays' name quickly disappeared from the Workingmen's ticket— most probably because he refused the nomination and insisted on support of the regular Democrats—and the Workingmen were forced to replace him. *Daily Picayune,* April 19, 1866; Vandal, *The New Orleans Riot,* 120; Lowrey, "The Political

Career of James Madison Wells," 1077; Caskey, *Secession and Restoration,* 202; *New Orleans Times,* May 6, 1866; *Daily Picayune,* May 6, 1866.

25. *New Orleans Times,* May 6, 1866; *Daily Picayune,* May 6, 1866.

26. The relationship between the small but rejuvenated Workman's Union and the new Workingmen's party remains obscure. According to P. J. K., a correspondent to the *Workingmen's Advocate,* economic depression was the reason for the lack of organization of the laboring classes in February 1868. The following month, however, witnessed the formation of a Workman's Union, which stressed its independence "of any party or politicians, its organization being for mutual benefit and self-protection" as a "combination of labor against capital;" its platform cautioned working men to "beware of the deception and intrigues of politicians and office seekers" and vowed to nominate no candidates for political office during the coming election. The new organization approved of the platform of the National Labor Reform party, protested "against the present unjust, oppressive and discriminating system of taxation, exempting the monopolist and compelling the poor to bear the whole burden of the national government," and advocated greenbacks and the eight-hour day. *Workingmen's Advocate,* January 11, February 15, March 24, April 4, 1868. In April 1868, a Workingmen's party nominated candidates for municipal office. P. J. K. disavowed any connection to the Workman's Union, and suggested that the aim of the party, "organized at the eleventh hour" might be to "spoil the Democratic ticket, and give a better chance to the extreme Radicals." *Workingmen's Advocate,* April 25, 1868. Also see: *Daily Picayune,* April 12, 14, 16, 17, 1868.

27. White parents who refused to "mix" their children, he further complained, found their children taken away from them and indoctrinated with "Radical 'religion' " in a children's asylum. "I feel the degradation inflicted upon us by this horde of adventurers,"—School Superintendent Conway and other "carpetbaggers"—who "have no intention of remaining long enough to hear their dusky offsprings call them father." *Workingmen's Advocate,* September 12, 1868.

28. *Workingmen's Advocate,* January 11, April 4, 1868.

29. *Workingmen's Advocate,* May 23, August 1, September 12, 1868.

30. *Daily Picayune,* April 19, 25, 27, 28, 1866.

31. *New Orleans Tribune,* December 29, 1864. For a discussion of class divisions in other black communities, see: Willard B. Gatewood, Jr., "Aristocrats of Color: South and North: The Black Elite, 1880–1920," *Journal of Southern History* LIV, No. 1 (February 1988), 3–20; Thomas Holt, *Black Over White: Negro Political Leadership in South Carolina During Reconstruction* (Urbana: University of Illinois Press, 1977); Eric Foner, "Reconstruction Revisited," in "The Promise of American History: Progress and Prospects," *Reviews in American History* 10, No. 4 (December 1982), 87–90.

32. Joseph Logsdon, "The Americanization of Black New Orleans, 1850–1900" (unpublished paper); Robert C. Reinders, "Slavery in New Orleans in the Decade Before the Civil War," in *Plantation, Town and Country: Essays on the Local History of American Slave Society,* ed. Elinor Miller and Eugene D. Genovese (Urbana: University of Illinois Press, 1974), 368–70, 374; John W. Blassingame, *Black New Orleans, 1860–1880* (Chicago: University of Chicago Press, 1973), 2; Joe Gray Taylor, *Negro Slavery in Louisiana* (Baton Rouge: The Louisiana Historical Association, 1963), 83–85; Claudia Dale Goldin, *Urban Slavery in the American South 1820–1860: A Quantitative History* (Chicago: University of Chicago Press, 1976), 19–21, 45–46; Robert S. Starobin, *Industrial Slavery in the Old South* (New York: Oxford University Press, 1970), 93–94; Thomas quoted in Loren Schweninger, "A Negro Sojourner in

Antebellum New Orleans," *Louisiana History* XX, No. 3 (Summer 1979), 311; Laurence J. Kotlikoff and Anton J. Rupert, "The Manumission of Slaves in New Orleans, 1827–1846," *Southern Studies* XIX, No. 2 (Summer 1980), 172–81; Judith Kelleher Schafer, "New Orleans Slavery in 1850 as Seen in Advertisements," *Journal of Southern History* XLVII, No. 1 (February 1981), 33–56.

33. Whites regularly complained about the autonomy of New Orleans blacks' social life. Olmsted noted the objections of the *New Orleans Crescent:* "The slave population of this city is already demoralized to a deplorable extent, all owing to the indiscriminate licence and indulgence extended them by masters . . . and to the practice of forging passes, which has now become a regular business in New Orleans . . . The consequence is that hundreds spend their nights drinking, carousing, gambling, and contracting the worst habits, which not only make them useless to their owners, but dangerous pests to society." But what whites found dangerous, slaves appreciated. One slave on a Louisiana sugar plantation informed Olmsted that he would "rather live in New Orleans than any other place in the world" because blacks there were "better off than those who lived in the country. Why? Because they make more money, and it is 'gayer' there, and there is more 'society.' " Olmsted, *The Cotton Kingdom*, 234–35, 260. On antebellum black life, see Richard C. Wade, *Slavery in the Cities: The South 1820–1860* (New York: Oxford University Press, 1964), 249–50; Blassingame, *Black New Orleans*, 10–14; David C. Rankin, "The Origins of Negro Leadership in New Orleans During Reconstruction," in *Southern Black Leaders of the Reconstruction Era*, ed. Howard N. Rabinowitz (Urbana: University of Illinois Press, 1982), 156–57; Reinders, *End of an Era*, 24; Leon Litwack, *Been in the Storm So Long: The Aftermath of Slavery* (New York: Vintage Books, 1979), 513; Geraldine Mary McTigue, "Forms of Racial Interaction in Louisiana, 1860–1880" (unpublished Ph.D. dissertation, Yale University, 1975), 18.

34. "New Orleans was captured," W.E.B. DuBois noted, "and the whole black population of Louisiana began streaming toward it." DuBois, *Black Reconstruction*, 82. Some of the former slaves "support themselves and their families in a miserable way by working now and then a day on the levies, but many appear to have no occupation whatever, nor any means of support," one Freedmen's Bureau official complained shortly after the war's end about the large number of ex-slaves congregated in the city. "Nothing would induce this class to go into the country and work on a plantation where they can have a comfortable home and good wages, they rather preferring to remain in a city and hire as best they can." See Report to General O.O. Howard, October 1866, "Annual Narrative Reports of Operations and Conditions, October 1866–October 1868," in *Bureau of Refugees, Freedmen and Abandoned Lands*, Louisiana, Reel 27, Frame 0010 (hereafter referred to as BRFAL, LA); William F. Messner, *Freedmen and the Ideology of Free Labor: Louisiana 1862–1865* (Lafayette: Center for Louisiana Studies, University of Southwestern Louisiana, 1978), 33–34; Blassingame, *Black New Orleans*, 50–55; Dan T. Carter, *When the War Was Over: The Failure of Self-Reconstruction in the South, 1865–1867* (Baton Rouge: Louisiana State University Press, 1985), 178; DuBois, *Black Reconstruction*, 68, 82.

35. "Education of the Colored Population of Louisiana," *Harper's New Monthly Magazine* XXXIII, No. CXCIV (July 1866), 246; Reid, *After the War*, 243–45; David C. Rankin, "The Impact of the Civil War on the Free Colored Community of New Orleans," *Perspectives in American History* XI, ed. by Donald Fleming (Cambridge: Charles Warren Center for Studies in American History, Harvard University, 1977–78), 385; Jean-Charles Houzeau, *My Passage at the New Orleans Tribune. A Memoir of*

the Civil War Era, ed. David C. Rankin (Baton Rouge: Louisiana State University Press, 1984), 72–75.

36. Laura Foner, "The Free People of Color in Louisiana and St. Domingue: A Comparative Portrait of Two Three-Caste Slave Societies," *Journal of Social History* (Summer 1970), 406–30; Ira Berlin, *Slaves Without Masters: The Free Negro in the Antebellum South* (New York: Vintage Books, 1974), 108–32.

37. "Table LXX—Occupations of Free Colored Males over Fifteen Years, Distinguishing Blacks and Mulattoes—1850," *Statistical View of the United States . . . Being a Compendium of the Seventh Census* (J.B.D. DeBow, Superintendent, 1854). Robert C. Reinders has warned that these figures are "woefully inadequate." They are used here only as rough estimates. See Reinders, "The Free Negro in the New Orleans Economy, 1850–1860," *Louisiana History* VI, No. 3 (Summer 1965), 273–75. On the place of free blacks in the occupational structure, see Leonard Curry, *The Free Black in Urban America, 1800–1850: The Shadow of the Dream* (Chicago: University of Chicago Press, 1981), 29; Reinders, *End of an Era,* 22–24; Blassingame, *Black New Orleans,* 10; Wade, *Slavery in the Cities,* 274–75; Berlin and Gutman, "Natives and Immigrants, Free Men and Slaves," 1192–94; Niehaus, *The Irish in New Orleans,* 46–54; Miller, "The Enemy Within," 34, 36, 44; Berlin, *Slaves Without Masters,* 231; Goldin, *Urban Slavery,* 21.

38. Manoj K. Joshi and Joseph P. Reidy, " 'To Come Forward and Aid in Putting Down This Unholy Rebellion': The Officers of Louisiana's Free Black Native Guard During the Civil War Era," *Southern Studies* XXI, No. 3 (Fall 1982), 326–42; Mary F. Berry, "Negro Troops in Blue and Gray: The Louisiana Native Guards, 1861–63," *Louisiana History* 8 (Spring 1967), 165–90; Charles Vincent, *Black Legislators in Louisiana During Reconstruction* (Baton Rouge: Louisiana State University Press, 1976), 1–2, 5–6, 13–14; Ted Tunnell, "Free Negroes and the Freedmen: Black Politics in New Orleans During the Civil War," *Southern Studies* XIX (Spring 1980), 8; Berlin, *Slaves Without Masters,* 126–28; Blassingame, *Black New Orleans,* 33–47; Litwack, *Been in the Storm So Long,* 42.

39. Rankin, "Origins of Negro Leadership," 158, 162, 167; *Daily Picayune,* July 16, 1859, quoted in Rankin, "Origins of Negro Leadership," 167; Rankin, "The Politics of Caste: Free Colored Leadership in New Orleans During the Civil War," *Louisiana's Black Heritage,* ed. by Robert R. Macdonald, John R. Kemp, and Edward F. Haas (New Orleans: Louisiana State Museum, 1979), 107–25; Vincent, *Black Legislators,* 1–15; Houzeau, *My Passage at the New Orleans Tribune,* 78–79.

40. *Grand Celebration in Honor of the Passage of the Ordinance of Emancipation, by the Free State Convention, on the Eleventh Day of May, 1864, Held in the Place D'Armes, New-Orleans, June 11th* (New Orleans: H.P. Lathrop, 1864).

41. Whitelaw Reid, *After the War,* 244; Donald E. Everett, "Demands of the New Orleans Free Colored Population for Political Equality, 1862–1865," *Louisiana Historical Quarterly* XXXVIII, No. 1 (January 1955), 43–64; McPherson, *The Negro's Civil War,* 278; Tunnell, *Crucible of Reconstruction,* 78; also see: Berlin and Gutman, "Natives and Immigrants, Free Men and Slaves," 1193–94; Charles Vincent, "Black Louisianians During the Civil War and Reconstruction: Aspects of Their Struggles and Achievements," *Louisiana's Black Heritage,* ed. by Robert R. Macdonald, John R. Kemp, and Edward F. Haas (New Orleans: Louisiana State Museum, 1979), 85–106; Logsdon, "The Americanization of Black New Orleans"; Joseph Logsdon, "Americans and Creoles in New Orleans: The Origins of Black Citizenship in the United States," *American Studies/Amerikastudien* 34 (1989), 192–95.

42. Reid, *After the War,* 244; *New Orleans Tribune,* January 24, 1865; Tunnell, "Free Negroes and the Freedmen, 16–17.

43. *New Orleans Tribune,* January 15, 1865.

44. "Education of the Colored Population of Louisiana," *Harper's New Monthly Magazine,* 247.

45. *New Orleans Tribune,* December 27, 29, 1864.

46. *New Orleans Tribune,* December 27, 29, 1864; January 4, 1865; June 6, 1867; Houzeau, *My Passage at the New Orleans Tribune,* 79. Also see Litwack, *Been in the Storm So Long,* 508–9, 513, 535–36; Tunnell, *Crucible of Reconstruction,* 112; Rodolphe Lucien Dusdunes, *Our People and Our History* (1911; rpt. Baton Rouge: Louisiana State University Press, 1973), 125–34.

47. *New Orleans Tribune,* November 21, 1865; William P. Connor, "Reconstruction Rebels: The New Orleans Tribune in Post-War Louisiana," *Louisiana History* XXI, No. 1 (Winter 1980), 159–81; Messner, *Freedmen and the Ideology of Free Labor,* 109–11; Blassingame, *Black New Orleans,* 57, 59; Vincent, *Black Legislators,* 35.

48. Messner, *Freedmen and the Ideology of Free Labor,* 50–60; Armstead L. Robinson, " 'Worser dan Jeff Davis': The Coming of Free Labor during the Civil War, 1861–1865," *Essays on the Postbellum Southern Economy,* ed. by Thavolia Glymph and John J. Kushma (College Station, Texas, 1985), 20–27. For more detailed discussions of postwar labor system, see Gerald David Jaynes, *Branches Without Roots: The Genesis of the Black Working Class* (New York: Oxford University Press, 1986); Eric Foner, *Nothing But Freedom: Emancipation and Its Legacy* (Baton Rouge: Louisiana State University Press, 1983); Roger L. Ransom and Richard Sutch, *One Kind of Freedom: The Economic Consequences of Emancipation* (New York: Cambridge University Press, 1977); Litwack, *Been in the Storm So Long,* 336–86.

49. *New Orleans Tribune,* December 8, 1864; December 17, 1865; November 28, December 20, 21, 1865; Tunnell, *Crucible of Reconstruction,* 85; Blassingame, *Black New Orleans,* 55.

50. The *Tribune* denounced urban harassment: "Why! A levee laborer going in the morning to his daily avocations, will be seized by a rebel police, taken from his family, and put to forced labor somewhere! Is Emancipation a by-word?" Superintendent T.W. Conway of the Freedmen's Bureau sustained the charge: "It has been the practice here to arrest as vagrants all colored laborers who were found on the streets in their working garments and not employed just at the moment when the police saw them . . . they may have their cotton hooks hanging to their belts, showing that they have proper employment; but still they have been arrested . . . Not a day passes without dozens of men being sent to me as vagrants, many of whom I release immediately upon ascertaining that they have been arrested unjustly. Those who are found to be vagrants are readily and effectively corrected at our colonies, where they are made to labor. . . ." *New Orleans Tribune,* October 10, 28, 1865. On hiring-out practices, see Wade, *Slavery in the Cities,* 38–54; Schweninger, "A Negro Sojourner in Antebellum New Orleans," 311; Starobin, *Industrial Slavery,* 135–37.

51. Parton, *General Butler in New Orleans,* 255; Marion Southwood, *"Beauty and Booty," The Watchword of New Orleans* (New York: M. Loolady, 1876), 61. A *Picayune* columnist supported these assessments in early 1864: "Looking up and down the wooden platform that crowns the Levee to the water's edge, we see acres of unoccupied space . . . There is no stir and bustle—none of that activity we were accustomed to see . . . All is dull and dreary where we have looked upon a picture of commercial activity unexcelled." *Daily Picayune,* February 13, 1864; also see February 17, March 13, 1864.

52. The "Levee itself is covered with bales of cotton and other produce," one journalist noted in 1858, "which hundreds of negroes, singing at their work, with here and there an Irishman among them, are busily engaged in rolling from the steamers and depositing in the places set apart for each consignee." "Transatlantic Sketches. 'The Crescent City' " *The Illustrated London News* XXXII, No. 912 (April 10, 1858), 278. Also see Henry A. Murray, *Lands of the Slave and the Free: or, Cuba, the United States, and Canada,* Vol. 1 (London: John W. Parker and Son, 1855), 251; William Howard Russell, *My Diary North and South* (Boston: T.O.H.P. Burnham, 1863), 249; Henry Didimus, *New Orleans as I Found It* (New York: Harper and Brother, 1845), 14; Marie Fontenay (Mme Manoel de Grandfort), *L'Autre Monde* (Paris: Librairie Nouvelle, 1855), 249. For assertions of white displacement of blacks in the waterfront labor force in the 1850s, see Reinders, "The Free Negro in the New Orleans Economy," 276–77; Shugg, *Origins of Class Struggle,* 118–19.

53. Russell, *My Diary,* 249, 231; Sylvia J. Pinner, "A History of the Irish Channel 1840–1860," 50; Capers, *Occupied City,* 78; *Daily Picayune,* July 15, 1865. On black migration to New Orleans, see *New Orleans Times,* December 30, 1865; Report to General O.O. Howard, October 1866, BRFAL, LA Reel 27, Frame 0010; Jaynes, *Branches Without Roots,* 264.

54. Charles H. Wesley, *Negro Labor in the United States 1850–1925: A Study in American Economic History* (New York: Vanguard Press, 1927), 71–73, 79–83, 99–100; Sterling D. Spero and Abram L. Harris, *The Black Worker: The Negro and the Labor Movement* (1931; rpt. New York: Atheneum Press, 1969), 8–9; Emerson David Fite, *Social and Industrial Conditions in the North During the Civil War* (New York: The Macmillan Co., 1910), 189–90; Richard B. Morris, "Labor Militancy in the Old South," *Labor and Nation* IV, No. 3 (May–June 1948), 34–35; *New Orleans True American,* August 31, 1835, quoted in *Niles' Weekly Register,* September 19, 1835, 34–5; Iver Charles Bernstein, "The New York City Draft Riots of 1863 and Class Relations on the Eve of Industrial Capitalism," I (unpublished Ph.D. dissertation, Yale University, 1985), 8, 43–48; Berlin, *Slaves Without Masters,* 229–31; Curry, *The Free Black in Urban American,* 96–97; Frederick Douglass, *Narrative of the Life of Frederick Douglass An American Slave, Written by Himself* (1845; rpt. New York: New American Library, 1968), 99–102; Eric Arnesen, " 'A Little Corner of Creation': Black Workers and the 1865 Baltimore Shipyard Strike" (unpublished essay, 1981); Olmsted, *The Cotton Kingdom,* 232; Ross, *Workers on the Edge,* 195, 197; Mandel, *Labor: Free and Slave,* 188–89; Golden, *Urban Slavery,* 28–31; DuBois, *Black Reconstruction,* 18–20; Lorenzo J. Greene and Carter G. Woodson, *The Negro Wage Earner* (New York: Russell & Russell, 1930), 15–18, 22–23; Starobin, *Industrial Slavery,* 211–12.

55. George P. Marks, III, "Notes and Documents: The New Orleans Screwmen's Benevolent Association, 1850–1861," *Labor History* 14, No. 2 (Spring 1973), 259–63; *Daily True Delta,* December 20, 21, 1865; *New Orleans Tribune,* December 20, 22, 29, 1865; *New Orleans Times,* December 28, 1865. The development of the screwmen's union is discussed in Chapter 2. Brief mention of the 1865 strike can be found in Arthur Raymond Pearce, "The Rise and Decline of Labor in New Orleans" (unpublished M.A. thesis, Tulane University, 1938), 12; F. Ray Marshall, *Organized Labor and the Negro* (New York: Harper and Brother Publishers, 1944), 149; Morris, "Labor Militancy in the Old South," 36; Dave Wells and Jim Stodder, "A Short History of New Orleans Dockworkers," *Radical America* 10, No. 1 (January–February 1976), 45–46; Litwack, *Been in the Storm So Long,* 441–42; Shugg, *Origins of Class Struggle,*

301; Jaynes, *Branches Without Roots,* 264. On the impressment of riverfront workers, see Parton, *General Butler in New Orleans,* 255.

56. *New Orleans Times,* December 21, 1865; *Daily True Delta,* December 22, 23, 1865.

57. *Daily True Delta,* December 21, 22, 23, 1865; *New Orleans Times,* December 22, 1865; *Daily Picayune,* December 22, 1865.

58. J.T. Trowbridge, *The South: A Tour of Its Battlefields and Ruined Cities* (Hartford: L. Stebbins, 1866), 405; Morris, "Labor Militancy in the Old South," 36; Jaynes, *Branches Without Roots,* 264.

59. *New Orleans Tribune,* December 20, 22, 29, 1865; *Daily True Delta,* December 21, 23 1865; *New Orleans Times,* December 22, 1865.

60. Trowbridge, *The South,* 405; *Daily True Delta,* December 21, 22, 1865; *New Orleans Tribune,* December 22, 1865.

61. *Daily True Delta,* December 28, 1865; *New Orleans Tribune,* December 29, 1863.

62. Trowbridge, *The South,* 405; *New Orleans Tribune,* December 23–25, 29, 1865; *Daily True Delta,* December 22, 1865; *Daily Picayune,* December 28, 29, 1865. For a discussion of Canby's activities in New Orleans politics, see Max L. Heyman, Jr., *Prudent Soldier: A Biography of E.R.S. Canby, 1817–1873* (Glendale, Ca.: The Arthur H. Clark Company, 1959), 271–94. Although Governor Wells signed the Black Codes into law on December 20 and 21, the first days of the strike, New Orleans authorities found no need to invoke these measures. The simple power of arrest on riot charges proved sufficient.

63. *New Orleans Tribune,* December 29, 1865; *Daily True Delta,* December 21, 1865, January 3, 1866; *Daily Picayune,* January 5, 1866.

64. *New Orleans Tribune,* December 22, 1865.

65. *New Orleans Tribune,* December 15, 17, 1865.

66. *Daily True Delta,* December 22, 1865; *New Orleans Tribune,* December 20, 29, 1865.

67. Philip Uzee, "The Beginnings of the Louisiana Republican Party," *Louisiana History* XII, No. 3 (Summer 1971), 197–211; Vandal, *The New Orleans Riot,* 79; *New Orleans Tribune,* September 29, 30, 1865; Taylor, *Louisiana Reconstructed,* 76.

68. Lowrey, "The Political Career of James Madison Wells," 1077–79; U.S. Congress, House Select Committee on New Orleans Riot, *Report of the Select Committee on the New Orleans Riot,* U.S. 39th Congress, 2nd Session, House Report 16 (Washington: Government Printing Office, 1867), 16; Altina L. Waller, "Community, Class and Race in the Memphis Riot of 1866," *Journal of Social History* (Winter 1984), 233–46; Carter, *When the War Was Over,* 248–53; Tunnell, *Crucible of Reconstruction,* 106–7; Vandal, *The New Orleans Riot;* Donald E. Reynolds, "The New Orleans Riot of 1866, Reconsidered," *Louisiana History* V, No. 1 (Winter 1964), 5–27.

69. Political opposition to civil rights issues came not just from white Democrats but from white Republicans as well. Convinced that the party's survival required broadening its appeal to attract white voters, some Republican politicians favored retreating from a commitment to equal rights for the state's black citizens. But despite hostility from some Republican party factions, in 1873 and 1875 black legislators succeeded in passing several civil rights bills. See *New Orleans Tribune,* April 16, 1867; Fischer, *Segregation Struggle in Louisiana,* 42, 47–49, 55, 61–87; Vincent, *Black Legislators,* 48–70, 71–112, 183–201; Tunnell, *Crucible of Reconstruction,* 107–18, 134.

70. *Daily Picayune,* March 6, 1875, quoted in Fischer, *Segregation Struggle in Louisiana,* 86–87; 40; Tunnell, *Crucible of Reconstruction,* 113–14; Dale Somers,

"Black and White in New Orleans: A Study in Urban Race Relations, 1865–1900," *Journal of Southern History* XL, No. 1 (February 1974), 22–26.

71. *New Orleans Tribune,* January 10, 1869, quoted in Fischer, *Segregation Struggle in Louisiana,* 86–87.

72. Richard C. Baylor, "Circular: An Appeal to the Colored Cotton Weighers, Cotton Pressmen, Generally, Levee Stevedores and Longshoremen," in *New Orleans Tribune,* October 22, 1865. Also see *Tribune,* October 26, 31, 1865; *Daily Picayune,* April 17, May 7, 1867; *New Orleans Crescent,* May 17, 1867; Taylor, *Louisiana Reconstructed,* 137.

73. *Daily Picayune,* May 12, 1867; *New Orleans Times,* May 17, 1867; *Commercial Bulletin,* May 17, 1867; *Tribune,* May 9, 11, 12, 17, 1867; *Charleston Mercury,* May 14, 1867; Charles Lewis Rast, III, "The Characteristics and Activities of Political Clubs in Reconstruction New Orleans, 1867–1872" (unpublished M.A. thesis, Tulane University, 1982), 27–29; Joe Gray Taylor, *Louisiana Reconstructed, 144.*

74. *Daily Picayune,* April 17, 1867; Joseph A. Mower, Assistant Commissioner, July 3, 1867, "Report of Operations in his District, of the Bureau for the Month of May, 1867," BRFAL, LA, Reel 27, Frames 0152–3.

75. Mower, "Report of Operations . . . May, 1867," Frames 0152–3; *New Orleans Tribune,* May 1, 3–5, 7, 8, 1867; *Daily Picayune,* May 2, 7, 8, 11, 17, 1867; *Charleston Mercury,* May 14, 1867; Fischer, *Segregation Struggle in Louisiana,* 30–38; Taylor, *Louisiana Reconstructed,* 141. In the same month, blacks in Richmond, Virginia, successfully challenged that city's segregated streetcar system through aggressive crowd actions. See Rachleff, *Black Labor in Richmond,* 42. Less successful were the efforts of a small number of New Orleans blacks to obtain service in white commercial establishments. There appeared to be little popular backing for the efforts of a small group of blacks to demand service in white-owned coffeehouses, theaters, and stores in early May 1867. The campaign failed. The city attorney ruled that as privately owned establishments, such businesses were only "subject to such rules and regulations as the proprietors may deem proper to adopt." Sustaining that judgment, Republican Mayor Heath issued a proclamation on May 15 that barred blacks from "any store, shop or other place of business conducted by private individuals, against the consent and wishes of the owners." Few rose up to protest this pro-segregation ruling.

76. *Daily Picayune,* May 16, 1867. On the issue of "star schools" and police segregation, see *New Orleans Tribune,* May 9, 10, 12, 25, 1867.

77. *New Orleans Crescent,* May 17, 1867; *Commercial Bulletin,* May 17, 1867; *Daily Picayune,* May 17, 1867; *New Orleans Times,* May 17, 1867; *New Orleans Tribune,* May 17, 1867.

78. *New Orleans Times,* May 17, 1867; *Commercial Bulletin,* May 17, 1867; *New Orleans Crescent,* May 17, 1867; *Springfield Republican* (Massachusetts), May 18, 1867.

79. *New Orleans Tribune,* May 24, 1867; *Springfield Republican,* May 20, 1867.

80. *Commercial Bulletin,* May 17, 1867; *Daily Picayune,* May 17, 1867.

81. *Commercial Bulletin,* May 18, 1867; *Daily Picayune,* May 18, 1867; *New Orleans Times,* May 18, 1867; *New Orleans Tribune,* May 18, 1867.

82. *Daily Picayune,* May 19, 1867; *New Orleans Tribune,* May 19, 1867; *New Orleans Times,* May 19, 1867; Mower, "Report of Operations . . . May, 1867," BRFAL, LA, Reel 27, Frame 0153. Mower's order did not permanently alleviate abuses of the payment system, however. In late May 1867 freedman Thomas Weitzel and twenty others filed a complaint with Freedmen's Bureau officials when the first mate of the *Sir*

Henry Ames failed to pay the agreed-upon wage of 30 cents an hour; freedmen on the docks brought similar charges against the steamboat *M.E. Allen* in mid-June. The following year, in March 1868, some 64 freedmen and 7 whites, "all laborers on the levee," charged contractors Archer Hinton (black) and John Chase (white) with cheating them of their wages. When this "practice of setting hands to work two or three days and then discharging them without remuneration" nearly resulted "in a riot," in the *Picayune*'s words, police arrested the two employers, apparently saving them from the fury of the dock workers who followed them to the station. At the same time, police arrested one worker on charges of inciting other laborers to riot. Less than a month later, black dock workers "rose in rebellion aginst their sable employer" who utilized "the old trick of unloading vessels at half price, and thus swindling the hands." Again police intervened by arresting "guilty parties." The Hinton and Chase incident led the Assistant Sub-Commissioner to conclude that in "cases of this nature, it would seem both proper and desirable that some means be devised rendering ships, steamboats, etc., directly responsible for all laboring done on them, either in loading, or, unloading them." L. Jolissaint to Captain Lucious H. Warner, March 31, 1868, Monthly and Trimonthly Reports, March–December 1868, Records of the Bureau of Refugees, Freedmen, and Abandoned Lands, Field Office Records Relating to New Orleans, La., Assistant Subassistant Commissioner for Orleans Parish, Left Bank, RG 105, Reel 1, Frames 0227–8; *Daily Picayune*, March 21, April 16, 1868.

83. *New Orleans Crescent*, May 17, 1867; *New Orleans Times*, May 17, 1867; *Daily Picayune*, May 17, 1867.

84. *New Orleans Tribune*, May 17, 18, 22, 24, 1867.

85. *Daily Picayune*, May 17, 1867. If the *Tribune*'s editors were unenthusiastic about the strike, they were appalled by Mower's statement. While heaping abuse on the ex-slaves, Mower had ignored the issue of segregated commercial facilities and failed to criticize an armed assault by whites on black streetcar riders the day prior to the strike or an attack on blacks attempting to desegregate a confectionary. *New Orleans Tribune*, May 18, 22, 24, 1867.

Chapter 2

1. *New Orleans Times*, September 11, 12, 1881; *Weekly Louisianian*, September 17, 1881; *Daily Picayune*, September 11, 1881; Savannah *Morning News*, September 12, 1881.

2. *Daily Picayune*, August 29, 1881; *Weekly Louisianian*, September 17, 1881; Savannah *Morning News*, September 2, 1881; Arthur Raymond Pearce, "The Rise and Decline of Labor in New Orleans" (unpublished M.A. thesis, Tulane University, 1938), 25.

3. *Daily Picayune*, August 29, 1881.

4. Edwin De Leon, "The New South," *Harper's New Monthly Magazine* XLIX, No. CCXCII (September 1874), 555; Paul M. Gaston, *The New South Creed: A Study in Southern Mythmaking* (Baton Rouge: Louisiana State University Press, 1970), 32–37.

5. *New Orleans Times*, August 28, 1881. Also see *Times*, December 27, 1865; Friedrich Ratzel, *Sketches of Urban and Cultural Life in North America*, translated and edited by Stewart A. Stehlin (1876; rpt. New Brunswick: Rutgers University Press, 1988), 197; Edward King, *The Great South*, ed. W. Magruder Drake and Robert R. Jones (1875; rpt. Baton Rouge: Louisiana State University Press, 1972), 50–51; Gary

Bolding, "Change, Continuity and Commercial Identity of a Southern City: New Orleans, 1850–1950," *Louisiana Studies* XIV, No. 2 (Summer 1975), 162–66.

6. Department of the Interior, U.S. Census Office, *Compendium of the Tenth Census* (June 1, 1880) Part II (Washington, D.C.: 1883), 1068–71; Norman Walker, "Manufactures," in *Standard History of New Orleans, Louisiana,* ed. Henry Rightor (Chicago: The Lewis Publishing Company, 1900), 511–37; Julian Ralph, "New Orleans, Our Southern Capital," *Harper's New Monthly Magazine* LXXXVI, No. DXIII (February 1893), 382–85; *The Industries of New Orleans, Her Rank, Resources, Advantages, Trade, Commerce and Manufactures* (New Orleans: J.M. Elstner & Co., 1885), 43; William Ivy Hair, *Bourbonism and Agrarian Protest: Louisiana Politics 1877–1900* (Baton Rouge: Louisiana State University Press, 1969), 57–59; Joy Jackson, *New Orleans in the Gilded Age: Politics and Urban Progress, 1880–1896* (Baton Rouge: Louisiana State University Press, 1969), 6. Also see Thomas Redard, "The Port of New Orleans: An Economic History, 1821–1860" I (unpublished Ph.D. dissertation, Louisiana State University, 1985), 39; Thomas L. Nichols, *Forty Years of American Life,* I (London: John Maxwell and Company, MDCCCLXIV), 184–85; David Taylor Kearns, "The Social Mobility of New Orleans Laborers 1870–1900" (unpublished Ph.D. dissertation, Tulane University, 1977), 10–14.

7. James S. Zacharie, *New Orleans Guide: With Descriptions of Routes to New Orleans, Sights of the City Arranged Alphabetically, and other Information Useful to Travelers; also, Outline of the History of Louisiana* (New Orleans, 1885), 81; *New Orleans: Her Relations to the New South,* 44. On skilled and factory labor in New Orleans, also see *Daily Picayune,* September 1, 1882; September 1, 1892; British Board of Trade, *Cost of Living in American Towns. Report of an Enquiry by the Board of Trade into Working Class Rents, Housing and Retail Prices, together with the Rates of Wages in Certain Occupations in the Principal Industrial Towns of the United States of America* (London: His Majesty's Stationery Office, 1911), 295.

8. Robert Somers, *The Southern States Since the War 1870–71* (1871; rpt. University: The University of Alabama Press, 1965), 191; B.M. Norman, *Norman's New Orleans and Environs: Containing a Brief Historical Sketch of the Territory and State of Louisiana, and the City of New Orleans* (New Orleans: B.M. Norman, 1845), 81; Alexander Mackay, *The Western World; or, Travels in the United States in 1846–47,* II (Philadelphia: Lea & Blanchard, 1849), 79–80; Henry Didimus, *New Orleans as I Found It* (New York: Harper & Brother, 1845), 13–14; "Transatlantic Sketches. The Cresent City,' " *The Illustrated London News* XXXII, No. 912 (April 10, 1858), 376; John B. Latrobe, *Southern Travels: Journal of John B. Latrobe,* ed. Samuel Wilson, Jr. (1834; rpt. New Orleans: The Historic New Orleans Collection, 1986), 39–40, 45; Carl David Arfwedson, *The United States and Canada in 1832, 1833, and 1834,* Vol. 2 (1834; rpt. New York: Johnson Reprint Corporation, 1969), 52–57; Henry A. Murray, *Lands of the Slave and the Free: Or, Cuba, The United States, and Canada,* Vol. 1 (London: John W. Parker and Sons, 1855), 250–51; C. A. Goodrich, *The Family Tourist. A Visit to the Principal Cities of the Western Continent* (Hartford: Case, Tiffany and Company, 1848), 429–31; A. Oakey Hall, *The Manhattaner in New Orleans, or, Phases of "Crescent City" Life* (1851; rpt. Baton Rouge: Louisiana State University Press, 1976), 4–6; Edward King, "The Great South. Old and New Louisiana.—II," *Scribner's Monthly* VII, No. 2 (December 1873).

9. Norman, *Norman's New Orleans,* 81; Didimus, *New Orleans As I Found It,* 5; George Augustus Sala, *America Revisited: From the Bay of New York to the Gulf of Mexico, and From Lake Michigan to the Pacific,* Vol. II (London: Vizetelly & Co.,

1883), 6; Belle Hunt, "New Orleans, Yesterday and To-Day," *Frank Leslie's Popular Monthly* XXXI, No. 6 (June 1891), 641; also see: Julian Ralph, "New Orleans, Our Southern Capital," 364; Lawrence H. Larsen, *The Rise of the Urban South* (Lexington: University Press of Kentucky, 1985), 43–44.

10. *The Weekly Louisianian,* February 4, 1882; Charles Dudley Warner, *Studies in the South and West, with Comments on Canada* (New York: Harper and Brothers, 1889), 39; *New Orleans Democrat,* October 9, 1881; *The Mascot,* February 18, 1882; *Tribune,* May 28, 1867; *Daily Picayune,* February 27, May 6, 1868; H.S. Herring, *History of the New Orleans Board of Trade, Ltd. 1880–1930* (New Orleans: 1930), 31; George W. Cable, "New Orleans in 1880," Tenth Census, *Report on the Social Statistics of Cities* XIX, Pt. 2 (Washington, 1887), 268, 272; William Ivy Hair, *Carnival of Fury: Robert Charles and the New Orleans Race Riot of 1900* (Baton Rouge: Louisiana State University Press, 1976), 87–88; Mary L. Dudziak, "The Social History of the Slaughter-House Cases: The Butchers of New Orleans and the Sacred Right of Labor" (unpublished essay, Yale University, 1983). Also see *Southwestern Christian Advocate,* February 20, 1879; Ratzel, *Sketches of Urban and Cultural Life,* 203, 207–9.

11. Joe Gray Taylor, *Louisiana Reconstructed 1863–1877* (Baton Rouge: Louisiana State University Press, 1974), 342–45, 350–51, 314–16, 319; Norman Walker, "Commercial and Mercantile Interests," in *Standard History of New Orleans, Louisiana,* ed. Henry Rightor (Chicago: The Lewis Publishing Company, 1900), 566–67; De Leon, "The New South," 557.

12. "Among the many striking changes wrought by the war through the extension of railroad lines," De Leon noted in 1874, "the diversion of trade from old routes is one of the most noticeable throughout the Southern country." De Leon, "The New South," 558. As an "outlet to the ocean for the grain trade of the West," a U.S. Census superintendent reported in 1860, the Mississippi River "has almost ceased to be depended upon by merchants." Quote in David R. Goldfield, *Cotton Fields and Skyscrapers: Southern City and Region, 1607–1980* (Baton Rouge: Louisiana State University Press, 1982), 60–61, 30–31; Walker, "Commercial and Mercantile Interests," 558–61, 572–73. On loss of trade, see *Southwestern Christian Advocate,* May 17, 1877; *Daily Picayune,* May 9, 1868.

13. *New Orleans: Her Relations to the New South,* 22–26; Walker, "Commercial and Mercantile Interests," 551–53, 572–75; *New Orleans Republican,* September 1, 1874; John Smith Kendall, *History of New Orleans,* II (New York: The Lewis Publishing Co., 1922), 630. On the antebellum era, see: Merl E. Reed, *New Orleans and the Railroads: The Struggle for Commercial Empire 1830–1860* (Baton Rouge: Louisiana State University Press, 1966).

14. De Leon, "The New South," 557; Ella Rightor, "Transportation," in *Standard History of New Orleans, Louisiana,* ed. Henry Rightor (Chicago: The Lewis Publishing Company, 1900), 310.

15. De Leon, "The New South," 557; *Preliminary Report of the Inland Waterways Commission: Message from the President of the United States, Transmitting a Preliminary Report of the Inland Waterways Commission,* U.S. Senate Document No. 325, 60th Cong., 1st sess. (Washington: Government Printing Office, 1908), 135, 132–36; Mark Twain, *Life on the Mississippi* (New York: Harper & Brothers, 1929 edition), 195; Warner, *Studies in the South and West,* 45. Also see "New Orleans Sketches. Alongshore," *Harlequin* 1, No. 16 (October 11, 1899), 9; *New Orleans Democrat,* October 6, 7, 12, 15, 1881; *New Orleans Times,* August 2, 1874; Walker, "Commercial and Mercantile Interests," 551–53, 572–75; Kendall, *History of New Orleans,* II, 630;

Lawrence H. Larsen, "New Orleans and the River Trade: Reinterpreting the Role of the Business Community," *Wisconsin Magazine of History* 61, No. 2 (Winter 1977–78), 114–18; British Board of Trade, *Cost of Living in American Towns,* 289; Taylor, *Louisiana Reconstructed,* 337; Redard, "The Port of New Orleans: An Economic History," 56–58, 178–82; J. Arthur White, "The Port of New Orleans" (unpublished M.A. thesis, Tulane University, 1924), 24; *New Orleans States,* December 1, 1883; "Analysis of Trade's Decline," *Harlequin* IX, No. 15 (October 24, 1907).

16. John Lovell, *Stevedores and Dockers: A Study of Trade Unionism in the Port of London, 1870–1914* (New York: Augustus M. Kelley, 1969), 34–35; Eric Hobsbawm, "National Unions on the Waterside," in *Labouring Men: Studies in the History of Labour* (London: Weidenfeld and Nicolson, 1964), 207–11. E.L. Taplin's description of the Liverpool waterfront as presenting "a job structure of bewildering complexity that probably only the dock labourer and his immediate employer could understand" could apply, with only slight exaggeration, to New Orleans as well. Taplin, *Liverpool Dockers and Seamen 1870–1900* (York: Occasional Papers in Economic and Social History No. 6, University of Hull, 1974), 9.

17. Zacharie, *New Orleans Guide,* 81–82; Edward King, *The Great South,* 54; "The Roustabouts of the Mississippi," *New Orleans Republican,* August 2, 1874; Sterling D. Spero and Abram L. Harris, *The Black Worker: The Negro and the Labor Movement* (1931; rpt. New York: Atheneum, 1968), 183–85.

18. General Committee of Arrangements, *A History of the Proceedings in the City of New Orleans, on the Occasion of the Funeral Ceremonies in Honor of James Abram Garfield, Late President of the United States, which took place on Monday, September 26th, 1881* (New Orleans: A.W. Hyatt, 1881), 251 (hereafter referred to as *Garfield Funeral Proceedings*).

19. Quote from W.E. Pedrick, *New Orleans As it Is with a Correct Guide to All Places of Interest* (Cleveland: W.E. Pedrick, 1885), 90; Edwin L. Jewell, ed., *Jewell's Crescent City Illustrated: The Commercial, Social, Political and General History of New Orleans, including Biographical Sketches of its Distinguished Citizens* (New Orleans, 1873), 316, 319; *Daily Picayune,* September 3, 8, 1880; *New Orleans: Her Relations to the New South,* 44; *Garfield Funeral Proceedings,* 251; *Norman's New Orleans,* 152–53; *Southwestern Christian Advocate,* April 19, 1883; Zacharie, *New Orleans Guide,* 81–82. The "clockwork manner" of cotton handling remained rather constant over the next several decades. The identical description appears in the 1902 edition of the guide. See: James S. Zacharie, *New Orleans Guide* (New Orleans: F.F. Hansell & Bro., 1902), 172.

20. *Harlequin* II, No. 46 (June 6, 1901), 6; Zacharie, *New Orleans Guide,* 81–82; *New Orleans: Her Relations to the New South,* 31–33.

21. See Chapter 3 for a discussion of freight handlers' work and unions.

22. W.H. Beveridge, *Unemployment: A Problem of Industry* (London: Longmans, Green and Co., 1909), 82–83; Hobsbawm, "National Unions on the Waterside," 205–7; David Montgomery, "Immigrant Workers and Managerial Reform" in *Workers' Control in America: Studies in the History of Work, Technology and Labor Struggles* (New York: Cambridge University Press, 1976), 36.

23. See Chapter 6.

24. Waterfront laborers were not the only group to experience seasonal unemployment. "There is not yet enough work in New Orleans to keep the workingmen busy all the year. In no other respect has [sic] the workingmen in New Orleans suffered more than in the short time of the busy season when there is something for them to do . . . Almost all the trades suffer from long periods when there is nothing doing in their

lines," the *Times-Democrat* observed in 1903. "Thus men employed in the hauling and loading trades are idle in the summer, while the carpenters, painters and bricklayers have little to do in the winter, when building is difficult and expensive and the weather against it." As late as 1888, the paper estimated, "the average working season in most of the industries in New Orleans was barely six and a half months, and the workingmen between seasons did anything they could to eke out a living." *Times-Democrat,* April 7, 1903. Also see *The Morning Star,* December 16, 1883.

25. On the shape-up and casual labor, see: Charles P. Larrow, *Shape-Up and Hiring Hall: A Comparison of Hiring Methods and Labor Relations on the New York and Seattle Waterfronts* (Berkeley: University of California Press, 1955), 249–60; Charles Barnes, *The Longshoremen* (New York: Russell Sage Foundation/Survey Associates, 1915), 59–68.

26. Hobsbawm, "National Unions on the Waterside," 205.

27. James V. Reese, "The Evolution of an Early Texas Union: The Screwmen's Benevolent Association of Galveston, 1866–1891," *Southwestern Historical Quarterly* LXXV, No. 2 (October 1971), 158–85; John P. Beck, "They Fought for Their Union: Upper Peninsula Iron Ore Trimmer Wars," *Michigan History* 73, No. 1 (January/February 1989), 24–31; William H. Sewell, Jr., "Uneven Development, the Autonomy of Politics, and the Dockworkers of Nineteenth-Century Marseille," *The American Historical Review* 93, No. 3 (June 1988), 604–37; Ian McKay, "Class Struggle and Mercantile Capitalism: Craftsmen and Labourers on the Halifax Waterfront, 1850–1902," in *Working Men Who Got Wet: Proceedings of the Fourth Conference of the Atlantic Canada Shipping Project,* ed. Rosemary Ommer and Gerald Panting (Maritime History Group/Memorial University of Newfoundland, 1980), 289–319.

28. The specialized dock worker, Hobsbawm noted, "had to have at least the qualities of the iron-puddler—strength and dexterity within a limited range, and very frequently the qualities of the all-round craftsman or supervisory worker—initiative, wide experience, the ability to make a variety of decisions to fit the necessities of loading and unloading the hundred and one non-standardized ships, the ability to supervise men. Even if not very highly skilled by conventional standards . . . their bargaining strength was considerable . . . For practical purposes the specialized man was as irreplaceable as a patternmaker." Hobsbawm, "National Unions on the Waterside," 207–8. On the "petty jealousies" produced by the "infinite variety of job specialisms," see Taplin, *Liverpool Dockers and Seamen,* 10.

29. George P. Marks, III, "The New Orleans Screwmen's Benevolent Association, 1850–1861," *Labor History* 14, No. 2 (Spring 1973), 259–63; Herbert R. Northrup, "The New Orleans Longshoremen," *Political Science Quarterly* 57, No. 4 (December 1942), 527; Carrol George Miller, "A Study of the New Orleans Longshoremen's Unions from 1850 to 1962" (unpublished M.A. thesis, Louisiana State University and Agriculture and Mechanical College, 1962), 10; Pearce, "The Rise and Decline of Labor in New Orleans," 12; *Garfield Funeral Proceedings,* 250–51. The best description of screwmen's work in Galveston, which at this time was much the same as in New Orleans, is found in Allen Clayton Taylor, "A History of the Screwmen's Benevolent Association from 1866 to 1924" (unpublished M.A. thesis, University of Texas at Austin, 1968), 25–39.

30. *Constitution and By-Laws of the Screwmen's Benevolent Association* (New Orleans: A.W. Hyatt, 1884), 13–16, 19–20.

31. *Constitution . . . of the Screwmen's Benevolent Association,* 13–16, 19–20. On the SBA's unequal division of work with its black counterpart, see *Weekly Louisianian,*

November 19, 1881. Racial tensions over the division of work exploded violently in the 1890s. See Chapter 4.

32. On New York, see *Fifth Annual Report of the Bureau of Statistics of Labor of the State of New York, for the Year 1887* (New York: Troy Press Co., 1888), 327–85; Barnes, *The Longshoremen*. According to German economist August Sartorius Freiherr von Waltershausen and labor economist Selig Perlman, New Orleans, Galveston, and Savannah all had trades assemblies in the 1880s that admitted black waterfront workers into membership on an equal basis with whites. See Selig Perlman, "Upheaval and Reorganization (since 1876)" in John R. Commons and Associates, *History of Labour in the United States,* Vol. 2 (New York: The Macmillan Company, 1918), 310; A. Sartorius Freiherr von Waltershausen, *Die nordamerikanischen Gewerkschaften unter dem Einfluss der fortschreitenden Productionstechnik* (Berlin: Hermann Bahr, 1886). Translation in the Abram Lincoln Harris Papers, Collection 43-1, Moorland-Spingarn Research Center, Manuscript Division, Howard University. This translation was produced by Rick Livingston, Yale University.

33. McKay, "Class Struggle and Mercantile Capitalism." On the British experience, see Hobsbawm, "National Unions on the Waterside," 214–15; Jonathan Schneer, *Ben Tillett: Portrait of a Labour Leader* (London: Croom Helm, 1981); Raymond Brown, *Waterfront Organisation in Hull 1870–1900* (Hull: Occasional Papers in Economic and Social History No. 5, University of Hull Publications, 1972); Eric Taplin, *Liverpool Dockers and Seamen 1870–1890* (Hull: University of Hull Publications, 1974); Eric Taplin, *The Dockers' Union: A Study of the National Union of Dock Labourers, 1889–1922* (New York: St. Martin's Press, 1985); John Lovell, *Stevedores and Dockers;* Philip J. Leng, *The Welsh Dockers* (Lancashire: G.W.&A. Hesketh, 1981); M.J. Daunton, *Coal Metropolis: Cardiff 1870–1914* (Bristol: Leicester University Press, 1977).

34. *Weekly Pelican,* June 4, 1887; *Garfield Funeral Proceedings,* 103, 249–51.

35. *Daily Picayune,* July 18, 1877. On the severity of the depression, see *Morning Star and Catholic Messenger,* October 12, 1873.

36. Taylor, *Louisiana Reconstructed,* 358–63. New Orleans had experienced economic hardship as a result of trade and political disruptions in the earlier Reconstruction era as well. In the winter and spring of 1868, the *Picayune* reported numerous "scenes of misery and distress never seen before in New Orleans. A city which a few years ago scarcely numbered a pauper within its limits, is now filled with poverty and distress. Everywhere scattered around one sees the idle and unemployed." See *Daily Picayune,* April 15, 1868; also see February 21, 23, May 6, 1868.

37. *Daily Picayune,* April 29, March 24, May 11, 1875; *New Orleans Republican,* May 4, 1875.

38. *New Orleans Republican,* May 4, 1875; *Daily Picayune,* March 16, 1875; Roger W. Shugg, *Origins of Class Struggle in Louisiana: A Social History of White Farmers and Laborers during Slavery and After, 1840–1875* (1939; rpt. Baton Rouge: Louisiana State University Press, 1968), 304. "God only knows what will become of our poor workingmen who have families," one impoverished man wrote in 1877, "if something soon don't turn up to give relief . . . [D]eath is staring us in the face . . . My family are, at this time, without the food necessary to keep life together." *Daily Picayune,* April 15, 1877, quoted in Philip Foner, *The Great Labor Uprising of 1877* (New York: Monad Press, 1977), 25.

39. See Herbert G. Gutman, "The Failure of the Movement by the Unemployed for Public Works in 1873," *Political Science Quarterly* 80, No. 2 (June 1965), 254–76;

Gutman, "The Tompkins Square 'Riot' in New York City on January 13, 1874: A Re-Examination of Its Causes and Its Aftermath," *Labor History* 6, No. 1 (Winter 1965), 44–70; Barbara Jeanne Fields, *Slavery and Freedom on the Middle Ground: Maryland during the Nineteenth Century* (New Haven: Yale University Press, 1985), 202–4.

40. *Daily Picayune,* July 19, 1874; March 16, 1875; Shugg, *Origins of Class Struggle,* 304.

41. *Daily Picayune,* March 24, April 29, 30, May 11–13, 27–29, 1875; *New Orleans Times,* April 20, 30, May 11, 27, 1875; *New Orleans Republican,* April 20, 24, 29, May 1, 1875.

42. *Daily Picayune,* May 12, June 3, 4, 1874. The employment situation temporarily improved by August 1874 when Captain James Eads began a major engineering feat of building jetties in the Mississippi, deepening the river to accommodate larger ships.

43. *New Orleans Republican,* March 12, 23, May 3, June 8, 21, 1876.

44. "If so many men were not watching for an opportunity to get employment," the *Picayune* noted, "a strike might be expected." Indeed, while cities linked by the nation's trunk railroads confronted massive strikes of unprecedented proportion in late July and early August 1877, New Orleans remained quiet. Despite rumors of strikes, only twenty disgruntled laboring men gathered near City Hall. Police stood ready to suppress any disturbance; the mayor temporarily banned union meetings; and the president of the workingmen's club was arrested for attempting to incite a disturbance. *New Orleans Republican,* January 3, June 9, 1877; *Daily Picayune,* July 3, 10, 14, 27, 28, 31, 1877.

45. *Daily Picayune,* May 5, 1875; April 1, 29, May 1, June 1, 9, 15, July 2, 3, 1875; July 15, August 7, 1877; *Weekly Louisianian,* July 11, 1874; also see Shugg, *Origins of Class Struggle,* 301–3.

46. *Daily Picayune,* April 30, May 7, 11, 27, June 1, 15, 25, 1875; *New Orleans Republican,* May 31, 1876. On relief, charity, social conflict, and depression, see Michael B. Katz, *In the Shadow of the Poorhouse: A Social History of Welfare in America* (New York: Basic Books, 1986), 66–68, 80–82; Alexander Keyssar, *Out of Work: The First Century of Unemployment in Massachusetts* (New York: Cambridge University Press, 1986), 150–55; Frances Fox Piven and Richard A. Cloward, *Regulating the Poor: The Functions of Public Welfare* (New York: Pantheon Books, 1971). Charity collection sometimes proved difficult during hard times. One Catholic newspaper criticized its readers and church-goers for stinginess in November 1873. "Here, with a Catholic population of at least 120,000, we have only 300 [in the Society of St. Vincent de Paul] and as to pecuniary assistance, the amounts contributed are so small in most parishes, that we would be ashamed to give a detailed statement of them." *Morning Star and Catholic Messenger,* November 9, 1873; November 15, 1874. On one occasion, "charitable and energetic gentlemen of the First District" sponsored an evening of "grand musical and variety entertainment," provided by the "Crescent City Serenaders" for the benefit of the poor and orphans of the Conferences of St. Patrick and St. Theresa." *Morning Star and Catholic Messenger,* November 16, 1873.

47. John Smith Kendall, *History of New Orleans,* I (New York: The Lewis Publishing Co., 1922), 342–58. For a more detailed discussion of Louisiana politics and the demise of the Reconstruction government, see: Taylor, *Louisiana Reconstructed;* Ted Tunnell, *Crucible of Reconstruction: War, Radicalism and Race in Louisiana, 1862–1877* (Baton Rouge: Louisiana State University Press, 1984); Roger A. Fischer, *The Segregation Struggle in Louisiana 1862–1877* (Urbana: University of Illinois Press, 1974); Agnes Smith Grosz, "The Political Career of Pinckney Benton Stewart Pinchback," *Louisiana Historical Quarterly* 27, No. 2 (April 1944), 527–612; Dennis

C. Rousey, "Black Policemen in New Orleans During Reconstruction," *The Historian* 49, No. 2 (February 1987), 223–43.

48. *New Orleans Republican,* March 23, 1876; Talyor, *Louisiana Reconstructed,* 238–52; Michael Perman, *The Road to Redemption: Southern Politics, 1869–1879* (Chapel Hill: University of North Carolina Press, 1984), 42; Melinda Meek Hennessey, "Race and Violence in Reconstruction New Orleans: The 1868 Riot," *Louisiana History* XX, No. 1 (Winter 1979), 77–91.

49. White League document quoted in Kendall, *History of New Orleans,* I, 360; *Southwestern Christian Advocate* editor the Rev. Mr. Hartzell quoted in *New Orleans Republican,* September 20, 1874; *New Orleans Republican,* September 4, 1874; *Birmingham Iron Age,* July 16, 1874; *New Orleans Times,* August 4, 1874; *Daily Picayune,* July 11, August 1, 4, 1874; *Morning Star and Catholic Messenger,* June 14, September 20, 1874; *New York Times,* August 8, September 16, 17, 1874. For an account of the battle of September 14, see Taylor, *Louisiana Reconstructed,* 279–96.

50. Taylor, *Louisiana Reconstructed,* 291–96; Kendall, *History of New Orleans,* I, 359–75; Jackson, *New Orleans in the Gilded Age,* 28–31; C. Vann Woodward, *Reunion and Reaction: The Compromise of 1877 and the End of Reconstruction* (1951; rpt. Boston: Little, Brown and Company, 1966).

51. New Orleans' black community also grew at a significantly slower rate during the 1870s. In contrast to the 1860s, in which the disruptions of the Civil War and Reconstruction period raised rural to urban black migration rates substantially, the decade of the 1870s, marked by a national economic depression and the rise of white rural terror, saw only a modest black growth rate of 14%, just above an 11% increase for whites. In 1880, 57,627 blacks out of a total of 158,367 people lived in the Crescent City; in both 1870 and 1880, the city's black population constituted just over a quarter of the total.

52. *Weekly Louisianian,* May 7, 1881. As Peter Rachleff has shown for Richmond, Virginia, these secret societies "provided an institutional framework through which black people could help each other—and seek help—beyond familial and churchly ties." See: Peter J. Rachleff, *Black Labor in Richmond 1865–1890* (1984; rpt. Urbana: University of Illinois Press, 1989), 25; also see: W. E. Burghardt DuBois, ed., *Some Efforts of American Negroes for Their Own Social Betterment: Report of an investigation under the direction of Atlanta University; together with the proceedings of the Third Conference for the Study of the Negro Problems, held at Atlanta University, May 25–26, 1898* (Atlanta: Atlanta University Press, 1898); Howard N. Rabinowitz, *Race Relations in the Urban South 1865–1890* (New York: Oxford University Press, 1978), 140–42, 227–28; Kenneth L. Kusmer, *A Ghetto Takes Shape: Black Cleveland, 1870–1930* (Urbana: University of Illinois Press, 1976), 96–98; David M. Katzman, *Before the Ghetto: Black Detroit in the Nineteenth Century* (Urbana: University of Illinois Press, 1973), 135–67; August Meier, *Negro Thought in America 1880–1915: Racial Ideologies in the Age of Booker T. Washington* (Ann Arbor: The University of Michigan Press, 1963), 130–38; Anne Firor Scott, "Most Invisible of All: Black Women's Voluntary Associations," *Journal of Southern History* LVI, No. 1 (February 1990), 3–22.

53. John W. Blassingame, *Black New Orleans 1860–1880* (Chicago: University of Chicago Press, 1973), 147; *Southwestern Christian Advocate,* December 6, 1883. Also see Harry J. Walker, "Negro Benevolent Societies in New Orleans: A Study of Their Structure, Function, and Membership" (unpublished essay, Department of Social Science, Fisk University, 1936; copy at Xavier University Archives, New Orleans); Claude F. Jacobs, "Benevolent Societies of New Orleans Blacks During the Late Nineteenth and Early Twentieth Centuries," *Louisiana History* XXIX, No. 1 (Winter

1988), 21–33; Dorothy Rose Eagleson, "Some Aspects of the Social Life of the New Orleans Negro in the 1880s" (unpublished M.A. thesis, Tulane University, 1961); *Weekly Louisianian,* May 7, 1881; *Daily Picayune,* December 30, 1913.

54. Blassingame, *Black New Orleans,* 156. A black lawyer, a descendant of an old free black family, told an investigator that "the peculiar end of the founding of these societies, is that they were first organized by free men of color, and among the free ones these societies would not admit men of a darker hue." Walker, *Negro Benevolent Societies in New Orleans,* 35–38. David Rankin, historian of New Orleans black elite, concluded that the black "Creole desire to remain distinct persisted long after the end of Reconstruction." The old community "forged before the war . . . was all that many free coloreds had left, and they eagerly sought refuge within its narrow confines, where neither blacks nor whites were admitted." "The Impact of the Civil War on the Free Colored Community of New Orleans," *Perspectives in American History* XXI, ed. Donald Fleming (Cambridge: Charles Warren Center for Studies in American History, 1977–78), 412. On separate associations of upper-class blacks and class divisions within the black community, see "'Color Lines' Among the Colored People," *The Literary Digest* 72, No. 11 (March 18, 1922); Willard B. Gatewood, Jr., "Aristocrats of Color: South and North: The Black Elite 1880–1920," *Journal of Southern History* LIV, No. 1 (February 1988), 3–20; Pops Foster, *Pops Foster: The Autobiography of a New Orleans Jazzman as told to Tom Stoddard* (Berkeley: University of California Press, 1971), 65; Loren Schweninger, "Prosperous Blacks in the South, 1790–1880," *American Historical Review* 95, No. 1 (February 1990), 31–56.

55. Blassingame, *Black New Orleans,* 64; *New Orleans Pelican,* June 4, 1887; *Registers of Signatures of Deposits in the Branches of the Freedmen's Savings and Trust Company, 1865–1874,* U.S. Department of the Treasury, RG 101, M816, Roll No. 12 (0174/#4634); *Garfield Funeral Proceedings,* 112–14, 248–51; "Longshoremen's Protective Union Benevolent Association," *Daily Picayune,* December 30, 1913 (part 6); Eagleson, "Some Aspects of the Social Life of the New Orleans Negro," 83. For a general discussion of black longshore unions, see Spero and Harris, *The Black Worker,* 183–85; Lorenzo J. Greene and Carter G. Woodson, *The Negro Wage Earner* (Washington, D.C.: The Association for the Study of Negro Life and History, Inc., 1930), 113–14; F. Ray Marshall, *Labor in the South* (Cambridge: Harvard University Press, 1967), 60–64; Herbert R. Northrup, *Organized Labor and the Negro* (New York: Harper & Brothers, 1944), 137–53.

56. *New Orleans Pelican,* June 4, 1887.

57. *Weekly Louisianian,* May 7, 1881; Judith Stein, *The World of Marcus Garvey: Race and Class in Modern Society* (Baton Rouge: Louisiana State University Press, 1986), 7–23; Nell Irvin Painter, *Exodusters: Black Migration to Kansas after Reconstruction* (New York: W.W. Norton and Co., 1976), 26–30. Addressing itself to both black benevolent societies and trade unions, the *Southwestern Christian Advocate* on numerous occasions pointed out the missed opportunities for investment and the development of black businesses, which would provide a firm basis for the economic uplift of the race. See *Southwestern Christian Advocate,* December 6, 1883; October 29, 1885; August 13, 1896; March 30, 1899.

58. *Weekly Louisianian,* October 1, November 19, 1881.

59. *Weekly Louisianian,* May 7, September 10, November 19, 1881. See below for details of the strike. See Chapter 4 for similar arguments in the aftermath of white union attacks against black dock workers in 1894–95.

60. *Weekly Louisianian,* May 7, 1881. For a broader discussion of African-

American ideology in the late nineteenth and early twentieth century, also see Meier, *Negro Thought in America,* John Brown Childs, *Leadership, Conflict, and Cooperation in Afro-American Social Thought* (Philadelphia: Temple University Press, 1989).

61. For treatments of the experiences of black dock workers in other southern cities during this period, see William C. Hine, "Black Organized Labor in Reconstruction Charleston," *Labor History* 25, No. 4 (Fall 1984), 504–17; George Brown Tindall, *South Carolina Negroes 1877–1900* (Columbia: University of South Carolina Press, 1952), 137–39; Robert E. Perdue, *The Negro in Savannah, 1865–1900* (New York: Exposition Press, 1973), 116; Jerrell H. Shofner, "The Pensacola Workingman's Association: A Militant Negro Labor Union During Reconstruction," *Labor History* 13, No. 4 (Fall 1972), 555–59; Jerrell H. Shofner, "Militant Negro Laborers in Reconstruction Florida," *Journal of Southern History* XXXIX, No. 3 (August 1973), 397–408.

62. *Daily Picayune,* October 21, 1872; *New Orleans Republican,* October 22, 1872; *New Orleans Times,* October 22, 1872; *New Orleans Pelican,* June 4, 1887. The LPUBA received a legislative charter of incorporation from the state government on April 23, 1872. See: "An Act to Incorporate the Longshoremen's Protective Union Benevolent Association of the city of New Orleans and State of Louisiana," and copy of the original charter in the William Lombard Collection, Amistad Research Center, Tulane University.

63. *Daily Picayune,* October 21, 1872.

64. *New Orleans Times,* October 19, 1872.

65. There is little information on this early white association. William Crotty, president of the United Laborers, was president of an association of the same name as early as 1866. In March of that year, the United Laborers celebrated its thirteenth anniversary. It was undoubtedly a benevolent society, described by the *Daily True Delta* as "one of the most estimable and respected charitable associations in the city, and has done much good in relieving sick and disabled workingmen, and widows and orphans, when they possessed no other source of assistance." *Daily True Delta,* March 8, 1866. According to a local account of organizations participating in the funeral ceremonies and procession honoring James Garfield in 1881, a white Longshoremen's Benevolent Association, whose object was "to help each other as fellow-men and fellow-laborers," was organized on October 5, 1873. This organization does not appear to have participated in any of the strikes in the 1870s. See *Garfield Funeral Proceedings,* 218; Reinders, "A Social History of New Orleans 1850–1860" (unpublished Ph.D. dissertation, University of Texas at Austin, 1957), 108; Shugg, *Origins of Class Struggle,* 115.

66. *New Orleans Republican,* October 18, 19, 1872; *Daily Picayune,* October 18, 19, 1872; *New Orleans Times,* October 18, 1872.

67. *New Orleans Republican,* October 18, 1872; *Daily Picayune,* October 21, 1872; *Mobile Register,* October 19, 1872.

68. *New Orleans Republican,* October 22, 1872.

69. *Daily Picayune,* October 18, 21, 1872; *Mobile Register,* October 19, 1872.

70. *New Orleans Times,* October 19–21, 1872; *Daily Picayune,* October 19–21, 1872; *New Orleans Republican,* October 20, 22, 1872.

71. *New Orleans Republican,* October 20, 22, 25, 26, 1872; *Daily Picayune,* October 23, 25, 29, 30, 1872; *New Orleans Times,* October 25–28, 1872.

72. *New Orleans Times,* October 18, 1872; *Daily Picayune,* October 21, 1872.

73. *New Orleans Republican,* October 26, 1872.

74. *Daily Picayune,* December 4, 1877.

75. *Daily Picayune,* October 15, 26, 1873.

76. *Daily Picayune,* October 15, 26, 30, 1873; September 2, 3, 1874; *New Orleans Republican,* September 3, 1874; *New Orleans Times,* September 2, 3, 1874; Taylor, *Louisiana Reconstructed,* 388.

77. *New Orleans Republican,* January 31, 1875; *New Orleans Times,* January 31, 1875; *Daily Picayune,* January 31, 1875.

78. *New Orleans Republican,* February 5, 1875; *New Orleans Times,* February 5, 1875; *Daily Picayune,* February 5, 1875.

79. *New Orleans Republican,* April 18, 20, 1875; *New Orleans Times,* April 18, 1875; *Daily Picayune,* April 18, 1875; November 4, December 2, 1877. It is possible that the Union Laboring Society was the Longshoremen's Protective Union Benevolent Society that struck unsuccessfully in 1872 and the Longshoremen's Protective Association that struck unsuccessfully in 1875. Joe Harris, president of the Union Laboring Society, may have been the same Joe Harris who was a marshal for and leader of the 1872 organization. In 1875, the *Republican* describes a Mr. Joseph Harris as a founder of the Longshoremen's Protective Association. Confusion stems from the inconsistency with which contemporary press accounts name individual organizations, as well as possible name changes of the organizations in question. See *New Orleans Times,* October 19, 1872; *New Orleans Republican,* January 31, 1875.

80. Eric Hobsbawm, "Economic Fluctuations and Some Social Movements since 1800," in *Labouring Men: Studies in the History of Labour* (London: Weidenfeld and Nicolson, 1964), 144. The "habit of solidarity, which is the foundation of effective trade unionism, takes time to learn . . . It takes even longer to become part of the unquestioned ethical code of the working-class." Hobsbawm, "The Machine Breakers," in *Labouring Men,* 9.

81. Sala, *America Revisited,* II, 49; Perlman, "Upheaval and Reorganization (since 1876)" 310–11; Steven Ross, *Workers on the Edge: Work, Leisure, and Politics in Industrializing Cincinnati, 1788–1890* (New York: Columbia University Press, 1985), 258–62; Leon Fink, *Workingmen's Democracy: The Knights of Labor and American Politics* (Urbana: University of Illinois Press, 1983), 25–26; Gregory S. Kealey and Bryan D. Palmer, *Dreaming of What Might Be: The Knights of Labor in Ontario, 1880–1900* (New York: Cambridge University Press, 1982).

82. George E. McNeill, ed., *The Labor Movement: The Problem of To-Day* (Boston: A.M. Bridgman & Co., 1887), 167–68; Dale Somers, "Black and White in New Orleans: A Study in Urban Race Relations, 1865–1900," *Journal of Southern History* XL, No. 1 (February 1974), 31; *Daily Picayune,* November 26, 1882, November 26, 1883.

83. *Garfield Funeral Proceedings,* 212, 248–49; *New Orleans Times,* September 7, 1880; *Daily Picayune,* September 2, 1880; November 26, 1883.

84. *The Weekly Pelican,* June 4, 1887. The unification process is discussed in Chapter 4.

85. *Daily Picayune,* February 10, 27, 1880.

86. The strike did prompt steamboat owners to address some of the underlying non-wage issues. Black longshoremen had complained bitterly of abuses "perpetrated by the contractors" who hired them. As they had in 1867, owners temporarily employed their ship's mates to direct the work of hiring labor for the loading and unloading of their boats. *Daily Picayune,* February 12, 14, 26, 27, March 9, 16, 17, 24, 1880.

87. *Daily Picayune,* July 7, September 2, 3, 5, 8, 1880; *New Orleans Times,* September 3, 4, 6, 1880; Pearce, "The Rise and Decline of Labor," 18–19; *Garfield Funeral Proceedings,* 248–49; also see *Weekly Louisianian,* January 20, 1881. The previous February, members of the "Cotton Rollers Benevolent Association"—the yardmen—

struck for one day demanding a wage increase. The rollers claimed that the "competition between the compresses has been carried so far that nothing is left to pay members." The following day, press owners granted a small increase. *New Orleans Times,* February 16, 1880.

88. *New Orleans Picayune,* September 7, 9, 1880; *New Orleans Times,* September 7, 8, 1880; Pearce, "The Rise and Decline of Labor," 19.

89. *Daily Picayune,* August 26, 1881; Savannah *Morning News,* September 2, 1881; *New Orleans Times,* January 27, 1881.

90. *New Orleans Times,* January 27, 28, 1881; *Daily Picayune,* January 27, 29, 1881.

91. A. Sartorius Freiherr von Waltershausen, *Die nordamerikanischen Gewerkschaften,* 146; Savannah *Morning News,* September 2, 1881.

92. *Daily Picayune,* August 26, 1881; Pearce, "The Rise and Decline of Labor," 25.

93. *Daily Picayune,* August 27, 30, 1881; *Weekly Louisianian,* August 13, 1881; *Appletons' Annual Cyclopaedia and Register of Important Events of the Year 1881,* New Series, VI (New York: D. Appleton and Co., 1885), 516; *The Mascot,* October 7, 1882.

94. *New Orleans Times,* August 30, 1881.

95. Kendall, *History of New Orleans,* I, 421–24, 428–36; *The Mascot,* May 27, 1882; Jackson, *New Orleans in the Gilded Age,* 56–66. For a more detailed discussion of political culture and class politics in New Orleans, see Chapter 4.

96. *Daily Picayune,* September 11, 1881; James E. Boyle, *Cotton and the New Orleans Cotton Exchange: A Century of Commercial Evolution* (Garden City, N.Y.: The Country Life Press, 1934), 65–94; Zacharie, *New Orleans Guide,* 85; L. Tuffly Ellis, "The New Orleans Cotton Exchange: The Formative Years, 1871–1880," *Journal of Southern History* XXXIX, No. 4 (November 1973), 545–64; De Leon, "The New South," 556; Herring, *History of the New Orleans Board of Trade,* 32; *Times-Democrat,* June 14, 1883.

97. *Daily Picayune,* July 2, August 27, 28, September 8, 1881.

98. *Daily Picayune,* September 7, 1881; *New Orleans Times,* September 8, 1881; Savannah *Morning News,* September 9, 1881.

99. *New Orleans Times,* August 28, September 8, 1881; *Galveston Daily News,* September 6, 8, 1881.

100. *Daily Picayune,* August 29, 30, September 10, 1881; Savannah *Morning News,* September 12, 1881; *Galveston Daily News,* August 28, 31, September 10, 1881.

101. *Daily Picayune,* August 31, September 1, 2, 1881; *Galveston Daily News,* September 2, 1881.

102. *New Orleans Times,* September 3, 13, 1881; *Daily Picayune,* September 9, 1881; *Galveston Daily News,* September 3, 4, 1881; Savannah *Morning News,* September 3, 1881. Unionized workers in Savannah publicly denounced the efforts of New Orleans employers to recruit scab labor in their city. The Workingmen's Benevolent Association, composed of cotton handlers, endorsed its New Orleans counterpart's "endeavor to obtain just, equitable, and living wages." It resolved that "the rights of laboring men, as portions of the body politic, are universal, and injuries done them in one section are always felt in another. The battle of our New Orleans brethren against combined capital is perhaps the same we should have to commence here tomorrow." The Savannah workers' association promised to expel permanently any member who accepted the offers of work of the New Orleans labor recruiter. When some city workers accepted the offer and boarded a train bound for New Orleans, crowds of Savannah blacks gathered to harass them. See Savannah *Morning News,* September 7, 8, 10, 1881; *Daily Picayune,* September 9, 10, 1881.

103. *Daily Picayune,* September 3, 6, 7, 10, 1881; *Weekly Louisianian,* October 10, 1881; *Galveston Daily News,* September 6, 7, 8, 10, 1881.

104. *Weekly Louisianian,* October 1, November 19, 1881. See Chapter 5 for discussion of the half-and-half division of work in the early twentieth century.

105. *New Orleans Times,* September 3, 1881; *Daily Picayune,* September 4, 6, 8, 9, 1881; *New York Times,* September 3, 1881.

106. *New Orleans Times,* September 11, 1881; *Daily Picayune,* September 12, 1881. Also see *Weekly Louisianian,* September 10, 17, 1881; *Southwestern Christian Advocate,* September 15, 1881; Savannah *Morning News,* September 12, 14, 1881; *New York Times,* September 12, 1881.

107. *Daily Picayune,* September 10 and 11, 1881; *New Orleans Times,* September 11, 1881; *Galveston Daily News,* September 11, 1881; *New York Times,* September 11, 1881.

108. *Daily Picayune,* September 11–13, 1881; *New Orleans Times,* September 12, 1881; *Galveston Daily News,* September 13, 1881; *New York Times,* September 11, 12, 14, 1881; *Atlanta Constitution,* September 14, 1881. Also see Joseph A. Shakspeare to Major General Behan, September 12, 1881; S.D. McEnery to Major General W.J. Behan, Commander, State National Guard, September 12, 1881; Proclamation of Mayor Shakspeare, September 13, 1881; Sam. D. McEnery, Gov., to British Consulate, September 13, 1881, in "New Orleans—Labor Riots, 1881," New Orleans Collection, Louisiana Militia and National Guard, First Division, General, September–December 1881, Box 3, Folder 5, 55-N, LHA, Tulane University; Shakspeare to Behan, September 11, 1881, in Washington Artillery Papers, Organizational Papers, 1877–92, Box 1, Folder 2, 55-W, LHA (in Manuscripts, Rare Books and University Archives, Howard-Tilton Memorial Library, Tulane University).

109. *New Orleans Times,* September 11, 12, 1881; *Daily Picayune,* September 12, 1881; *Galveston Daily News,* September 13, 1881.

110. *New Orleans Times,* September 13, 1881; *Daily Picayune,* September 13, 14, 1881; *Galveston Daily News,* September 13, 14, 1881.

111. *Daily Picayune,* September 7, 8, 15, 1881. Major General W.J. Behan, whom the governor placed in charge of the state militia in New Orleans, chaired the arbitration committee. On September 13, Behan, John Glynn, Jr., commander of the First Brigade of the First Division of the militia, and Adolphe Meyer, commander of the Second Brigade, met with Acting Governor McEnery and members of his staff and decided to call a conference of the laborers' and employers' associations. City Administrator Patrick Mealey of the white Cotton Yardmen's Benevolent Association and four white and four black union leaders represented the unions at the conference. Mayor Shakspeare, several other city commissioners, and Acting Governor McEnery observed at least some of the hearings. *New Orleans Times,* September 8, 1881; *Daily Picayune,* September 8, 14, 1881.

112. *New Orleans Times,* September 15, 1881; *Daily Picayune,* September 15, 1881; *Galveston Daily News,* September 15, 1881; *New York Times,* September 15, 1881; *Southwestern Christian Advocate,* September 22, 1881; Savannah *Morning News,* September 15, 1881.

113. *Daily Picayune,* September 15, 1881. See Chapter 4 for a discussion of municipal affairs and class politics. Also see Jackson, *New Orleans in the Gilded Age;* George M. Reynolds, *Machine Politics in New Orleans, 1897–1926* (1936; rpt. New York: AMS Press, 1968); Raymond O. Nussbaum, " 'The Ring Is Smashed!': The New Orleans Municipal Election of 1896," *Louisiana History* XVII, No. 3 (Summer 1976), 283–97; Matthew James Schott, "John M. Parker of Louisiana and the Varieties of American

Progressivism" (unpublished Ph.D. dissertation, Vanderbilt University, 1969); Schott, "The New Orleans Machines and Progressivism," *Louisiana History* XXIV, No. 2 (Spring 1983), 141–53; Schott, "Progressives Against Democracy: Electoral Reform in Louisiana, 1894–1921," *Louisiana History* XX, No. 3 (Summer 1979), 247–60; Edward F. Haas, "John Fitzpatrick and Political Continuity in New Orleans," *Louisiana History* XXII, No. 1 (Winter 1981), 7–29; Brian Gary Ettinger, "John Fitzpatrick and the Limits of Working-Class Politics in New Orleans, 1892–1896," *Louisiana History* XXVI, No. 4 (Fall 1985), 545–64; Edward F. Haas, *Political Leadership in a Southern City: New Orleans in the Progressive Era 1896–1902* (Ruston, La.: McGinty Publications, 1988).

Chapter 3

1. James T. Newman, "The Future of the South—V," *Southwestern Christian Advocate,* April 9, 1885.

2. Charles Dudley Warner, *Studies in the South and West, with Comments on Canada* (New York: Harper and Brothers, 1889), 61; Alan Brinkley, *Huey Long, Father Coughlin and the Great Depression* (New York: Alfred A. Knopf, 1982), 15; Matthew James Schott, "John M. Parker of Louisiana and the Varieties of American Progressivism" (unpublished Ph.D. dissertation, Vanderbilt University, 1969), 14–15; Edward F. Haas, "John Fitzpatrick and Political Continuity in New Orleans, 1896–1900," *Louisiana History* XXII, No. 1 (Winter 1981), 7–29; Brian Gary Ettinger, "John Fitzpatrick and the Limits of Working-Class Politics in New Orleans," *Louisiana History* XXVI, No. 4 (Fall 1985), 341–67.

3. William O. Scroggs, "Commission Government in the South," *The Annals of the American Academy of Political and Social Science* XXXVIII, No. 3 (November 1911), 13–14; William W. Howe, "Municipal History of New Orleans," *Johns Hopkins University Studies in Historical and Political Science,* Seventh Series, IV (April 1889), 18–19; Joy J. Jackson, *New Orleans in the Gilded Age: Politics and Urban Progress 1880–1896* (Baton Rouge: Louisiana Historical Association/Louisiana State University Press, 1969), 56; John Smith Kendall, *History of New Orleans,* Vol. I (Chicago: The Lewis Publishing Company, 1922), 444–45; William Saunders, *Through the Light Continent; or, The United States in 1877–8* (London: Cassell, Petter, and Galpin, 1879), 68; Norman Walker, "Municipal Government," *Standard History of New Orleans, Louisiana,* ed. Henry Rightor (Chicago: The Lewis Publishing Company, 1900), 102–5.

4. *The Mascot,* March 27, 1886. Historians have shared this nineteenth-century assessment of New Orleans politics. James Schott describes the Ring as an "alliance of colorful, self-aggrandizing, aldermen-assessors . . . [who were] unmistakably cynical, corrupt, and selfish . . ." In her overview of New Orleans politics in the Gilded Age, Joy Jackson referred only slightly less harshly to the "corrupt or power-usurping Ring forces." Schott, "John M. Parker," 14–15; Jackson, *New Orleans in the Gilded Age,* 55. For contemporary denunciations of Ring practices, see *Morning Star,* July 19, 1874; April 27, May 4, June 8, 1884; *Southwestern Christian Advocate,* March 15, September 6, November 15, 1883; *New York Times,* October 12, 1880; October 17, November 24, December 1, 1882; December 2, 1894.

5. *The Mascot,* July 15, 1882; March 29, August 30, 1884. Also see Raymond O. Nussbaum, " 'The Ring is Smashed!': The New Orleans Municipal Election of 1896," *Louisiana History* XVII, No. 3 (Summer 1976), 284; Reynolds, *Machine Politics in New Orleans,* 14.

6. Kendall, *History of New Orleans,* II, 506; Covington Hall, "Labor Struggles in

the Deep South," Part I, Section I (unpublished manuscript, n.d., Manuscripts, Rare Books and University Archives, Howard-Tilton Memorial Library, Tulane University), 12; *The Mascot,* January 2, 7, 1888; Ettinger, "John Fitzpatrick and the Limits of Working-Class Politics," 347–48; "Biographies of the Mayors of New Orleans," Works Progress Administration Project 665-64-3-112, May 1939, 142–43 (in New Orleans Public Library); *Daily Item,* April 7, 1919; *New Orleans States,* April 7, 1919; *Times-Picayune,* April 8, 1919.

7. *Daily Picayune,* January 2, 3, 1888; *The Mascot,* January 2, 7, 1888.

8. *The Mascot,* July 1, 1882.

9. *The Mascot,* July 15, 1882. "The chief objection" to Fitzpatrick, the satirical journal *Figaro* similarly complained, "refers to the gang of vicious rowdies and bummers that he is compelled to carry with him wherever he goes in political life." *Figaro,* March 15, 1884. On the machine's use of patronage, also see *The Mascot,* February 25, March 25, July 1, 8, 22, 29, August 26, November 4, 11, 1882; May 10, 17, September 20, 1884; May 9, September 26, October 10, 1885; April 10, 1886.

10. See Section VI below for a discussion of the general strike of 1892 and Chapter 5 for a discussion of the riverfront racial conflicts of 1894–95. Also see Ettinger, "John Fitzpatrick and the Limits of Working-Class Politics," 341–67; Haas, "John Fitzpatrick and Political Continuity," 7–29. The lack of police interference in labor disturbances was evident during a late 1884 streetcar strike as well. "While the rioting was going on, and hundreds of people were gathered in the streets in the various neighborhoods of violence, not a policeman could be seen in any direction or within any distance. And this thing was going on for a portion of two days." *The Morning Star,* January 4, 1885; also see November 9, 1884.

11. On the Greenback party in New Orleans, see *Times-Democrat,* September 3, October 3, 18, 1882; *Daily Picayune,* August 15, 1882; *Southwestern Christian Advocate,* September 25, 1884.

12. *The Mascot,* October 21, 28, 1882; April 17, 1886; *Daily Picayune,* October 11, 18, November 8, 1882.

13. Delaney's presence on the municipal reform ticket led some businessmen to refuse to endorse it. "Perhaps it occurred to the merchants to wonder," the *Figaro* noted, "if this Mr. John Delaney, who, some four years ago, figured as the leader and organizer of a strike which for days paralyzed the business of New Orleans, and threatened to saturate our streets with blood. No doubt they asked themselves whether it were wise to put into an important office a man who recognized, before his duty as a citizen, his obligations as a member of the Trade Union—a man who would connive at—nay! order and approve—the commission of violence, the infraction of the law, in pursuit of his society's policy." *Figaro,* April 12, 1884; *The Mascot,* August 30, 1884. Two years earlier, Delaney turned down a nomination from the city's small Greenback party, stating that his leadership in the Trades Assembly would not permit him to take such a step, since "the public would consider that his object in uniting the trades was for the purpose of advancing himself politically." *Times-Democrat,* September 3, 1882. By the early 1890s, Delaney had left the labor movement altogether to become a contracting stevedore. *Daily Picayune,* March 8, 1892. On the candidacy of the Dock and Cotton Council leader in the twentieth century, see *Harlequin,* November 10, 1904.

14. *Daily Picayune,* April 8, 1888; January 22, 1901; *States,* January 21, 1901; *Times-Democrat,* January 22, 1901.

15. *Times-Democrat,* September 3, 7, October 3, 1882; *New Orleans Times,* August 28, 1881; *Daily Picayune,* October 21, 1877; August 15, 1882; William Ivy Hair, *Bourbonism and Agrarian Protest: Louisiana Politics 1877–1900* (Baton Rouge: Louisi-

ana State University Press, 1969), 62–63, 71–74. For a broader discussion of politics and class, see: Martin Shefter, "Trade Unions and Political Machines: The Organization and Disorganization of the American Working Class in the Late Nineteenth Century," *Working-Class Formation: Nineteenth-Century Patterns in Western Europe and the United States,* ed. Ira Katznelson and Aristide R. Zolberg (Princeton: Princeton University Press, 1986), 197–276; Richard Oestreicher, "Urban Working-Class Political Behavior and Theories of American Electoral Politics, 1870–1940," *Journal of American History* 74, No. 4 (March 1988), 1257–86; Leon Fink, *Workingmen's Democracy: The Knights of Labor and American Politics* (Urbana: University of Illinois Press, 1983); Ira Katznelson, *City Trenches: Urban Politics and the Patterning of Class in the United States* (Chicago: University of Chicago Press, 1981), 5–21.

16. Kendall, *History of New Orleans,* I, 409; Nussbaum, " 'The Ring is Smashed!' " 283–85; *The Morning Star,* April 27, May 4, June 8, 1884. On the lack of business interest in reform and the inability of businessmen to sustain their reform organizations, see *Daily Picayune,* June 6, 1875; *The Morning Star,* April 20, 1884.

17. *Daily Picayune,* April 18, 1884, cited in Kendall, *History of New Orleans,* I, 447–48. Also see "To the People of New Orleans," *Daily Picayune,* April 10, 1884; *Report of the United States Senate Committee to Inquire into Alleged Frauds and Violence in the Elections of 1878, with the Testimony and Documentary Evidence,* Vol. 1. *Louisiana* (Washington: Government Printing Office, 1879), 45th Cong., 3rd Sess., Senate Report No. 855), xxiii–xxiv, 433–47.

18. Samuel P. Hays, "The Politics of Reform in Municipal Government in the Progressive Era," *Pacific Northwest Quarterly* 55, No. 4 (October 1964), 157–69; Schott, "John M. Parker of Louisiana," 13; Katz, *In the Shadow of the Poorhouse,* 152–53. See Chapter 4 for discussion of disfranchisement and the Constitutional Convention of 1898.

19. Jackson, *New Orleans in the Gilded Age,* 46.

20. Howe, "Municipal History of New Orleans," 18–20; Kendall, *History of New Orleans* I, 436–37; Jackson, *New Orleans in the Gilded Age,* 46–47; Schott, "John M. Parker," 13; Kendall, *History of New Orleans,* I, 455; Walker, "Municipal Government," 102–3. Despite the enactment of new charters in 1896 and 1912, a new political machine, the Choctaw Club, dominated municipal government for the first two decades of the twentieth century. Although far more receptive to the interests of business and based far less on fraud and violence than its predecessor, the new machine continued to give the city's elite much about which to complain.

21. *Times-Democrat,* September 9, 1883.

22. Kendall, *History of New Orleans* I, 448–49, 451, 454; *The Mascot,* April 12, 1884.

23. Howe, "Municipal History of New Orleans," 32–33; Kendall, *History of New Orleans* I, 455–56; Jackson, *New Orleans in the Gilded Age,* 93–94.

24. *Daily Picayune,* March 29, 1888, quoted in Kendall, *History of New Orleans* I, 470–71.

25. Schott, "John M. Parker of Louisiana," 16–18; Jackson, *New Orleans in the Gilded Age,* 94–96; Kendall, *History of New Orleans* I, 470–73; *The Mascot,* April 7, 14, 21, 1888; Howe, "Municipal History of New Orleans," 33.

26. Jackson, *New Orleans in the Gilded Age,* 96–110; Kendall, *History of New Orleans* 1, 473–82; W. R. Parkerson, "Memoirs, 1857–1912" (in Manuscripts, Rare Books and University Archives, Howard-Tilton Memorial Library, Tulane University).

27. John W. Blassingame, *Black New Orleans 1860–1880* (Chicago: University of Chicago Press, 1973), 210.

28. *Southwestern Christian Advocate,* March 31, 1881.

29. Warner, *Studies in the South and West,* 13, mentioned in C. Vann Woodward, *The Strange Career of Jim Crow,* 3rd revised ed. (New York: Oxford University Press, 1974), 42.

30. Dale Somers, *The Rise of Sports in New Orleans, 1850–1900* (Baton Rouge: Louisiana State University Press, 1968), 286; Dolores Egger Labbe, *Jim Crow Comes to Church: The Establishment of Segregated Catholic Parishes in South Louisiana* (1971, rpt. New York: Arno Press, 1978), 17–26; *Southwestern Christian Advocate,* July 21, 1887.

31. British Board of Trade, *Cost of Living in American Towns. Report of an Enquiry by the Board of Trade* (1911), 288–89, 295. The British investigators offered an assessment of residential patterns that revealed the widespread persistence of both class segregation and relative ethnic and racial mixture in some wards: "The poor streets, as might be expected, are found in the neighbourhood of the river, the banks of which are lined with wharves and railway tracks. The streets gradually improve as one goes inland from the river, until St. Charles Avenue is reached. Further inland still are districts occupied mainly by the middle classes and some of the upper working classes. There are areas of some magnitude in which no coloured people live, and some districts which are almost exclusively occupied by them. It is only in this inland district that the two races are separated to any considerable extent. All along the belt between Magazine Street and the river bank, white and coloured people live in close proximity. Beyond the old French quarter, going further east, is another large working class district, in which there are both white and coloured people, but rather more separated than is the case in the districts near the river on the west side."

32. Howard N. Rabinowitz, *Race Relations in the Urban South, 1865–1890* (New York: Oxford University Press, 1978), 226; Fink, *Workingmen's Democracy,* 150; John W. Blassingame, "Before the Ghetto: The Making of the Black Community in Savannah, Georgia, 1865–1880," *Journal of Social History* 6, No. 4 (Summer 1973), 462–88. On separate religious orders, see *Southwestern Christian Advocate,* November 26, 1885; July 21, 1887.

33. *City Item,* quoted in the New Orleans *Weekly Pelican,* July 23, 1887; Warner, *Studies in the South and West,* 13; Eagleson, "Some Aspects of the Social Life of the New Orleans Negro," 114.

34. For examples of the numerous events sponsored by black labor organizations, see *Weekly Pelican,* March 5, April 23, June 1, July 30, August 6, 13, October 15, 1887; July 20, 1889; Eagleson, "Some Aspects of the Social Life of the New Orleans Negro," 68–69, 82. No adequate, scholarly study of African-American social, political and economic life in late nineteenth-century New Orleans currently exists. See Edward F. Haas, "Black Louisianians," in *A Guide to the History of Louisiana,* ed. Light Townsend Cummins and Glen Jeansonne (Westport, Conn.: Greenwood Press, 1982), 80.

35. *Weekly Pelican,* February 5, 26, March 12, April 2, May 21, 1887; *Southwestern Christian Advocate,* April 7, 1892.

36. *Weekly Pelican,* October 1, 22, 29, 1887; September 7, 1889.

37. *Weekly Pelican,* July 4, 1887; *Daily Picayune,* November 10, 1892; *New Orleans Republican Courier,* December 2, 1899; *Weekly Louisianian,* January 28, 1882.

38. *Weekly Pelican,* June 4, July 23, 1887; *Southwestern Christian Advocate,* January 18, 1894; "Obituary: James E. Porter. Memorial Address by Thomas P. Woodland, President of the Central Labor Union of the A.F. of L., New Orleans, La., Nov. 26, 1916," *The Longshoreman* 8, No. 3 (January 1917), 3.

39. *Weekly Pelican,* June 4, 1887. In 1900, a census enumerator recorded the occupation of Swann as "day laborer" and that of his wife, Mary, as "laundress." In 1887, the Swanns, who lived at 383 First Street between Howard and South Liberty streets, were neighbors of Alice Bolden and her family (which included young Buddy Bolden, the future jazz musician). See Twelfth Census of the United States, Schedule No. 1— Population, New Orleans, National Archives, Washington, D.C.; Donald M. Marquis, *In Search of Buddy Bolden: First Man of Jazz* (Baton Rouge: Louisiana State University Press, 1978), 24.

40. *New Orleans Republican Courier,* December 2, 1899; *Daily Picayune,* December 24, 26, 1902; *Southwestern Christian Advocate,* January 1, 1903.

41. *Weekly Pelican,* May 14, 1887; *Southwestern Christian Advocate,* January 1, 1903; *Daily Picayune,* December 24, 26, 1902. Other black labor leaders shared similar organizational affiliations. Sumpter J. Watt, a leader of the black freight handlers in 1883, was a prominent member of the Odd Fellows during the 1880s. Initiated into the O.J. Dunn Lodge 1548 in March of 1883, he became permanent secretary by August, holding that post through the decade; he was also elected secretary of the Louisiana District Lodge No. 21 in 1885. See *Weekly Pelican,* May 14, 1887. I.G. Wynn, president of the black Cotton Yardmen in the early twentieth century, was elected delegate to the national BMC convention by the Israel Lodge No. 1971 in 1890; he was Master Mason for Degury Lodge in 1903. See *The Crusader,* July 19, 1890; *Proceedings of the Thirty-Ninth Annual Communication of the MW Eureka Grand Lodge, Free and Accepted Masons, for the State of Louisiana and Jurisdiction, February 11, 1902 to February 13, 1902* (New Orleans: Merchants Printing Co., 1903), 49. Penn and Wynn served on the board of directors for an industrial school. *Southwestern Christian Advocate,* October 19, 1899. Black longshore leader Albert Workman was the Master Mason of the Richmond Lodge, No. 1, in 1903; T.P. Woodland was Master Mason for the Gilbert Lodge, No. 6, while E.S. Swann was Master Mason for the Richmond Lodge in 1893. See: *Proceedings of the Twenty-Ninth Annual Communication of the MW Eureka Grand Lodge, Free and Accepted Masons, for the State of Louisiana and Jurisdiction . . . 1892; Proceedings of the Thirtieth Annual Communication of the MW Eureka Grand Lodge, Free and Accepted Masons for the State of Louisiana and Jurisdiction . . . 1893; Proceedings of the Thirty-Nine Annual Communication. . .1902.* These proceedings are in Masonic Proceedings, Box 9, George Longe Jr. Papers, Amistad Research Center, Tulane University.

42. *Weekly Pelican,* August 20, 1887; *Birmingham Age,* August 29, 1887; *Daily Picayune,* April 12, 13, 1888.

43. *Weekly Pelican,* February 19, August 20, September 3, October 15, 29, 1887; *Southwestern Christian Advocate,* November 5, 1896; *Birmingham Age,* August 29, 1887.

44. New Orleans Republicans attempted, without great success, to broaden their appeal to other social and racial groups. "Nineteen-twentieths of the Republicans of this State are Negroes," the *Pelican* asserted in 1887. "We heartily desire an accession of the whites into our ranks; we believe it to be in the interests of every planter, every mechanic, every laborer to ally himself with the Republican party, for therein lies his safest protection in combatting the free trade heresies of the Democratic party, and in competing with the pauper labor of the Old World." *Weekly Pelican,* August 6, 1887.

45. *Weekly Pelican,* September 10, 1887; May 14, 1887.

46. *Weekly Pelican,* August 13, 27, September 3, 1887.

47. *Weekly Pelican,* August 13, 1887. On earlier support for reformers, see *Southwestern Christian Advocate,* September 20, 1883.

48. *Weekly Mascot,* June 2, 1888. On the 1888 election, also see *Southwestern Christian Advocate,* March 15, April 5, May 3, June 4, 1888.

49. *Weekly Pelican,* August 13, 1887; June 1, 1889; *Southwestern Christian Advocate,* July 10, 1890; March 13, 1890. "The Young Men's Democracy, after securing 12,000 Negro votes last year for their city ticket treated them so shamefully after the election that they failed to command their support. They gave some of their colored allies forty-five days work in cleaning the streets and then discharged every one of them." *Southwestern Christian Advocate,* June 18, 1889.

50. *Weekly Pelican,* August 6, 1887; Stanley P. Hirshson, *Farewell to the Bloody Shirt: Northern Republicans and the Southern Negro, 1877–1893* (Bloomington: Indiana University Press, 1962), 180–82.

51. *Weekly Pelican,* August 20, 1887; *Southwestern Christian Advocate,* March 24, 1887. On separate rail cars, see *Southwestern Christian Advocate,* July 11, 1889.

52. *Southwestern Christian Advocate,* June 6, 1895; *Daily Picayune,* May 20, 1895; *Crusader,* February 28, 1895 (in Crusader File, Folder P37, Xavier University, New Orleans); Labbe, *Jim Crow Comes to Church,* 27–62; Charles Barthelemy Rousseve, *The Negro in Louisiana: Aspects of His History and His Literature* (New Orleans: The Xavier University Press, 1937), 138–39; Roger Bauder, *The Catholic Church in Louisiana* (New Orleans: 1939), 489; Fischer, *Segregation Struggle in Louisiana,* 148–49. In the realm of sports, historian Dale Somers has noted that racial distinctions emerged sharply in the mid-1880s. Some white baseball players threatened to boycott white teams that competed with black players; in the early 1890s, white bicyclists quit their national body, the League of American Wheelmen, which enrolled blacks as members in northern states. By the turn of the century, he found that "interracial sporting events had all but disappeared." Somers, *The Rise of Sports in New Orleans,* 286; *Southwestern Christian Advocate,* March 8, 1894. In economic life, some unskilled black workers confronted growing hostility from white workers. The New York Trust Company's Planters' Crescent Oil Mills and the Union Oil Company in Gretna each employed at least 200 non-union workers; in 1887 one worker reported that "there is no discrimination on account of color in the employment of men" and that both mills relied on black foremen. Yet two years later, white oil mill "hoodlums" attempted to pressure the oil mill proprietors into dismissing their black workforces. Threatening to close the mills early for the summer, the firms responded by temporarily stopping their operations. Black workers retained their employment when the mills reopened days later. *Weekly Pelican,* February 5, 1887; June 8, 1889; *Southwestern Christian Advocate,* June 13, 1889; *Times-Democrat,* June 3–5, 7, 1889.

53. *Southwestern Christian Advocate,* August 2, 1894; Henry C. Dethloff and Robert R. Jones, "Race Relations in Louisiana, 1877–98," *Louisiana History* IX, No. 4 (Fall 1968), 315–16; Germaine A. Reed, "Race Legislation in Louisiana, 1864–1920," *Louisiana History* VI, No. 4 (Fall 1965), 380–81; Fischer, *The Segregation Struggle in Louisiana,* 151–52. Also see: *Southwestern Christian Advocate,* September 30, 1886; November 14, 1889; January 9, March 20, June 19, September 11, 18, 1890; December 17, 1891, January 28, 1892, March 1, 1894; July 5, 26, 1900; *The Crisis* 4, No. 5 (September 1912), 219; *The Crisis* 4, No. 3 (July 1912), 124; "Education in New Orleans," *Baltimore Sun,* September 8, 1900.

54. Hair, *Bourbonism and Agrarian Protest,* 113–17, 185; Hirshson, *Farewell to the Bloody Shirt,* 181; Perry H. Howard, *Political Tendencies in Louisiana* (Baton Rouge: Louisiana State University Press, 1971), 154–56; *Weekly Pelican,* February 2, 1889. On election fraud and political violence against blacks, see *Southwestern Christian Advo-*

cate, April 19, October 18, 25, November 1, 22, 1888; May 16, August 8, September 19, 1889; *Savannah Tribune,* February 2, 1889.

55. Wallis quoted in Jeffrey Gould, "The Strike of 1887: Louisiana Sugar War," *Southern Exposure* XII, No. 6 (November–December 1984), 53. Also see *Southwestern Christian Advocate,* November 10, 17, December 8, 1887; April 19, May 10, 1888; March 28, 1889; Hair, *Bourbonism and Agrarian Protest,* 177–84; Thomas Becnel, *Labor, Church, and the Sugar Establishment in Louisiana, 1887–1976* (Baton Rouge: Louisiana State University Press, 1980), 5–8; Hall, "Labor Struggles in the Deep South," Section II, Pt. VI, 30–41. Gould's "The Strike of 1887" is the best account of the activities of rural Knights of Labor in that period.

56. *Weekly Pelican,* November 19, 1887.

57. *Daily Picayune,* December 3, 1887, also cited in Philip S. Foner and Ronald L. Lewis, eds., *The Black Worker: A Documentary History from Colonial Times to the Present,* Vol. III: *The Black Worker During the Era of the Knights of Labor* (Philadelphia: Temple University Press, 1978), 208–9.

58. *National Republican,* reprinted in *Weekly Pelican,* November 19, 1887.

59. *The Mascot,* November 5, 1887, reprinted in *Weekly Pelican,* November 12, 1887.

60. C. Vann Woodward, *Origins of the New South 1877–1913* (1951; rpt. Baton Rouge: Louisiana State University Press, 1971), 229.

61. Sterling D. Spero and Abram L. Harris, *The Black Worker: The Negro and the Labor Movement* (1931; rpt. New York: Atheneum, 1969), 44. Also see Woodward, *Origins of the New South,* 321.

62. For discussions of the Knights of Labor and southern workers, see Fink, *Workingmen's Democracy;* Spero and Harris, *The Black Worker;* Sidney H. Kessler, "The Organization of Negroes in the Knights of Labor," *Journal of Negro History* 37 (1952), 248–76; Sidney H. Kessler, "The Negro in the Knights of Labor" (M.A. thesis, Columbia University, 1950); Melton A. McLaurin, *The Knights of Labor in the South* (Westport, Conn.: Greenwood Press, 1978); Dominic J. Parisi, "The Knights of Labor in the Black Press" (unpublished essay, Afro-American Studies Program, Yale University, 1979).

63. Spero and Harris, *The Black Worker,* 43; *Weekly Pelican,* November 19, 1887. No single pattern can account for the regional variations in the configuration of labor alliances. For example, in Richmond's highly segmented labor market, blacks were largely excluded from artisanal crafts, and many fields of employment were segregated by race. Black and white workers did not compete directly for the same jobs, and autonomous union locals entered into complex and short-lived political alliances. In contrast, blacks and whites directly competed for work on the docks of most southern cities. In at least Galveston, Savannah, and New Orleans, workers established federations of waterfront unions which functioned, to some degree, to regulate that job competition and to provide institutional frameworks for the resolution of racial issues. See: Fink, *Workingmen's Democracy,* 149–77; Peter J. Rachleff, *Black Labor in Richmond, Virginia, 1865–1890* (1984; rpt. Urbana: University of Illinois Press, 1989); Spero and Harris, *The Black Worker,* 43–44; Woodward, *Origins of the New South,* 229.

64. Quote from *Times-Democrat,* November 26, 1887. Following a tour of the South, Knights of Labor Master Workman Terrence Powderly noted that in May 1884, New Orleans had one Knights assembly. The editor of *John Swinton's Paper* found four assemblies there in May 1885. By the fall of 1885, New Orleans Knights formed a

district assembly. *John Swinton's Paper*, February 1, May 17, November 8, 1885. For a discussion of the Knights in New Orleans, see David Paul Bennetts, "Black and White Workers: New Orleans 1880–1900" (unpublished Ph.D. dissertation, University of Illinois at Urbana-Champaign, 1972), 335–56; *The Mascot*, August 4, 1883; *Daily Picayune*, July 25, 1882; July 27, 1883; December 31, 1886; January 15, September 19, 1887; July 8, November 11, 1894.

65. *Weekly Pelican*, June 4, 1887; also see June 8, 1889. German economist August Sartorius Freiherr von Waltershausen offered a similar assessment: "In the large cities, especially those exporting cotton, where a lively commerce exists and the mutual aversion between the races is often reduced by business interests, the greatest rise in the standard of living of the Negro is to be found." A. Sartorius Freiherr von Waltershausen, *Die nordamerikanischen Gewerkschaften*, 93.

66. George E. McNeill, ed., *The Labor Movement: The Problem of Today* (Boston: A.M. Bridgman and Co., 1887), 168; A. Sartorius Freih. von Waltershausen, *Die nordamerikanischen Gewerkschaften*, 47; *Southwestern Christian Advocate*, December 7, 1882. On labor day parades, see *Times-Democrat*, November 26, 1883; November 26, 1884; November 26, 1885; November 26, 1886; November 26, 1887; November 26, 1888. Obviously not everyone saw the development of black unions and interracial collaboration as a positive thing. The *Figaro*, a journal of political satire, offered this racist interpretation: "The most instructive and exhilarating feature of the Trades-Unions of this city, whether manifested in strikes or only in street parades, is the colored man and brother . . . When it comes to strikes . . . Sambo is the most excited, the most carnivorous of them all. He seems to acquire fervor by the sense of multitude. He feels that great forces are at his back and he is violent to a surprising degree. Verily, the Trades-Unions have done wonders for Sambo! They have changed him from a peaceful, bucolic hind to a spluttering fire-brand." *Figaro*, December 14, 1883.

67. *Daily Picayune*, July 28, November 26, 1882; November 26, 1883; November 26, 1884; December 3, 1883; November 26, 1886; *The Morning Star*, December 9, 1883; *Times-Democrat*, November 26, 1885; November 25, 26, 1886 November 26, 1887; November 26, 1888.

68. *Daily Picayune*, July 28, October 3, 17, 31, November 4, 1882; Bennetts, "Black and White Workers," 314–15. It "is a fact of peculiar interest and importance to those who are studying the race and labor questions to know that in New Orleans a very large proportion of skilled labor in the house-building trades, such as carpentry, bricklaying, plastering, and painting, is performed by colored mechanics. Many of the most important public and private buildings in this city have been chiefly if not wholly erected by the labor of colored men." *Daily Picayune*, December 4, 1886.

69. *Daily Picayune*, March 17, April 5–10, 12, 13, 15–20, 1884; *Southwestern Christian Advocate*, September 8, 1904; *British Board of Trade, Cost of Living in American Towns*, 293; Bennetts, "Black and White Workers," 357–62.

70. *Conference Rules of the Stevedores and Longshoremen's Benevolent Association and Longshoremen's Protective Union Benevolent Association* (New Orleans: Paragon, 1892), in Department of Archives and Manuscripts, Screwmen's Benevolent Association Records, Folder 31, Louisiana State University, Baton Rouge, Louisiana. In the Vertical File (Benevolent Societies), Louisiana Collection, Howard-Tilton Memorial Library, Tulane University, see *Constitution and By-Laws of the Cotton Yard Men's Benevolent Association* (New Orleans: Henry Powers, Printer, 1896), and *Constitution and By-Laws of the Screwmen's Benevolent Association* (New Orleans: A.W. Hyatt, Printer, 1884).

71. *Daily Picayune,* October 30, 1879; October 18, 1882; February 23, 1886; *Weekly Pelican,* June 4, 1887; *Times-Democrat,* February 23, 1886.

72. *Daily Picayune,* February 22, July 30, 1886; *Times-Democrat,* February 23, 27, 1886. Also see *Daily Picayune,* May 6, 1887; June 30, 1891; *The Mascot,* July 4, 1891.

73. *Weekly Pelican,* June 4, 1887; *Daily Picayune,* February 23, 27, July 30, August 6, 1886; *Times-Democrat,* February 23, 27, 1886. Tensions persisted between the branches, however, and during the Cotton Councils crisis in early 1887 some members of these unions again temporarily withdrew from the LPUBA. The secession was short-lived.

74. *Times-Democract,* September 8, 1887; *States,* September 7, 11, 1887.

75. *Times-Democrat,* September 8, 1887.

76. *Daily Picayune,* September 6, 1887; *Times-Democrat,* September 8, 1887.

77. *Daily Picayune,* September 6–12, 1887; *Times-Democrat,* September 7–10, 1887; *States,* September 8, 9, 11, 1887.

78. *Daily Picayune,* September 14, 1887; *States,* September 14, 1887. The text of these rules also can be found in *Conference Rules of the Stevedores and Longshoremen's Benevolent Association and Longshoremen's Protective Union Benevolent Association.*

79. *Times-Democrat,* November 26, 1883; *Daily Picayune,* November 27, 1883.

80. *Times-Democrat,* November 29, 1883; *The Mascot,* February 18, 1882; December 8, 1888; *States,* November 23, 1883. The struggle between the freight handlers and railroads, the Catholic *Morning Star* argued, "rises above mere considerations, for it involves a question of *human right . . .* Capital, thus concentrating, by all possible methods, is daily strengthening its hands, and so becoming daily more aggressive and more unscrupulous . . ." *The Morning Star,* December 2, 1883; also see December 9, 1883. On the business history of Morgan's Louisiana and Texas Railroad and Steamship Company, see James P. Baughman, *Charles Morgan and the Development of Southern Transportation* (Nashville: Vanderbilt University Press, 1968), 208–35.

81. *Daily Picayune,* November 19, 25, 1883; *Times-Democrat,* November 25, 26, 1883. On strikes of freight handlers in New York and New Jersey in the same era, see Shelton Stromquist, *A Generation of Boomers: The Pattern of Railroad Labor Conflict in Nineteenth-Century America* (Urbana: University of Illinois Press, 1987), 26–28.

82. *Times-Democrat,* November 25, 1883.

83. *Daily Picayune,* November 25–27, 1883; *Times-Democrat,* November 25–27, 1883; *L'Abeille de la Nouvelle-Orleans,* November 28, 1883.

84. *Times-Democrat,* November 28, 1883; *Daily Picayune,* November 28, 30, 1883; *L'Abeille de la Nouvelle-Orleans,* November 28–30, 1883.

85. *Times-Democrat,* November 28, 29, December 1, 1883; *Daily Picayune,* November 28, 29, 1883.

86. *Times-Democrat,* November 20, December 1, 2, 1883; *Daily Picayune,* November 29, 30, 1883; *States,* December 1, 1883.

87. *Times-Democrat,* December 3, 4, 5, 1883; *Daily Picayune,* December 4, 1883; *L'Abeille de la Nouvelle-Orleans,* December 4, 1883; *The Mascot,* December 8, 1883.

88. The *Mascot* alone recognized that a stronger expression of solidarity might alter the strike's outcome. "If the Trades Assembly come promptly to the aid of the freight handlers, the bosses will have to relinquish the unfair and unreasonable war they are waging against legitimate labor . . . The staple by which this community exists will rot at the depots, and the small upstarts who are imitating our Goulds and Vanderbilts, will have to return to work for a living. These fellows seem to forget that the men

whom they are trying to subdue made them, and that they can as readily unmake them." *The Mascot,* December 1, 1883.

89. Just over two years later in February 1886, a strike of freight handlers against the Southern Pacific Railroad was ended by arbitration, producing slightly better results: wages rose from 25 to 30 cents an hour. See *Times-Democrat,* February 23, 1886; *Daily Picayune,* February 22, 23, 1886; *John Swinton's Paper,* April 12, 1885; February 21, 1886.

Coal wheelers were another group of workers along the Mississippi River that fell outside of the Cotton Men's Executive Council's protective umbrella. Their membership in the Central Trades and Labor Assembly proved inadequate to improve their own bargaining position vis-à-vis their employers. Taking advantage of the presence of more than 27 steamers in port, many of which required immediate coaling, and frustrated by fruitless talks with employers over the past months, as many as 1200 black coal wheelers and white boat pumpers struck in January 1884 to increase daily wages from $3 to $4. The Central Trades and Labor Assembly refused to sanction the strike, however, because the wheelers' union had violated a promise to the Assembly to seek arbitration of its differences. Without the Assembly's backing, strikers watched helplessly as employers hired several hundred whites at $2.50 a day under police protection to replace black union members. When the wheelers agreed to submit to its authority, the Assembly appointed a mediation committee composed of key labor leaders to investigate the wheelers' claims and demands. Unpersuaded by the wheelers' strategy, the committee rejected a recommendation to involve the Assembly in the strike, relegating the entire matter back to the strikers "to settle it in their own way." Without allies in the labor movement, the wheelers resorted to violence; in late January and early February groups of black strikers threw stones and fired shots at various groups of white and black strikebreakers until the police had arrested enough union leaders to break the strike. The Trades Assembly had not entirely abandoned the wheelers. While it ordered no general sympathetic strike, its arbitration committee did press coal operators to reinstate the strikers, acquiesce to union demands, and hire only union men. Although these efforts failed, the Assembly refused to admit to membership a newly organized coal wheelers union composed of the white strikebreakers. *Times-Democrat,* January 15–17, 19, 26, 29, February 4, 5, 8, 13, 1884; *Daily Picayune,* January 15, 17, 22, 23, February 4, 8, March 8, 1884; Kessler, "The Organization of Negroes in the Knights of Labor," 251; Bennetts, "Black and White Labor," 323–26. Also see *Daily Picayune,* April 14, 1886.

90. Valcour Chapman, "Roustabouts Are Treated Worse Than Brutes," *Southwestern Christian Advocate,* November 29, 1900; cited in Hampton University Newspaper Clipping File (Item 295, Labor: Overview: Negro Labor by State, Frame 43).

91. "The Roustabouts of the Mississippi," *New Orleans Republican,* August 2, 1874.

92. Julian Ralph, "The Old Way to Dixie," *Harper's New Monthly Magazine* LXXXVI, No. DXII (January 1893), 175; also see Ernest Peixotto, "The Charm of New Orleans," *Scribner's Magazine* LIX, No. 4 (April 1916), 461.

93. Ralph, "The Old Way to Dixie," 175; *Daily Picayune,* November 9, 1877, November 2, 1882; "The Roustabouts of the Mississippi," *New Orleans Republican,* August 2, 1874; *New Orleans Times,* January 16, 1875. "The roustabout is a very peculiar individual. He does not care about saving money or having a home, and does not care much about a family; he has no moral obligations, hardly, of any kind. All he cares about is playing craps. They are very improvident: they like to change about and go on one boat one time and then another the next time." Testimony of John W.

Bryant, Secretary of Steamboat Captains and Owners' Exchange, New Orleans, February 6, 1901, in *Report of the Industrial Commission on Transportation,* IX (Washington: Government Printing Office, 1901), 396.

94. *Times-Democrat,* December 21, 1883; *Daily Picayune,* November 4, 11, 1901; *Times-Democrat,* March 18, 1908; William Ivy Hair, *Carnival of Fury: Robert Charles and the New Orleans Race Riot of 1900* (Baton Rouge: Louisiana State University Press, 1976), 76–77.

95. *The Mascot,* October 19, 1889; *Daily Picayune,* October 27, 29, 1892; August 23, 1892.

96. *Southwestern Christian Advocate,* November 29, 1900; *The Mascot,* October 19, 1889; Charles B. Spahr, "America's Working People IV. The Negro as an Industrial Factor," *The Outlook* 62, No. 1 (6 May 1899), 35. Also reprinted in Spahr, *America's Working People* (New York: Longmans, Green, and Co., 1900), 83. "The negro roustabout's importance is based on two apparently insurmountable facts: First, the boat cannot be operated without him; and second, his place cannot be filled by white men." Undated, untitled newspaper clipping, in Hampton University Newspaper Clipping File (Item 295, Labor: Overview: Negro Labor by State, Frame 38). "There are very few white men employed as roustabouts, because they can not stand the work . . . it is just about the hardest work . . . and about the only laborer we find that can stand it is the negro," testified J.W. Bryant in 1901. *Report of the Industrial Commission on Transportation,* 395.

97. *Daily Picayune,* September 9, 1880.

98. *Daily Picayune,* October 24, 1901; November 2, 1900; September 1, 1897; *Times-Democrat,* November 3, 1882; *Birmingham News,* October 12, 1904; *Southwestern Christian Advocate,* November 8, 1900; "The Roustabout Question," February 7, 1901, no paper indicated, in Hampton University Newspaper Clipping File (Item 295, Labor: Overview: Negro Labor by State, Frame 44); "Negro Roustabouts on the Mississippi," *Brooklyn Citizen,* June 7, 1903, in Hampton University Newspaper Clipping File (Item 295, Frame 47).

99. *The Mascot,* October 19, 1889; *Daily Picayune,* October 23, 24, 30, November 4–6, 8, 9, 11, 1901; October 12, 13, 18, 1904; *Birmingham News,* October 12, 1904; *Times-Democrat,* March 18, 1908; "To Use White Roustabouts," *Transcript,* October 14, 1904, in Hampton Newspaper Clipping File (Item 297, Frame 26); "Negro Roustabout Displaced," *Brooklyn Citizen,* November 9, 1903, in Hampton Newspaper Clipping File (Item 295, Frame 48); untitled article, *Macon Telegraph,* February 22, 1901, in Hampton Institute Clipping File (Item 295, Frame 45); *Times-Picayune,* August 30, 1919; Raymond A. Pearce, "The Rise and Decline of Labor in New Orleans" (unpublished M.A. thesis, Tulane University, 1938), 46–47.

100. *Chicago Inter-Ocean,* May 5, 1887, also quoted in *Weekly Pelican,* May 14, 1887. On New York waterfront strike of 1887, see Barnes, *The Longshoremen,* 102–8; *John Swinton's Paper,* January 23, 30, February 6, 13, 20, 1887; *Fifth Annual Report of the Bureau of Statistics of Labor of the State of New York for the Year 1887* (Troy: Troy Press Company, 1888), 327–85; on the 1886 San Francisco strike, see Bruce Nelson, *Workers on the Waterfront: Seamen, Longshoremen, and Unionism in the 1930s* (Urbana: University of Illinois Press, 1988), 40–41.

101. A. Sartorius Freiherr von Waltershausen, *Die nordamerikanischen Gewerkschaften,* 146.

102. *Daily Picayune,* September 11, 1886.

103. *Daily Picayune,* May 28, July 22, August 30, 1886.

104. *Daily Picayune,* August 30, 1886.

105. *Daily Picayune,* September 2–4, 6, 7, 11, 1886; *Times-Democrat,* January 15, 1887.

106. *The Mascot,* September 4, 1886.

107. *Daily Picayune,* November 30, 1886.

108. *The Mascot,* November 20, 1886; *Times-Democrat,* November 16–25, 27, December 4, 8, 1886; *Weekly Pelican,* December·4, 1886; *Daily Picayune,* November 17, 26, 30, 1886. On earlier complaints of high port charges, see *The Mascot,* September 2, 1882; January 2, November 20, 1886; September 10, 1887; October 27, 1888; May 25, 1889; *New Orleans Democrat,* October 6, 8, 12, 15, 1881; *Daily Picayune,* June 9, 1875; December 17, 18, 1879; August 28, 1880; *Times-Democrat,* January 10, 1884; *Weekly Louisianian,* December 4, 1886.

109. *Times-Democrat,* November 22, 1886.

110. *Daily Picayune,* November 22, 1886; *States,* November 22, 1886.

111. *Times-Democrat,* November 23, 1886.

112. *Times-Democrat,* November 18–20, 22, December 8, 1886; April 2, 1887; *States,* November 17, 18, 1886. Specifically, several articles in the new Council's constitution provided the basis for disunity by prohibiting certain practices of important groups of workers. One rule, directed at longshoremen, stated that "no member of any association represented in this council shall be supported in a demand for remuneration for any labor not actually performed. The tariff, if found extortionate or objectionable, shall be rejected by a majority vote of the council."

113. Leading the new Council were president Robert Brough (white screwmen) and vice president J.A. Allen (black screwmen). The old Council was headed by white longshoreman Hugh Reiley as president, and James Porter of the black longshoremen served as secretary and later treasurer. See *Daily Picayune,* December 6, 11, 23, 1886; January 12, 14–16, February 25, March 25, 1887; *Times-Democrat,* December 6, 23, 1886; January 15, 20, 28, 1887; *Weekly Pelican,* December 11, 25, 1886; January 15, 29, February 19, 1887.

114. *Weekly Pelican,* March 12, April 2, 1887; *Times-Democrat,* March 25, 30, April 1–3, 1887; *Daily Picayune,* March 11, 25–27, 29, 30, April 1–3, 1887; *States,* March 24, 25, 1887.

115. *Times-Democrat,* March 30, April 1, 2, 1887; *Daily Picayune,* March 27, 29, 30, April 1, 2, 1887; *States,* March 27, 28, April 1, 1887.

116. *The Mascot,* April 2, 1887; *Times-Democrat,* April 2, 1887.

117. *Times-Democrat,* April 2–4, 6–9, 1887; *Daily Picayune,* April 2, 3, 6–9, 1887; *States,* March 28–31, April 4, 7, 11, 1887; *Weekly Pelican,* April 9, 1887.

118. *States,* March 24–26, 1887; *Times-Democrat,* November 15, 1886.

119. *The Mascot,* April 2, 1887. The editors of the *Pelican,* a black paper, also worried about the racial dimension of the conflict: "As long as the black and white laborer worked alongside of each other contentedly, as long as they were jealous of each other's interests, as long, in fact, as they were a united band, working for a common cause and with the same object in view—self-betterment—they were a power for good. But when disintegration set in, when petty jealousies were allowed to overweigh common sense and reason, when the question of color was considered in connection with labor, therein was the danger, therein lies the germ of future trouble." *Weekly Pelican,* April 2, 1887.

120. *Weekly Pelican,* April 9, 1887; *Times-Democrat,* April 6, 7, 1887; *Daily Picayune,* April 6, 7, 1887.

121. *Weekly Pelican,* April 2, 16, 1887; *Times-Democrat,* April 9–11, 1887; *Daily Picayune,* April 9–11, 13, 14, 1887. Four proposals received widespread attention. The

first suggested that the old Council join the new one in exchange for equal representation on its executive committee. The second advised that both councils dissolve and that a new body be organized on "a basis different from either of the present councils." A third requested the new Council dissolve and the old Council be reorganized to embrace most of the contending elements. The last, and most strenuously opposed by the white members of the new Council, was that half of the entire work of handling cotton in the city, including screwing, classing, sampling, rolling, and longshore work be given to the black workers of the old Council. None of these proposals was acceptable to the screwmen. *Daily Picayune,* April 9, 11, 1887; *Times-Democrat,* April 9–11, 1887; *States,* April 9, 11–13, 1887.

122. *Weekly Pelican,* May 14, 1887, August 27, 1887. This did not preclude occasional dealings with other riverfront unions. During the September 1887 longshoremen's strike, discussed above, white and black screwmen rolled their own bales of cotton—an action which would have been unthinkable during the Council's height—but refused to handle any other goods. While opposed to the longshoremen's walkout, they held joint meetings with the strikers to encourage negotiations and a settlement.

123. In a strike reminiscent of the September 1886 cotton press conflict, black and white screwmen, and black teamsters and loaders struck in Sept. 1891 on behalf of the boss draymen, who demanded an increase in their tariff from the cotton presses. Although the rump Cotton Men's Executive Council of screwmen and teamsters eventually ruled that boss draymen were not true workers, for several days they halted the flow of cotton on and off the docks. Longshoremen and yardmen, outside of the Council, took no interest in the conflict. See *Times-Democrat,* September 4–12, 1891; *Daily Picayune,* September 3–8, 1891.

Craft divisions also remained pronounced over the next five years. In late February and early March 1892, the black and white longshoremen's unions struck for a reform of the quarter system, whereby workers received payment for every fifteen minutes over the standard day at the rate of 10 cents per quarter, or 25-cents per half-hour. Since standard wages were 50 cents an hour, longshoremen objected to being sometimes shortchanged by the system. Moreover, they claimed that foremen—even union foremen in the stevedores' employ—sometimes miscalculated the time the men worked in favor of the employer. ("Many of the longshoremen are too ignorant to correctly estimate the time, and twenty or twenty-five minutes are not infrequently booked against them as quarters, for which they get but ten cents," argued James Porter, secretary of the unions' joint executive committee.) The longshoremen had few allies outside their own ranks. Opposing them were contracting stevedores and cotton screwmen, both members of the Cotton Men's Executive Council; longshoremen were not members. Thus, screwmen supported the stevedores in their refusal to acquiesce to the longshoremen's demand, continuing to stow cotton in ships' holds. Council rules specified that any stevedore who broke ranks and signed with the longshoremen would be subject to expulsion from the organization. In the end, longshoremen and stevedores agreed to a compromise that seemed to favor the unions. Employers abolished the quarter system, agreeing to pay longshoremen a full half-hour rate for any fraction of time worked up to 30 minutes. See *Daily Picayune,* February 24–26, March 4–6 8, 9, 1892; *Louisiana Review,* March 9, 1892; *Birmingham Age-Herald,* February 27, 1892.

124. *United Mine Workers Journal,* November 10, 1892; Testimony of Samuel Gompers, April 18, 1899, *Report of the Industrial Commission on the Relations and Conditions of Labor and Capital Employed in Manufactures and General Business,* VII (Washington: Government Printing Office, 1901), 647; also in Samuel Gompers, "The

'Color Line' in Labor," *Labor and the Employers,* compiled and edited by Hayes Robbins (New York: E. P. Dutton and Co., 1920), 166–67.

125. *Savannah Tribune,* November 19, 1892. On the May streetcar strike, see *Daily Picayune,* May 20–27, June 1, 1892; *Louisiana Review,* May 25, June 1, 8, 1892. On the Workingmen's Amalgamated Council and the summer and fall strike wave, see *Louisiana Review,* April 20, May 11, 25, June 1, July 6, 20, September 7, 1892; *Daily Picayune,* July 18, 24, 26–29, August 22, 23, 25, 30, September 2, 12, 17, 26, October 4, 5, 11, 1892; Birmingham *Labor Advocate,* June 4, October 1, 1892; *United Mine Workers Journal,* August 25, 1892.

126. For accounts of the 1892 general strike, see Roger Wallace Shugg, "The New Orleans General Strike of 1892," *Louisiana Historical Quarterly* 21, No. 2 (April 1938), 547–60; Covington Hall, "Labor Struggles in the Deep South," Part I, Section II, 10–25, (unpublished ms., n.d., Tulane University); Jeremy Brecher, *Strike!* (San Francisco: Straight Arrow Books, 1972), 64–66; Philip S. Foner, *History of the Labor Movement in the United States,* Vol. 2: *From the Founding of the A.F. of L. to the Emergence of American Imperialism* (New York: International Publishers, 1955), 200–204; Bernard A. Cook, "The Typographical Union and the New Orleans General Strike of 1892," *Louisiana History* XXIV, No. 4 (Fall 1983), 377–88; Jackson, *New Orleans in the Gilded Age,* 228–29; Dave Wells and Jim Stodder, "A Short History of New Orleans Dockworkers," *Radical America* 10, No. 1 (January–February 1976), 47–48; Bernard A. Cook and James R. Watson, *Louisiana Labor: From Slavery to "Right-to-Work"* (Lanham, Md.: University Press of America, 1985), 71–76; Bennetts, "Black and White Labor," 394–438; Christopher Tomlins, *The State and the Unions: Labor Relations, Law, and the Organized Labor Movement in America, 1880–1960* (New York: Cambridge University Press, 1985), 61; Daniel Rosenberg, *New Orleans Dockworkers: Race, Labor and Unionism* (Albany: State University of New York Press, 1988), 33–36. For a day-to-day account, see *Daily Picayune,* October 20, 22, 23, 25–31, November 1–13, 16, 1892; January 1, 1893; *Times-Democrat,* October 23–29, 31, November 2–12, 14, 1892; *Louisiana Review,* November 2, 9, 16, 1892; *Birmingham Age-Herald,* October 26, 1892.

127. *Daily Picayune,* October 25, 1892; *Louisiana Review,* November 2, 1892; *Savannah Tribune,* November 16, 1892; *United Mine Workers Journal,* November 10, 1892.

128. Breen, Porter, and Kier had worked together earlier in the year in a somewhat similar situation. During the May 1892 streetcar strike, they were members of another committee of five representing the carmen's union before the committee of the presidents of the street railway companies. They also served as labor's official delegates on the arbitration committee that resolved the strike. *Daily Picayune,* May 26, 27, 1892.

129. *Daily Picayune,* November 5, 1892; *Louisiana Review,* November 9, 1892; Shugg, "The General Strike of 1892"; W.R. Parkerson Memoirs, 1857–1912 (Special Collections, Tulane University), 4. While the Triple Alliance received the backing of many of the city's labor organizations, a significant number did not participate in the general strike. Master Workman Walter S. Crawford of the Knights of Labor informed all the assemblies in District Assembly 102 that "owing to the stringency" of the order's by-laws, the Knights could not participate as an organization. "This order," he informed his members, "is opposed to strikes." Many of its members, on the other hand, were not, and groups of Knights did join the strike. Although the Typographical Union voted narrowly to support the Triple Alliance, a significant faction of the city's second oldest labor association adhered to a conservative, " 'pragmatic' vision of the labor movement" and strongly opposed sympathetic action. *Daily Picayune,* Novem-

ber 10, 1892; *Times-Democrat,* November 10, 1892; *Galveston Daily News,* November 5, 1892; Birmingham *Labor Advocate,* November 12, 1892; Cook, "The Typographical Union and the New Orleans General Strike," 388.

130. *Times-Democrat,* November 4, 10, 1892; October 14, 1894; *Daily Picayune,* November 10, 1892; Mobile *Daily Register,* November 9, 11, 1892; Birmingham *Labor Advocate,* November 12, 1892.

131. *Times-Democrat,* November 7, 10, 1892; Mobile *Daily Register,* November 9, 1892. In contrast, waterfront workers not in the Cotton Men's Executive Council joined the strikers. Both the city's freight handlers and roustabout deck hands walked off the job in support of the Triple Alliance. See *Daily Picayune,* October 29, November 8, 9, 1892. Hestor quote from *United Mine Workers Journal,* November 10, 1892; also cited in Shugg, "The New Orleans General Strike," 555.

132. Sidney James Romero, Jr., "The Political Career of Murphy James Foster, Governor of Louisiana, 1892–1900,"*Louisiana Historical Quarterly* 28, No. 4 (October 1945), 1165–66; *Times-Democrat,* November 11, 12, 1892; *Daily Picayune,* November 11, 1892; Mobile *Daily Register,* November 12, 1892; *Galveston Daily News,* November 11, 1892; *Savannah Tribune,* November 16, 1892.

133. *Times-Democrat,* November 12, 1892; *Daily Picayune,* November 13, 1892; *Galveston Daily News,* November 11, 1892; *Savannah Tribune,* November 16, 26, 1892.

134. *Times-Democrat,* November 14, 1892. For other criticism of Fitzpatrick, see *Times-Democrat,* November 5, 1892; *Galveston Daily News,* November 11, 1892; *Louisiana Review,* November 9, 1892; Mobile *Daily Register,* November 12, 1892.

135. *Savannah Tribune,* November 19, 1892; Mobile *Daily Register,* November 12, 1892; *Louisiana Review,* December 14, 1892.

Chapter 4

1. *Weekly Pelican,* June 8, 1889; *Times-Democrat,* June 3, 1889. On Gould's activities in Galveston as president of the Independent Colored Organization in the twentieth century, see speech by Gould, *Proceedings of the Seventeenth Convention of the International Longshoremen's Association, Galveston, Texas, July 12th to 17th inclusive, 1909* (Detroit: A.W. Brooks, 1909), 105.

2. C. Vann Woodward, *Origins of the New South 1877–1913* (1951; rpt. Baton Rouge: Louisiana State University Press, 1971), 321–49.

3. *The Freeman* (Indianapolis), March 23, 1895.

4. "Longshoremen Draw No Line," *New York Age,* August 14, 1913.

5. Young Men's Business League, *New Orleans of 1894: Its Advantages, Its Prospects, Its Conditions, as shown by a Résumé of a Year's Record, 1893–1894* (New Orleans: L.Graham and Son, 1894), 4, 21–23. "Almost every department of our vast commercial and industrial interests have felt the effects of the unusual commotion in our great financial system," complained one black paper. *Southwestern Christian Advocate,* August 17, 1893. On the depression, see: Nell Irvin Painter, *Standing at Armageddon: The United States, 1877–1919* (New York: W.W. Norton & Co., 1987), 116–26; Alexander Keyssar, *Out of Work: The First Century of Unemployment in Massachusetts* (New York: Cambridge University Press, 1986), 47; Michael Katz, *In the Shadow of the Poorhouse: A Social History of Welfare in America* (New York: Basic Books, 1986), 147; Carlos A. Schwantes, *Coxey's Army: An American Odyssey* (Lincoln: University of Nebraska Press, 1985), 13–17; David Montgomery, *The Fall of the House of Labor:*

The Workplace, the State, and American Labor Activism (New York: Cambridge University Press, 1987), 171–74.

6. *Times-Democrat,* November 20, 1886; November 2, 1894; Mobile *Daily Register,* November 4, 1894; *Daily Picayune,* July 22, August 6, 1894.

7. *Daily Picayune,* October 28, 1894.

8. *Daily Picayune,* July 22, August 6, October 28, November 11, 1894; *Times-Democrat,* December 4, 1886; November 2, 20, 1894; John Lovell, "Sail, Steam and Emergent Dockers' Unionism in Britain, 1850–1914," *International Review of Social History* XXXII (1987), 237–42. See Chapter 5 for a detailed discussion of the technological changes in the maritime industry.

9. *Southwestern Christian Advocate,* August 4, 1892; July 12, 1894. On the *Advocate*'s stance toward unions and strikes, also see *Southwestern Christian Advocate,* September 15, 1881.

10. *Southwestern Christian Advocate,* May 20, 1886. The paper continued: "If he united with the strikers they would drop him as soon as they carried their point, and should he unite with the capitalist he would go down when the troubles were over. Choosing either horn of the dilemma the benefits would be given him only by necessity, and denied him so soon as it became convenient to do so. Therefore . . . the Negro remains passive spectators of this far-reaching revolution."

11. *Southwestern Christian Advocate,* July 12, 1894.

12. Booker T. Washington, *Up from Slavery: An Autobiography* (1901; rpt. New York: Bantam Books, 1977), 157, 155; Booker T. Washington, "The Negro and the Labor Unions," *Atlantic Monthly* III (June 1913), 756–67; Louis R. Harlan, *Booker T. Washington: The Making of a Black Leader, 1865–1901* (New York: Oxford University Press, 1972); August Meier, *Negro Thought in America 1880–1915: Racial Ideologies in the Age of Booker T. Washington* (Ann Arbor: University of Michigan Press, 1963), 100–18; John Brown Childs, "Concepts of Culture in Afro-American Political Thought, 1890–1920," *Social Text* 4 (1981), 29.

13. *Daily Picayune,* January 14, 1893; *Birmingham Age-Herald,* January 16, 1893; *Times-Democrat,* October 14, 1894. On membership figures, see *Weekly Pelican,* June 8, 1889. Black resentment over racial restrictions and practices first emerged publicly more than a decade before when black and white unions struck briefly on behalf of white screwmen against employers of black screwmen, demanding that the black wage rate be raised to match that of the whites. During the 1881 strike, two groups of black workers broke ranks with the Cotton Men's Executive Council and independently contracted with employers. The Cotton Men's Executive Council forced the longshoremen's groups to merge with the larger Longshoremen's Protective Union Benevolent Association in 1886. If the black screwmen chafed at the enduring restrictions on the number of members who could work, they nonetheless sided with the more powerful whites against the largely black, old Cotton Men's Executive Council during the Cotton Councils war in 1886–87. Race, craft identification, considerations of power, and the dynamics of group politics produced complicated alliances in this era. See Chapter 3.

14. "Until last year," Stoddart explained in October 1894 "I always employed white labor, but when I was offered considerably reduced rates I thought it only fair to give the other people a trial." *Daily Picayune,* October 10, 14, 1894; *New York Times,* October 28, 1894.

15. *Daily Picayune,* January 14, 1893; July 22, 1894; *Galveston Daily News,* October 27, 1894. The black and white screwmen, the Galveston paper explained, "some years ago were on terms of the utmost harmony, working together and parading together and belonged to the same council. Trouble has been brewing between them

for some time . . . The appearance along the river front of firms of negro stevedores hiring negro screwmen has served to accentuate the trouble, and the white screwmen have been charging that negroes have been cutting rates and that the whites are losing ground [due to the] influx of negro labor."

16. Edna Bonacich, "A Theory of Ethnic Antagonism: The Split Labor Market," *American Sociological Review* 37, No. 5 (October 1972), 547; George Frederickson, *White Supremacy: A Comparative Study in American and South African History* (New York: Oxford University Press, 1981), 212.

17. *Daily Picayune,* July 22, August 6, October 7, 10, 15, 1894; *Galveston Daily News,* October 27, 1894.

18. *Daily Picayune,* October 14, 15, 1894; *Times-Democrat,* October 14, 1894; *Sunday States,* October 14, 1894; *Southwestern Christian Advocate,* November 29, 1894.

19. *Southwestern Christian Advocate,* June 29, 1893.

20. *Times-Democrat,* October 14, 1894; *Daily Picayune,* October 14, 15, 16, November 4, 1894; also see William Ivy Hair, *Carnival of Fury: Robert Charles and the New Orleans Race Riot of 1900* (Baton Rouge: Louisiana State University Press, 1976), 94; *Galveston Daily News,* October 27, 1894; *New York Times,* October 27, 1894.

21. *Daily Picayune,* December 20, 1894; *Sunday States,* October 14, 1894.

22. *Times-Democrat,* October 14, 1894; *Daily Picayune,* October 14, 15, 1894.

23. *Times-Democrat,* October 14, 15, 1894; *Daily Picayune,* October 15, 16, 1894; *Sunday States,* October 14, 1894; John Smith Kendall, *History of New Orleans,* II (New York: The Lewis Publishing Co., 1922), 515.

24. *Daily Picayune,* October 17, 18, 1894.

25. *Times-Democrat,* October 27, 1894; *Daily Picayune,* October 27, 28, 1894; *Birmingham Age–Herald,* October 27, 1894; *Appletons' Annual Cyclopaedia and Register of Important Events of the Year 1894,* New Series, XIX (New York: D. Appleton and Company, 1895), 443; F. Ray Marshal, *Labor in the South* (Cambridge: Harvard University Press, 1967), 63; Carroll George Miller, "A Study of the New Orleans Longshoremen's Unions from 1850 to 1962" (unpublished M.A. thesis, Louisiana State University and Agriculture and Mechanical College, 1962), 17–19.

26. *Times-Democrat,* October 27, 1894; *Daily Picayune,* October 27, 1894; *Daily States,* October 27, 1894; *Galveston Daily News,* October 27, 1894; *Washington Bee,* November 10, 1894; *New York Times,* October 27, 1894; British Foreign Office, *Diplomatic and Consular Reports on Trade and Finance. United States. Report for the Year 1894 of the Trade of the Consular District of New Orleans,* Annual Series No. 1551, May 1895, 15. For general accounts or mention of the 1894–95 riots, see Woodward, *Origins of the New South,* 267; Jackson, *New Orleans in the Gilded Age,* 230–31; Raymond Arthur Pearce, "The Rise and Decline of Labor in New Orleans" (unpublished M.A. thesis, Tulane University, 1938), 31–37; *Appleton's Annual Cyclopaedia . . . 1894,* 443; Kendall, *History of New Orleans,* II, 515.

27. *Times-Democrat,* October 28, 29, 1894; *Daily Picayune,* October 28, 29, 1894; March 20, 1908; *Daily States,* October 28, 29, 1894; *Southwestern Christian Advocate,* November 1, 1894; *Washington Bee,* November 10, 1894.

28. *Daily Picayune,* October 30, 1894; *Times-Democrat,* October 30, 1894.

29. *Times-Democrat,* October 30, 1894.

30. *Times-Democrat,* October 28, 30, 31, November 1, 1894; *Daily Picayune,* October 30, 31, November 1, 4, 1894; Meeting of Board of Directors, November 1, 1894, New Orleans Cotton Exchange Minute Book, H (Manuscripts, Rare Books and University Archives, Howard-Tilton Memorial Library, Tulane University), 81.

31. *Times-Democrat,* October 28, 1894; *Daily Picayune,* October 28, 1894; *Daily States,* October 27, 28, 1894; Meeting of Board of Directors, October 29, 1894, New Orleans Cotton Exchange Minute Book, H, 72.

32. *Daily Picayune,* November 3, 1894. "Neither race conflicts nor any other sort of conflicts must be allowed to destroy a great shipping industry," the *Picayune* editorialized. "The labor of both whites and blacks is needed. Laborers of all classes and both races have their rights, and the protection of these laborers in their rights must be maintained without any failure or default." *Daily Picayune,* October 30, 1894.

33. *Times-Democrat,* November 5, 1894; *Daily Picayune,* November 4–7, 11, 1894; *Daily States,* November 5, 1894; Minutes of Meeting of Board of Directors, November 7, 1894, New Orleans Cotton Exchange Minute Book, H, 83.

34. Following the blaze at the West India and Pacific Steamship wharves, black workers threatened to seek their own injunction and to demand federal intervention to prevent further harassment by whites. If the president of the United States "did not allow them the troops," they noted, "he would have to apologize for the action he took in the late railroad strike, headed by Mr. Debs." Federal troops, however, were unnecessary, as white longshoremen and screwmen abided by the terms of the injunction received by the West India line. *Times-Democrat,* November 9, 11, 1894; *Daily Picayune,* December 2, 9, 10, 1894; *Daily States,* November 9, 11, 1894; *Birmingham Age-Herald,* November 9, 18, 1894.

35. *Times-Democrat,* October 28, 29, 31, 1894; *Daily Picayune,* October 28, 29, November 1, 2, 5, 1894; *Southwestern Christian Advocate,* November 1, 8, 1894. The *Times-Democrat* declared: "It may as well be understood now as later that if the force furnished to-day shall prove inadequate that then the Governor of the State will be called upon for troops, and that if the State should unfortunately prove not to have sufficient men at command to enforce the law and disperse the rioters, then the Federal government will be asked to take a hand in the affair." *Times-Democrat,* November 13, 1894.

36. *Daily Picayune,* November 9, 1894; *Southwestern Christian Advocate,* November 15, 1894; *Birmingham Age-Herald,* November 9, 1894.

37. *Daily Picayune,* November 7, 1894.

38. *Times-Democrat,* November 8, 1894; *Daily Picayune,* November 7, 11, December 17, 1894; *Daily States,* November 7, 1894.

39. *Daily Picayune,* November 2, 1894.

40. The demise of the interracial alliance and the return to the pre-1885 rates must have been particularly difficult for Swann, one of the architects of the alliance. "I have been doing all I could to help the white laboring men ever since the formation of the union," he recalled to his members, "and I can say that I have done all I could for the white men, as well as the colored men." *Daily Picayune,* December 16, 1894. Not all black union officials felt this way, however. Black screwmen always had had a far more problematic relationship than black longshoremen to the federation of riverfront labor. One leader of the black screwmen's Association No. 1 reflected in November 1894: "I lay the cause of this trouble to a much older strife, and date its origin to the very establishment of a labor council in this city. As colored men, when the council was first organized, bowing in humble submission to the dictates of the white screwmen, and acting under the advice of some colored men, who were members of other laboring organizations, even though we feared our chances for justice, blindfolded to self, we became members of said council, then known as the Cotton Men's Executive Council. We upheld the price . . . and in order to preserve what was then called the union, we made the best out of a bad bargain." *Daily Picayune,* November 2, 1894.

41. *Daily Picayune,* December 17–19, 1894; *Times-Democrat,* December 21, 1894; *Daily States,* December 17, 18, 1894; Minutes of Meeting of Board of Directors, December 19, 1894, New Orleans Cotton Exchange Minute Book, H, 115–16.

42. *Daily Picayune,* December 20, 1894; *Daily States,* December 19, 20, 1894; *Birmingham Age-Herald,* December 21, 1894.

43. *Times-Democrat,* December 21, 1894; *Daily Picayune,* December 21, 1894; *Daily States,* December 20, 1894.

44. *Daily Picayune,* December 28, 1894.

45. British Foreign Office, *Diplomatic and Consular Reports . . . 1894,* 16–17. For accounts of waterside unionism in Great Britain during this period, see E.L. Taplin, *Liverpool Dockers and Seamen, 1870–1890* (York: William Sessions Ltd., 1974); Raymond Brown, *Waterfront Organization in Hull, 1870–1900* (Hull: University of Hull Publications, 1972); Philip J. Leng, *The Welsh Dockers* (Lancashire: G.W.& A. Hesketh, 1981); E.J. Hobsbawm, "National Unions on the Waterside," in *Labouring Men: Studies in the History of Labour* (London: Weidenfeld and Nicolson, 1964), 204–30; John Lovell, *Stevedores and Dockers: A Study of Trade Unionism in the Port of London, 1870–1914* (London: Macmillan and Co., 1969); Jonathan Schneer, *Ben Tillett: Portrait of a Labour Leader* (London: Croom Helm, 1982).

46. *Daily Picayune,* February 1, 5, 7, 8, March 3, 1895; *States,* February 5, 1895; British Foreign Office, *Diplomatic and Consular Reports,* 7. Also see: *Daily Picayune,* February 24, 1886.

47. The unemployed Gretna whites, the *Picayune* noted, "thought it a little hard that a crowd of unskilled negroes should be allowed to cross the river and literally take away their means of making a living. All the mills had closed, and the men employed there had nothing to do, and were forced to work on the levee or seek employment on this side of the river." *Daily Picayune,* March 3, 5–8, 1895. On the use of injunctions, see British Foreign Office, *Diplomatic and Consular Reports,* 17; *Sunday States,* February 10, 1895; *Daily Picayune,* March 6, 1895.

48. *Daily Picayune,* March 13, 1895.

49. *Times-Democrat,* March 22, 24, 26, 1895.

50. *Times-Democrat,* February 11, 1895; *Daily States,* February 11, 1895.

51. *Daily Picayune,* February 9, 1895.

52. *Daily Picayune,* January 25, 1895.

53. *Times-Democrat,* February 11, 1895; *Daily Picayune,* February 11, 1895; *Daily States,* February 11, 1895.

54. *Times-Democrat,* February 11, 1895; *Daily Picayune,* February 11, 1895; *Daily States,* February 11, 1895.

55. *Daily Picayune,* February 9, 1895. The *States* agreed with this interpretation: when the Harrison line agent Alfred LeBlanc dismissed his white labor force and instead contracted with black stevedore George Geddes for some 300 black screwmen and longshoremen, the paper predicted trouble, "for it is more than flesh and blood can stand to see work taken out of the mouth of a man by another who is affecting his purpose by a dead 'cut' in the rate of wages that prevailed." *Daily States,* February 5, 1895.

56. *Daily Picayune,* March 3, 1895.

57. *Daily Picayune,* February 5, 8, 9, 1895.

58. *Daily Picayune,* January 24, 25, February 11–14, March 1, 3, 7, 1895; *Times-Democrat,* March 12, 1895. A third approach that never materialized was the reconstitution of a Cotton Men's Executive Council. Although the details are sketchy, accounts suggest that four organizations—the black and white yardmen and the black

and white longshoremen—formed a new cotton council at some point in January, electing on a temporary basis white longshoreman H.O. Hassinger president and black longshoreman James Porter vice president. In February, the *Picayune* reported rumors that "a union of the cotton organizations"—although in a "crude state"—would "be brought about [soon] and a general strike would follow." Although teamsters allied with the new organization, the white screwmen remained apart. The move toward unification, however, abruptly ended on February 12. "The stumbling block seemed to be Sanders' men [—black screwmen and longshoremen hired by stevedore Geddes—] and as they could not be induced to join the council, the matter was for the time dropped." See *Daily Picayune,* February 1, 8, 11–13, 1895; *Twentieth Century* XIV, No. 14 (April 4, 1895), 4.

59. *Daily Picayune,* March 3, 5, 7, 14, 1895.

60. *Times-Democrat,* March 9, 1895; *Daily Picayune,* March 9, 10, 1895; Mobile *Daily Register,* March 10, 1895; Savannah *Morning News,* March 10, 1895. For a discussion of Cuney's role as a contracting stevedore in Galveston and for a history of Galveston dock workers, see Maud Cuney Hare, *Norris Wright Cuney: A Tribune of the Black People* (New York: The Crisis Publishing Co., 1913); Virginia Neal Hine, "Norris Wright Cuney" (unpublished M.A. thesis, Rice University, 1965); Lawrence D. Rice, *The Negro in Texas* (Baton Rouge: Louisiana State University Press, 1971); Ruth Allen, *Chapters in the History of Organized Labor in Texas* (Austin: 1941); James V. Reese, "The Evolution of an Early Texas Union: The Screwmen's Benevolent Association of Galveston, 1866–1891," *Southwestern Historical Quarterly* LXXXV, No. 2 (October 1971), 158–85; Allen Clayton Taylor, "A History of the Screwmen's Benevolent Association from 1866 to 1924" (unpublished M.A. thesis, University of Texas, 1968); Kenneth Kann, "The Knights of Labor and the Southern Black Worker," *Labor History* 18, No. 1 (Winter 1977), 47–70.

61. *Daily Picayune,* March 10, 11, 1895; *States,* March 10, 1895; *Times-Democrat,* March 10, 1895; Birmingham *Age-Herald,* March 10, 1895; *Atlanta Constitution,* March 11, 1895; Mobile *Daily Register,* March 12, 1895; Savannah *Morning News,* March 11, 1895. George A. Patrick, a black Galveston screwman, explained upon his return that he and his fellow men had been misled by stevedore Lincoln, who told them that New Orleans suffered from a serious scarcity of labor. *Galveston Daily News,* March 12, 1895.

62. *Times-Democrat,* March 10–12, 1895; *Daily Picayune,* March 10–12, 1895; *Daily States,* March 11, 1895; Mobile *Daily Register,* March 10, 12, 1895; *Galveston Daily News,* March 12, 1895; *Appletons' Annual Cyclopaedia and Register of Important Events of the Year 1895,* New Series XX (New York: D. Appleton and Company, 1896), 427–28; *Southwestern Christian Advocate,* March 14, 1895; St. Louis *Post-Dispatch,* March 12, 1895; Louisville *Courier-Journal,* March 12, 1895; Benjamin Brawley, *A Social History of the American Negro* (New York: The Macmillan Company, 1921), 321. For a collection of articles on the 1895 riots, also see Charles C. Titcomb Collection, 1895–1900 (Newspaper Clippings: Labor and Levee Riots, March, May, June 1895), Louisiana State University, Baton Rouge.

63. *Times-Democrat,* March 13, 1895; *Daily Picayune,* March 13, 1895; *Daily States,* March 12, 1895; Mobile *Daily Register,* March 13, 1895; *New York Times,* March 13, 1895; Chicago *Daily Inter-Ocean,* March 13, 1895; *Galveston Daily News,* March 13, 1895; Jacksonville *Florida Times-Union,* March 13, 1895; *Southwestern Christian Advocate,* March 14, 21, 1895; *Boston Evening Transcript,* March 13, 1895; *The Nation,* March 21, 1895; *Harper's Weekly* XXXIX, No. 1997 (March 30, 1895) 295;

Minutes of Meeting of Board of Directors, March 12, 1895, Cotton Exchange Minute Books, H, 144.

64. *Daily Picayune*, March 12, 1895; *Daily States*, March 12, 1895; *Times-Democrat*, March 12, 1895; *Southwestern Christian Advocate*, March 21, 1895; *Galveston Daily News*, March 12, 1895; Mobile *Daily Register*, March 12, 13, 1895; *Boston Evening Transcript*, March 12, 1895; Baltimore *Sun*, March 13, 1895; *Appletons' Annual Cyclopaedia . . . 1895*, 427–28.

65. The federal government's role in crushing the recent American Railway Union strike stood as a powerful example for employers, who petitioned Governor Murphy Foster to request federal troops to quell the disturbances. Refusing to work without ample protection, black union leaders also announced that they "did not care to risk the protection of the police again, but wanted State militia or Federal soldiers." Promising to cooperate with city merchants to end the lawlessness, Governor Foster rejected the demand for federal troops and instead issued a proclamation barring crowds from assembling and ordered the state militia to the waterfront. On the military occupation, see *Times-Democrat*, March 14, 1895; *Daily Picayune*, March 13, 1895; *Daily States*, March 13, 1895; *Southwestern Christian Advocate*, March 21, 1895; *New York Times*, March 14, 1895; *Galveston Daily News*, March 14, 1895; Mobile *Daily Register*, March 14, 1895; Baltimore *Sun*, March 14, 1895.

66. *Times-Democrat*, March 15–17, 20, 21, 1895; *Daily States*, March 14, 1895; *Daily Picayune*, March 14, 15, 16, 17, 19, 21, 1895; Chicago *Daily Inter-Ocean*, March 15, 16, 1895; *Florida Times-Union*, March 15, 16, 1895; Savannah *Morning News*, March 15, 16, 1895; *Galveston Daily News*, March 15, 1895; Mobile *Daily Register*, March 14, 16, 1895; *New York Times*, March 15, 1895; Baltimore *Sun*, March 15, 1895; *Baltimore American*, March 15, 1895; *Birmingham Age-Herald*, March 15, 16, 1895.

67. *Times-Democrat*, March 22–26, 29, 1895; *Daily Picayune*, March 22, 23, 25, 26, 1895; *Sunday States*, March 17, 1895; *New York Times*, March 21, 22, 26, 27, 1895; Mobile *Daily Register*, March 21, 1895; *Birmingham Age-Herald*, March 22, 23, 26, 27, 1895. Although peace reigned on the New Orleans side of the river, there were several incidents of racial harassment and violence in Gretna. Unemployed Gretna whites—who were not members of the New Orleans longshore unions—carried out a small-scale "labor riot" in mid-May against black workers employed by stevedore Geddes on a West India and Pacific steamship. The "riot" was not a white effort to drive all blacks off the docks; whites demanded half of the available work. The violence began when the stevedore attempted to work a number of white men under the supervision of a black foreman. See *Daily Picayune*, May 17, 22, 1895; *Southwestern Christian Advocate*, May 23, 1895; *New York Times*, May 17, 1895; *Birmingham Age-Herald*, May 17, 1895.

68. *Times-Democrat*, March 23, 26, 29, 1895; *Daily Picayune*, March 17, 26, 29, April 9, 1895.

69. *Daily Picayune*, April 9, 11, 18, May 17, 22, 29, July 27, October 12, 15, 1895; *New York Times*, May 17, 1895; *Southwestern Christian Advocate*, May 23, 1895.

70. *Daily Picayune*, October 15, 29, November 15, December 6, 1895; *Birmingham Age-Herald*, November 15, December 11, 1895.

71. *Florida Times-Union*, March 14, 1895. Northern newspapers saw the matter differently. The riots were not born of "a short-lived fit of popular madness," the *Boston Evening Transcript* noted. The origins lay in the simple fact that the screwmen had decided that cotton screwing was "exclusively a white man's business, and that the colored men must leave it." *Boston Evening Transcript*, March 14, 1895. The *New York*

Tribune saw the source of the riot in labor competition and in the heritage of violence that marked the state: "The New Orleans riot is a labor conspiracy rather than a race outbreak. The negroes were shot down on the wharves not because they were blacks, but because they were substitutes for strikers. . . . The white mobs of longshoremen are profiting by the education in lawlessness which they have received in the politics of Louisiana. There were white leagues riding roughshod over political rights guaranteed by the Constitution, and there were negroes coerced, intimidated and shot down in order to secure the establishment of Democratic ascendancy in the State . . . There is nothing which the longshoremen have done in the way of intimidation, murder and massacre which was not anticipated during the political period immediately following reconstruction times. With a population trained in the use of bowie-knife and revolver, and with police and militia in sympathy with 'nigger-hunting,' New Orleans is to-day the most turbulent and lawless city in the South." *New York Tribune,* March 14, 1895. Days later, the *Tribune* expressed approval of the city's commercial elite for standing up for the rights of blacks to work: "The declaration of New-Orleans business men and leading citizens that they will henceforth uphold and defend, with armed force if necessary, the right of men to work if they wish, without regard to race, color or previous condition, is only astonishing because it is the first time the controlling influences in the State have been enlisted to defend the rights of negroes formerly enslaved. Self-interest in this case opens the eyes blinded so long and so completely by race prejudice, and by the habits of the 'peculiar institution.' " *New York Tribune,* March 16, 1895. Also see *Twentieth Century* XIV, No. 12 (March 21, 1895), 2.

72. Baltimore *Sun,* March 13, 1895. Both the *Florida Times-Union* and the Baltimore *Sun* were referring to the Chicago-centered 1894 American Railway Union strike on behalf of Pullman Car Company workers and the January 1895 strike of trolley car workers in Brooklyn. See Nick Salvatore, *Eugene V. Debs: Citizen and Socialist* (Urbana: University of Illinois Press, 1982), 126–38; Joshua B. Freeman, *In Transit: The Transport Workers Union in New York City, 1933–1966* (New York: Oxford University Press, 1989), 16.

73. Mobile *Daily Register,* March 13, 1895.

74. *Washington Bee,* November 10, 1894; *Southwestern Christian Advocate,* March 14, 1895.

75. *Southwestern Christian Advocate,* November 29, 1894.

76. *The Freeman* (Indianapolis), March 23, 1895.

77. *Daily Picayune,* March 13, 1895; *Times-Democrat,* March 13, 1895; Birmingham *Labor Advocate,* November 19, 1892. For further discussion of Parker's views of labor, see Matthew J. Schott, "John M. Parker of Louisiana and the Varieties of American Progressivism" (unpublished Ph.D. dissertation, Vanderbilt University, 1969), 64–67. Fitzpatrick received widespread press condemnation as well. See Savannah *Morning News,* March 14, 15, 1895; Charleston *News and Courier,* March 14, 1895; Baltimore *Sun,* March 13, 1895; *Daily Picayune,* March 18, 1895; Chicago *Inter-Ocean,* March 14, 1895; *Atlanta Constitution,* March 14, 16, 1895. Also see statements of Produce Exchange President E.S. Stoddard and Cotton Exchange secretary Henry Hester on poor government and business prospects, in Young Men's Business League, *New Orleans of 1894,* 4–5.

78. Schott, "John M. Parker of Louisiana," 57; George E. Cunningham, "The Italians, A Hindrance to White Solidarity in Louisiana, 1890–1898," *Journal of Negro History* L, No. 1 (January 1965), 25–28; Henry C. Dethloff, "The Alliance and the Lottery: Farmers Try for the Sweepstakes," *Louisiana History* VI, No. 2 (Spring 1965), 141–59; Kendall, *History of New Orleans,* II, 483–501, 509–11.

79. *Southwestern Christian Advocate,* September 19, 1895; Edward F. Haas, *Political Leadership in a Southern City: New Orleans in the Progressive Era 1896–1902* (Ruston, La.: McGinty Publications, 1988), 39–55; Kendall, *History of New Orleans,* II, 517–23; Raymond O. Nussbaum, " 'The Ring is Smashed!': The New Orleans Municipal Election of 1896;" *Louisiana History* XVII, No. 3 (Summer 1976), 283–97; Philip D. Uzee, "The Republican Party in the Louisiana Election of 1896," *Louisiana History* II, No. 3 (Summer 1961), 332–44; George M. Reynolds, *Machine Politics in New Orleans, 1897–1926* (1936; rpt. New York, AMS Press, 1968), 26–27; Schott, "John M. Parker of Louisiana," 70–77; Schott, "Progressives Against Democracy: Electoral Reform in Louisiana, 1894–1921," *Louisiana History* XX, No. 3 (1979), 253–54; Lucia Elizabeth Daniel, "The Louisiana People's Party," *Louisiana Historical Quarterly* 26, No. 4 (October 1943), 1109. On black attitudes toward the machine and the reformers, also see The *Crusader,* no date, *Crusader* Clipping File, Folder P58, Xavier University; The *Crusader* clipping file n.d. 1895, Folder P101, Xavier University; *Southwestern Christian Advocate,* May 16, 1895.

80. Reynolds, *Machine Politics in New Orleans,* 32–34; Kendall, *History of New Orleans,* II, 517–31; Edward F. Haas, "John Fitzpatrick and Political Continuity in New Orleans, 1869–1899," *Louisiana History* XXII, No. 1 (Winter 1981), 7–29.

81. Hair, *Bourbonism and Agrarian Protest,* 234–67; Uzee, "The Republican Party," 332–44; Reynolds, *Machine Politics in New Orleans,* 27–30; Daniel, "The Louisiana People's Party," 1107–11. By mid-decade, Louisiana politics had reached a crisis point. Incumbent Democrats faced a twin threat: the Populist upheaval (always a rural phenomenon in Louisiana) and the emergence of "sugar Republicans" (pro-sugar tariff planters who bolted the anti-tariff Democratic party to form a faction within the Republican party). When an alliance between Populists and the sugar Republicans threatened to unseat Governor Foster in the 1896 state election, Democrats crushed the insurgency through widespread intimidation and violence.

82. J. Morgan Kousser, *The Shaping of Southern Politics: Suffrage Restriction and the Establishment of the One-Party South, 1880–1910* (New Haven: Yale University Press, 1974), 152–65; Hair, *Bourbonism and Agrarian Protest,* 268–79; Reynolds, *Machine Politics in New Orleans,* 26–30; Charles Barthelemy Rousseve, *The Negro in Louisiana: Aspects of his History and his Literature* (New Orleans: The Xavier University Press, 1937), 132. On black reactions to disfranchisement, see *Southwestern Christian Advocate,* April 2, 1896; December 16, 1897; February 24, 1898.

83. *The Harlequin,* June 23, 1900; *Southwestern Christian Advocate,* July 19, November 1, 1900; June 5, 26, November 6, 1902; February 26, March 26, April 23, 1903; A.R. Holcombe, "The Separate Street-Car Law in New Orleans," *The Outlook* 72, No. 13 (November 29, 1902); *Times-Democrat,* November 4, 1902; *Daily Picayune,* July 31, 1902; *Galveston Daily News,* August 7, 1903.

84. *Southwestern Christian Advocate,* September 21, 1899.

85. *Southwestern Christian Advocate,* October 19, 1899.

86. *Southwestern Christian Advocate,* July 25, 1895.

87. Hair, *Carnival of Fury;* Joel Williamson, *The Crucible of Race: Black-White Relations in the American South Since Emancipation* (New York: Oxford University Press, 1984), 201–9; Ida B. Wells-Barnett, *Mob Rule in New Orleans. Robert Charles and His Fight to the Death* (1900), reprinted in Wells-Barnett, *On Lynchings* (New York: Arno Press, 1969); *The Harlequin,* July 28, August 4, 18, 1900; *Southwestern Christian Advocate,* August 2, 9, 16, 1900; Herbert Shapiro, *White Violence and Black Response: From Reconstruction to Montgomery* (Amherst: The University of Massa-

chusetts Press, 1988), 61–63; "The New Orleans Mob," *The Outlook* 65, No. 14 (August 4, 1900), 760–61.

88. Hair, *Carnival of Fury,* xiv; Dale A. Somers, "Black and White in New Orleans: A Study in Urban Race Relations, 1865–1900," *Journal of Southern History* XL, No. 1 (February 1974), 42; Williamson, *The Crucible of Race,* 201.

89. Gompers to Leonard, March 18, 1899, Samuel Gompers Letterbooks, Volume 27 (January 27 to April 12, 1899), Reel 18, Frame 633, Library of Congress; *Daily Picayune,* September 1, 1897.

90. *United Labor Council Directory of New Orleans, Louisiana* (New Orleans: United Labor Council Publishing Committee, 1895), in Louisiana Collection, Howard-Tilton Memorial Library, Tulane University; *Southern Economist and Trade Unionist* 1, No. 1 (April 17, 1897); *Union Advocate,* April 27, 1903. On the growth of the AFL, see Leo Wolman, *The Growth of American Trade Unions 1880–1923* (New York: National Bureau of Economic Research, 1924), 33–34.

91. "The New South," *American Federationist* VI, No. 3 (May 1899), 57–58; Gompers to Leonard, March 18, 1899, Samuel Gompers Letterbooks, Volume 27 (January 27 to April 12, 1899), Reel 18, Frame 633. Correspondence from Winn, Greene, and McGruder can be found in the Gompers letterbooks for 1899.

92. *American Federationist* VI, No. 3 (June 1899); *Daily Picayune,* June 28, 1899. From the Samuel Gompers Letterbooks, Reel 20 (July 22–September 26, 1899), see Gompers to Winn, August 3, 1899, frames 120–21; Gompers to Leonard, August 9, 1899, frame 267; Gompers to Leonard, August 17, 1899, frame 4457; Gompers to Leonard, August 25, 1899, Frame 538; Gompers to Winn, September 1, 1899, frame 691.

93. *Daily Picayune,* August 14, September 11, 1899; *Times-Democrat,* August 14, September 11, 1899. Also see Minutes of the Central Trades and Labor Council, October 8, 1899–April 26, 1901, in Stoddard Labor Collection, Archives and Special Collections, University of New Orleans.

94. Bernard Cook, "The Typographical Union and the New Orleans General Strike of 1892," *Louisiana History* XXIV, No. 4 (Fall 1983), 382–88.

95. *Daily Picayune,* September 21, 28, 30, October 19, November 1, 2, 30, 1894; October 23, 1895; *Daily States,* October 29, 1894; *Crusader,* January 10, 1895 (date unclear), Folder P 23, in *Crusader* Clipping File, Xavier University; Minutes of Special Meetings on November 1, 4, 5, 7, 9, December 11, 1892, in Records of the New Orleans Typographical Union No. 17, in Archives and Special Collections, University of New Orleans. On some Populists' repudiation of interracialism and adoption of white supremacy, see Woodward, *Origins of the New South,* 323.

96. Gompers to John Callahan, May 17, 1892, in Samuel Gompers Letterbooks, Reel 6 (Vol. 7), frame 419; also cited in Philip Taft, *The A.F. of L. in the Time of Gompers* (New York: Harper and Brothers, 1957), 311. Gompers maintained this line of reasoning, arguing that white New Orleans workers accept the formation of a separate black city union council at the end of the century: "I have always insisted on the right to organize being according to colored men. And if we do not give them this opportunity and thus make friends of them, they would of necessity be our enemies and utilized by our opponents in every struggle and whenever opportunity presents itself." Gompers to Leonard, May 23, 1900, Samuel Gompers Letterbooks, Reel 24 (Vol. 34), frame 344.

97. Foner, *History of the Labor Movement,* Vol. 2, 350; John Stephens Durham, "The Labor Unions and the Negro," *The Atlantic Monthly* LXXXI, No. CCCCLXXXIV (January 1898), 226; Taft, *The A.F. of L. in the Time of Gompers,* 308–13.

98. *Southwestern Christian Advocate,* August 2, 1900.

99. *New Orleans Republican Courier,* December 2, 1899; W.E.B. DuBois, ed., *The Negro Artisan: Report of a Social Study Made Under the Direction of Atlanta University; Together with the Proceedings of the Seventh Conference for the Study of the Negro Problems, Held at Atlanta University, on May 27th, 1902* (Atlanta: Atlanta University Press, 1902), 127–28; *Daily Picayune,* September 8, 1903; *United Mine Workers Journal,* June 9, 1904.

100. "Report of Vice President James E. Porter," *Proceedings of the Eleventh Annual Convention of the International Longshoremen, Marine and Transportworkers' Association held at Chicago, Illinois, July 14–19,1902,* 73–74; *American Federationist* VIII, No. 6 (June 1901), 224; *American Federationist* VIII, No. 10 (October 1901), 438; *American Federationist* VIII, No. 11 (November 1901), 489. On Porter, other black longshore leaders, and the International Longshoremen's Association, see *New York Age,* July 13, August 7, 1913.

101. The National Longshoremen's Association of the United States was founded in 1892 by representatives of ten lumber handlers unions from the Great Lakes. It joined the AFL the following year, eventually changing its name to the International Longshoremen's Association. By 1901, the organization claimed as many as 250 locals with 40,000 members in its ranks. The International's strength centered on the Great Lakes, though by the 1910s, a Southern Gulf District of the ILA joined together Galveston, New Orleans, Mobile, and a number of smaller ports. From the ILA's formation through 1911, some New Orleans longshore workers valued their institutional independence, maintaining a rocky relationship with the ILA. Some years locals affiliated with the ILA, other years they broke with it. Not until the First World War did membership in the larger organization have much direct bearing on the struggles of the New Orleans men. On the ILA's early history, see: John R. Commons, "The Longshoremen of the Great Lakes," *Labor and Administration* (New York: The Macmillan Company, 1913), 267–68; Maud Russell, *Men Along the Shore: The I.L.A. and Its History* (New York: Brussell & Brussell, Inc., 1966), 62–74; Lloyd G. Reynolds and Charles C. Killingsworth, *Trade Union Publications: The Official Journals, Convention Proceedings and Constitutions of International Unions and Federations, 1850–1941,* Vol. 1, *Description and Bibliography* (Baltimore: The Johns Hopkins Press, 1944), 95–100; "International Longshoremen's Association," *Report of the Industrial Commission on Labor Organizations, Labor Disputes, and Arbitration, and on Railway Labor,* Vol. XVII (Washington: Government Printing Office, 1901), 264–65.

102. In the Samuel Gompers Letterbooks, see Gompers to Leonard, March 9, 1900 (Reel 22, Vol. 32), frame 908; Gompers to Porter, March 9, 1900 (Reel 24, Vol. 34), frame 907; Gompers to Porter, April 25, 1900 (Reel 24, Vol. 34), frame 754; Gompers to Leonard, May 23, 1900 (Reel 24, Vol. 34), frame 344.

103. Leonard to Gompers, June 29, 1900, cited in Philip S. Foner and Ronald L. Lewis, eds., *The Black Worker: A Documentary History from Colonial Times to the Present,* Vol. V: *The Black Worker from 1900 to 1919* (Philadelphia: Temple University Press, 1980), 121; also see Foner, *History of the Labor Movement,* II, 351–52; Morrison to L. B. Lauwdry, March 16, 1901, in Central Trades and Labor Council, New Orleans, Louisiana, Correspondence, in Stoddard Labor Collection, Archives and Special Collections, University of New Orleans.

104. Porter to Gompers, June 14, 1900, cited in Foner and Lewis, *The Black Worker: A Documentary History,* V, 120.

105. American Federation of Labor, *Report of Proceedings of the Twentieth Annual Convention held at Louisville, Kentucky, December 6th to 15th, inclusive, 1900,*

12–13. "Separate charters may be issued to Central Labor Unions, Local Unions, or Federal Labor Unions, composed exclusively of colored members, where, in the judgement of the Executive Council, it appears advisable and to the best interest of the Trade Union movement to do so," read the amended article to the Federation's constitution. AFL, *Report of Proceedings of the Twentieth Annual Convention,* xiii. For short documents on the CLU, see Porter to Gompers, May 20, 1901; and Application for Certificate of Affiliation to the American Federation of Labor, May 20, 1901, in Charter File, The George Meany Memorial Archives, Silver Spring, Md. Also see *American Federationist* VIII, No. 6 (June 1901), 224; *American Federationist* VIII, No. 10 (October 1901), 438; *American Federationist* VIII, No. 11 (November 1901), 489; *American Federationist* VIII, No. 12 (December 1901), 562; David Paul Bennetts, "Black and White Workers: New Orleans, 1880–1900" (unpublished Ph.D. dissertation, University of Illinois at Urbana-Champaign, 1972), 527–30; *Official Roster of the Central Labor Union of New Orleans* (New Orleans: Creighton Bro., 1907), in Vertical File (Trade Unions), Louisiana Collection, Tilton Memorial Library, Tulane University.

106. "There is no apparent discrimination in wages in this city and the trade unions are open to Negroes in most cases," the DuBois study noted. "On the whole the Negro artisans seem better organized and more aggressive in this state than in any other." DuBois, ed., *The Negro Artisan,* 127–28.

107. Board of Trade, *Cost of Living in American Towns. Report of an Enquiry by the Board of Trade into Working Class Rents, Housing and Retail Prices, Together with the Rates of Wages in Certain Occupations in the Principal Industrial Towns of the United States of America* (London: His Majesty's Stationery Office, 1911).

108. Board of Trade, *Cost of Living in American Towns,* 290, 292.

109. DuBois, ed., *The Negro Artisan,* 160, 128; "Report of the Joint Committee of the Senate and House of Representatives of the State of Louisiana, Appointed to Investigate the Port of New Orleans," *Official Journal of the Proceedings of House of Representatives of the State of Louisiana at the First Regular Session of the Third General Assembly,* May 28, 1908, p. 200.

110. *Times-Democrat,* July 28, 1900; *Daily Picayune,* July 27, 1900.

111. *Daily Picayune,* June 27, 1900.

112. *Harlequin,* May 12, 19, July 28, October 13, 1900; November 28, 1901; *Daily Picayune,* November 17, 1901; *Baltimore Sun,* September 26, 1900.

113. *Times-Democrat,* September 11, 12, 1901; *Daily Picayune,* September 11, 1901; Pearce, "The Rise and Decline of Labor," 43; Daniel Rosenberg, *New Orleans Dockworkers: Race, Labor, and Unionism 1892–1923* (Albany: State University of New York Press, 1988), 72–73.

114. *Daily Picayune,* September 19, 1901.

115. *Daily Picayune,* September 11–14, 17, 19, 22, 1901; *Times-Democrat,* September 13, 1901; *New York Tribune,* September 13, 1901; *First Annual Report of the Bureau of Labor Statistics for the State of Louisiana 1901,* Thomas Harrison, Commissioner (Baton Rouge: The Advocate, 1902), 191. Labor's widespread backing for the port's grain trimmers extended beyond the city's limits. Unable to unload in New Orleans because of the strike, the German ship *Pointers* sailed for Galveston. Texas union screwmen, in support of the New Orleans longshoremen, refused to handle the cargo, forcing the ship to return to the Crescent City. Its agent reluctantly signed the trimmers' tariff. This action, the *Picayune* observed, "effectively stopped any attempt that the railroads or anyone else might have made to divert commerce to that port." *Daily Picayune,* September 19, 23, 1901.

116. *Daily Picayune,* September 19, 1901.

117. *Daily Picayune,* September 20–24, 1901; *Times-Democrat,* September 23, 1901; Gompers to Porter, September 13, 1901, Samuel Gompers Letterbooks, Reel 33 (vol. 46), frame 199.

118. *Daily Picayune,* September 26, 1901.

119. *Daily Picayune,* September 26, 1901.

120. *Daily Picayune,* October 6, 8, November 30, 1901; September 21, 1903; April 29, 1904; Pearce, "The Rise and Decline of Labor," 49.

121. On other examples of late nineteenth- and early twentieth-century interracial labor collaboration that qualify the earlier historiographical portrait of total black exclusion from trade unions, see: Paul B. Worthman, "Black Workers and Labor Unions in Birmingham, Alabama, 1894–1904," *Labor History* 10, No. 3 (Summer 1969), 375–407; Herbert Gutman, "The Negro and the United Mine Workers of America: The Career and Letters of Richard L. Davis and Something of Their Meaning," *Work, Culture and Society in Industrializing America: Essays in American Working Class and Social History* (New York: Vintage Books, 1977), 121–208; James Green, *Grassroots Socialism: Radical Movements in the Southwest, 1895–1943* (Baton Rouge: Louisiana State University Press, 1978), 176–227; Karin A. Shapiro, "The Convicts Must Go! The East Tennessee Coal Miners' Rebellion of 1891–2" (unpublished M.A. thesis, Yale University, 1983); Paul B. Worthman and James Green, "Black Workers in the New South," in *Key Issues in the Afro-American Experience,* Vol. 2, ed. Nathan Huggins, Martin Kilson, and Daniel Fox (New York: Harcourt Brace Jovanovich, 1971), 47–69.

122. Covington Hall to Abram L. Harris, August 27, 1929, in Abram L. Harris Papers, Collection 43-1, Moorland-Spingarn Research Center, Howard University.

Chapter 5

1. *Daily Picayune,* March 28, April 25, 1908; Minutes of the Louisiana Port Investigation Commission, in the Samuel L. Gilmore Papers, City Attorney, Box 3, in Manuscripts, Rare Books and University Archives, Howard-Tilton Memorial Library, Tulane University; *Official Journal of the Proceedings of House of Representatives of the State of Louisiana at the First Regular Session of the Third General Assembly,* May 28, 1908, 200.

2. John R. Commons, ed., *Regulation and Restriction of Output: Eleventh Special Report of the Commissioner of Labor* (Washington: Government Printing Office, 1904), 13–14. Also see David Montgomery, "Workers' Control of Machine Production in the Nineteenth Century," in *Workers' Control in America: Studies in the History of Work, Technology, and Labor Struggles* (New York: Cambridge University Press, 1979), 9–31; David M. Gordon, Richard Edwards, and Michael Reich, *Segmented Work, Divided Workers: The Historical Transformations of Labor in the United States* (New York: Cambridge University Press, 1982), 100–164. Susan Porter Benson has explored the work culture of urban department stores in the late nineteenth and early twentieth century that involved an impressive array of tactics by non-union saleswomen. See Benson, *Counter Cultures: Saleswomen, Managers, and Customers in American Department Stores 1890–1940* (Urbana: University of Illinois Press, 1986), 227–82.

3. On unskilled workers and managerial control, see David Montgomery, *The Fall of the House of Labor: The Workplace, the State, and American Labor Activism, 1865–1925* (New York: Cambridge University Press, 1987), 90–96; James Barrett, *Work and*

Community in the Jungle: Chicago's Packinghouse Workers 1894–1922 (Urbana: University of Illinois Press, 1987), 54–58. The work rules of the Joint Screwmen's Conference, the black and white longshoremen, and the black and white cotton yardmen are listed in *Report of the Bureau of Statistics of Labor for the State of Louisiana 1904–1905,* ed. Robert E. Lee, Commissioner (New Orleans: Miller and Brandao, 1906), 73–87.

    4. Francis E. Hyde, *Liverpool and the Mersey: An Economic History of a Port 1700–1970* (Newton Abbot: David & Charles, 1971), 95–103, 111–14; Hyde, *Shipping Enterprise and Management 1830–1939: The Harrisons of Liverpool* (Liverpool: Liverpool University Press, 1967), 24–33; Hyde, *Cunard and the North Atlantic 1840–1973: A History of Shipping and Financial Management* (London: Macmillan Press, Ltd., 1975), 90–93. John Lovell has described well the major changes in the structure of British dock employment ushered in by the gradual advent of steamships and the concomitant maritime technical revolution. New economic pressures, based upon the need to reduce time in port, led steamship companies to assume "more or less direct control over loading and discharging operations on their vessels," and to "wage an intensive campaign against organized labor." Lovell, "Sail, Steam and Emergent Dockers' Unionism in Britain, 1850–1914," *International Review of Social History* XXXII (1987), 237–40. Also see R. Bean, "Employers' Associations in the Port of Liverpool, 1890–1914," *International Review of Social History* XXI (1976), 360. During the 1890s, the expansion of the Elder-Demster Line's New Orleans business also offered serious challenges to established shipping firms. In the early twentieth century, New Orleans-based steamship companies conducted their "battle for cargoes" by negotiating contracts for cotton and grain on as favorable terms as possible with railroad companies—the Southern Pacific, Illinois Central, and Texas and Pacific—that increasingly dominated the inland transportation of cotton. See: Hyde, *Liverpool and the Mersey,* 113–14; Hyde, *Shipping Enterprise and Management,* 27–33.

    5. *Daily Picayune,* May 8, 1903; November 4, 1902; April 19, 21, 1903; *Times-Democrat,* November 4, 1902; May 10, 1903; also see *Daily Picayune,* March 20, 1908. For a discussion of the roots of these pressures, see Chapter 4; Lovell, "Sail, Steam and Emergent Dockers' Unionism in Britain."

    6. *Daily Picayune,* November 4, 1902. For similar observations, see *Times-Democrat,* November 4, 1902; *Daily Picayune,* April 19, 21, July 9, 1903.

    7. Levee laborers feared the consequences of such "parity." In his testimony before the Port Investigation Commission in 1908, black longshore leader E.S. Swann noted that laborers in New Orleans were "free" men, while in Galveston, Savannah, Mobile, and Pensacola, dock workers were "little better than slaves, because the white men and the blacks are fighting each other, and each side strives to load more than the other." *Daily Picayune,* November 5, 1902, March 14, 1908; *Times-Democrat,* March 14, 1908. Outside of Galveston, railroad companies dominated most important southern ports. The Port Investigation Committee found Savannah "completely under the control of the railroads." Local unions proved no match for the rail lines; Savannah did "not know what serious labor trouble" was, the press reported, "and ever since the great wharf strike of fifteen years ago, when the backbone of unionism was broken, she has had peace on her levee. . . ." In its final report the Commission found the labor performed by Savannah's black freight handlers to be inferior to the quality of New Orleans labor. In Pensacola, railroads and steamship companies dispensed with stevedores and hired their own gangs directly. Pensacola unions exercised little power; the *Picayune* noted approvingly that Pensacola had "not had a serious levee strike in years." The Port Investigation Commission stated that Savannah "laborers are kept down by fear of

dismissal, and every attempt to organize a union is checked by wholesale discharging of the guilty men, and those suspected of favoring them." See *Times-Democrat*, February 5, 6, 1908; *Daily Picayune*, February 5, 10, 1908; Testimony of Thomas W. Kent, Minutes of Investigation had in the city of Pensacola, Florida, as to the conditions prevailing at that port, February 7, 1908; and Testimony of Mr. Harris, ship broker, Minutes of Investigation had in the city of Savannah, Georgia, as to Conditions prevailing at the Port, February 4th and 5th, 1908, in Minutes of Louisiana Port Investigation Commission, Box 3, Samuel L. Gilmore Papers; *Official Journal of the Proceedings of House of Representatives of the State of Louisiana at the First Regular Session of the Third General Assembly, May 28, 1908,* 195–200. On Savannah dock workers, also see Mercer Griffin Evans, "The History of the Organized Labor Movement in Georgia" (unpublished Ph.D. dissertation, University of Chicago, 1920), 185–89; Mark V. Wetherington, "The Savannah Negro Laborers' Strike of 1891," *Southern Workers and Their Unions, 1880–1975: Selected Papers, the Second Southern Labor History Conference, 1978,* ed. Merl E. Reed, Leslie S. Hough, and Gary M. Fink (Westport: Greenwood Press, 1981), 4–21; *Times-Democrat*, October 10, 1903; *Atlanta Constitution*, October 6, 10, 21, November 14, 1903; *Savannah Tribune*, September 15, October 27, 1894; October 31, 1903.

8. *Daily Picayune*, September 23, October 4, 22, 1902. Black workers' contract with agents stipulated they would receive half of the loading work available, but agents "didn't give the black man half and half," stevedore J.B. Honor recalled in 1908. Therefore, "the black man goes to his white brother and says 'take me in.' It was just simply a case of self- preservation with the screwmen; they were forced to amalgamate." Testimony of J.B. Honor, Port Investigation Commission, 1908, 21–22, in Samuel L. Gilmore Papers. Also see Testimony of James Byrnes, President of Screwmen's Benevolent Association, March 10th, 1908, Minutes of Session Held in the City of New Orleans, Port Investigation Commission, 52, in Samuel L. Gilmore Papers.

9. *Daily Picayune*, April 4, 1903. Stevedore J.B. Honor would have concurred with Ellis' assessment. In 1908, he testified that steamship agents who had pitted blacks against whites for a number of years were "the cause of all the friction on the river front." See Testimony of J.B. Honor, Minutes of the Louisiana Port Investigation Commission, 1908, 19, 22, 55, in Samuel L. Gilmore Papers.

10. *Daily Picayune*, April 17, 1902; James Porter, "Report of Screwmen's Lockout, New Orleans," *Proceedings of the Twelfth Annual Convention of the International Longshoremen and Marine Transport Workers' Association, Bay City, Michigan, July 13–18, 1903,* 86. At some point during the 1890s, the two black screwmen's unions, Associations No. 1 and 2, merged, assuming the name Screwmen's Benevolent Association No. 2.

11. Ellis quoted in *Daily Picayune*, April 4, 1903; Porter, "Report of Screwmen's Lockout," *Proceedings,* 86; *Times-Democrat*, March 30, October 30, 1902; March 12, 1908; *Daily Picayune*, September 22, 25, 26, 28, 30, October 1, 4, 22, 27, 28, 30, November 6, 1902; March 12, 1908; "No Labor-Saving Devices," October 23, No Title, (Item 295, Labor: Overview: Negro Labor by State), in Hampton University Newspaper Clipping File; Herbert Northrup, *Organized Labor and the Negro* (New York: Harper and Brothers, 1944), 149; Rosenberg, *New Orleans Dockworkers,* 82–85. E.S. Swann offered a description of the pre-amalgamation era similar to Ellis' when he testified before the Port Investigation Commission in 1908. The *Picayune* reported him as saying: "The negroes had the forward hatches and the whites the aft hatches, and the bosses used to go to the niggers and say, 'Niggers, them white men is beating you two to one; if you don't do better, we'll give all the work to the whites.'

Then they'd tell the same thing to the whites, and they kept war to the knife, and knife to the hilt, between the two races, and a riot was likely to break out at any time. Then the two races amalgamated." *Daily Picayune*, March 14, 1908.

12. *Daily Picayune*, November 1, 5, 1902; *Times-Democrat*, November 1, 6, 1902; "Special Meeting of the Board of Directors, April 29, 1903," New Orleans Cotton Exchange Minute Book, J (in Manuscripts, Rare Books and University Archives, Howard-Tilton Memorial Library, Tulane University), 232.

13. *Daily Picayune*, November 3, 4, 1902.

14. *Daily Picayune*, November 3, 4, 1902; *Times-Democrat*, November 4, 1902.

15. *Daily Picayune*, November 4–6, 1902; *Times-Democrat*, November 6, 7, 1902. "There is no trouble where the agents have a union foreman. If they did not . . . then there was trouble." *Daily Picayune*, November 4, 1902. On the capacity of waterfront employers to mount effective anti-union campaigns in New York and on the West coast, see Howard Kimeldorf, *Reds or Rackets? The Making of Radical and Conservative Unions on the Waterfront* (Berkeley: University of California Press, 1988), 51–79.

16. *Daily Picayune*, November 4–7,9, December 25, 26, 1902; *Labour Gazette*, May 1903. Several weeks later another agent concluded that while there had been no fixed figure of a day's work, the "old number of bales have never been handled" since. One stevedore then proposed that screwmen be paid by the piecework system; that way, he reasoned, gangs could load as many or as few bales as they desired. The unions, predictably, rejected the suggestion and promised a major confrontation if employers attempted to change the current system. Nothing more was heard of this plan.

17. *Report of the Bureau of Statistics of Labor for the State of Louisiana 1902–1903* (Baton Rouge: *The Advocate*, Official Journal of Louisana, 1904), 36; *Daily Picayune*, April 7, 1903; *Times-Democrat*, April 6, May 10, 1903.

18. For discussion of the open shop drive of the early twentieth century, see Selig Perlman, *A History of Trade Unionism in the United States* (New York: The Macmillan Company, 1922), 193–96; Clarence E. Bonnett, *Employers' Associations in the United States: A Study of Typical Associations* (New York: The Macmillan Company, 1922); Leo Wolman, *The Growth of American Trade Unions 1880–1923* (New York: National Bureau of Economic Research, 1924), 33; Foster Rhea Dulles and Melvyn Dubofsky, *Labor in America: A History*, 4th ed. (Illinois: Harlan Davidson, 1984), 184–86; Gordon, Edwards, and Reich, *Segmented Work, Divided Workers*, 111. On the New Orleans meeting of the National Association of Manufacturers, see *Daily Picayune*, April 14–17, 1903. In the spring and summer of 1904, the New Orleans Employers Association, composed of businessmen and contractors in the building trades, declared war on the carpenters and other building tradesmen by insisting upon the open shop. See *Daily Picayune*, April 11, 12, 22, 26–30, May 3–5, 10, 12, 20, 21, 23–29, June 1, 3–5, 8–11, 14, 17–19, 1904; *Birmingham News*, April 27, 1904.

19. *Report of the Bureau of . . . Labor . . . 1902–1903*, 35–36; Porter, "Report of Screwmen's Lockout," 86–87; *Daily Picayune*, April 4–8, 1903; *Times-Democrat*, April 6, 1903; *Union Advocate*, April 13, 1903.

20. *Daily Picayune*, April 9–11, 20, 21, 29, June 10, 12–17, 23, July 10, October 1, 4, 9, 1903; *Union Advocate*, April 13, 1903; *Times-Democrat*, April 10, 21, 1903; Porter, "Report of Screwmen's Lockout," 86.

21. *Report of the Bureau of . . . Labor . . . 1902–1903*, 35–39; Porter, "Report of Screwmen's Lockout," 86–87; *Daily Picayune*, April 14, 15, 20, 1903; *Times-Democrat*, April 19, 20, 1903. For a detailed discussion of other mediation efforts, see Eric Arnesen, "To Rule or Ruin: New Orleans Dock Workers' Struggle for Control 1902–

1903," *Labor History* 28, No. 2 (Spring 1987), 157–59; *Report of the Bureau . . . of Labor . . . 1902–1903,* 39–40; *Times-Democrat,* April 21–24, 26, 27, 29–30, May 2–3, 1903; *Daily Picayune,* April 25–26, 30, May 1–3, 5–8, 10, 13, 15, 1903; Porter, "Report of Screwmen's Lockout," 86–87; *Union Advocate,* April 20, 27, May 11, June 1, 15, 1903.

22. *Report of the Bureau of . . . Labor . . . 1902–1903,* 47; *Daily Picayune,* September 16, 18–23 1903; *Atlanta Constitution,* September 19, 1903; *Savannah Tribune,* September 26, 1903; *Galveston Daily News,* September 18, 26, 29, 1903; *Mobile Register,* September 26, 1903. When negotiations resumed in earnest in September, numerous proposals and counterproposals centered on precisely how many bales of cotton loaded constituted a fair day's work. While screwmen renewed their earlier compromise offer to stow 150 bales by hand and 85 to 90 with screws, the same position they took before the Cotton Exchange Mediation Committee in April, employers put forth yet another compromise offer: 85–95 bales screwed, 175–225 hand stowed, *or* submit the entire matter to arbitration by former New Orleans Mayor John Fitzpatrick. A mass meeting of black and white screwmen overwhelmingly rejected this proposal and voted to maintain their position. *Daily Picayune,* September 24, 25, 26, 30, October 2, 1903; *Times-Democrat,* October 1, 1903; *Report of the Bureau . . . Labor . . . 1902–1903,* 48; *Union Advocate,* September 28, October 5, 19, 1903.

23. *States,* October 5, 1903; *Mobile Register,* October 2, 1903; also see *Galveston Daily News,* August 27, September 25, 26, 29, 1903; *Mobile Register,* September 26, 1903.

24. The events of the week-and-a-half-long lockout/strike can be followed in *Daily Picayune,* October 1–11, 1903; *Times-Democrat,* October 1–11, 1903. Also see Monthly Meeting of Board of Directors, October 7, 1903, New Orleans Cotton Exchange Minute Book, J, 307; *States,* October 10, 1903; *Galveston Daily News,* October 2, 4, 7, 9–13, 1903; *Mobile Register,* October 2, 3, 6, 8–11, 13, 1903; Rosenberg, *New Orleans Dockworkers,* 88–90. The Dock and Cotton Council approached the conflict with caution. Rejecting motions for a general strike, it instead named a committee of five to meet with the screwmen and agents to work out an agreement as well as to investigate the situation of the 200 strikebreakers quartered, or imprisoned, on a British steamship anchored in the river. Individual longshoremen, however, did not wait for official instructions. On most ships along the riverfront, they simply failed to report for work. While there was no order to strike, one stevedore noted, "it seems impossible to keep the men at work when the other associations are in trouble and are continually urging sympathy action."

Machine mayor Paul Capdeveille's response was measured. Breaking a vow that he would never involve himself in labor disputes (made after a streetcar strike), he promised protection to employers but outlawed the private armed "strikebreakers"— that is, detectives—in the employ of the agents. *Daily Picayune,* October 10, 1903; *Labour Gazette,* November 1903; *Daily States,* October 5, 6, 8–10, 1903; *Times-Democrat,* October 5–11, 1903; *Mobile Register,* October 9, 1903.

25. *Daily Picayune,* August 30, 1903. Divisions within the employers' ranks were evident during the summer as well, when the city's stevedores attempted to use the crisis to advance their own position. By the early twentieth century, the largest steamship lines that were most adamant about repealing the screwmen's rules no longer employed stevedores as middlemen, and many stevedores felt no great love for these agents. The deadlock offered them new opportunities: an alliance of stevedores and screwmen, some felt, might force the "big four" to employ them once again as middlemen. The screwmen, however, spurned the alliance; they did not even bother to

respond to a request for a conference. The talk of the stevedores' plan was not without adverse effect. An article in the *Sunday Item* reported that the white screwmen were indeed conspiring to form a combination with the stevedores "to down the colored brother at the last moment, when the September rush comes on." The white union men vehemently denied that they would draw the color line, and the two unions held a mass meeting in late June to determine who was responsible for the false story. Affirming that the "affiliation between the two organizations is as strong as ever, and [that] nothing can sever it," the meeting declared that "the whites cannot do without the blacks and the colored must have the white." *Daily Picayune*, May 13, 14, 20, 29, 30, June 10, 20, 21, 25, 1903; Porter, "Report of Screwmen's Lockout," 86–87; *Times-Democrat*, July 7, 1903.

26. *Times-Democrat*, October 2–4, 1903; *Daily Picayune*, October 2, 4, 6, 1903; *Galveston Daily News*, October 4, 1903; *Report of the Bureau of . . . Labor . . . 1902–1903*, 48; *Daily States*, October 5, 1903.

27. *Report of the Bureau of . . . Labor . . . 1902–1903*, 50; *Daily Picayune*, September 26, October 3, 4, 6, 9, 13, 1903; *Times-Democrat*, October 1–4, 12–14, 21, 1903; *Galveston Daily News*, October 3, 12, 13, 1903; *Atlanta Constitution*, October 12, 13, 1903; *The Labour Gazette* XI, No. II (November 1903), 305; *Union Advocate*, October 19, 1903; *Daily States* October 12, 13, 1903; *Mobile Register*, October 13, 14, 1903. Agents and stevedores remained obsessed by union work rules and the restriction of output. Months later, in March 1904, the stevedoring firm of John Honor and Company sought some way to "obtain quicker dispatch in loading vessels" and proposed to the screwmen that they stow 105 bales with screws and 220 bales by hand in exchange for $30 per day per gang of men. A joint meeting of black and white screwmen—equally represented—rejected the proposition after animated discussion. Not only would it be impossible for a gang "under existing conditions" to perform that much work, they argued, but the offer failed to compensate them proportionately for the proposed increase in work. Agents and stevedores found new reason to complain later that month when the screwmen unilaterally implemented a new work rule that prohibited members from working "quarters" in the afternoon after two p.m.—that is, if a ship was not ready to be loaded or unloaded before that time, screwmen would not work until five p.m., at which hour "double time" begins. Little came of the offer of three smaller shipping lines in August 1904 to stabilize the situation by entering into a five-year contract with the screwmen. Although black workers reportedly favored the plan, whites who preferred that "matters drift along as now" had the final say. *Daily Picayune*, March 9, 15, 23, August 13, 1904.

28. Testimony of William J. Kearney, Minutes of the Port Investigation Commission, 118–20, 125–27, in Samuel L. Gilmore Papers; *Times-Democrat*, March 28, 1908; *Daily Picayune*, March 28, 1908. Screwmen's Benevolent Association president and business agent James Byrnes explained the nature of the employment relation: A stevedore seeking to employ a screwmen would "hire a foreman. . . . Some ship agents have superintendents they hire, a 'walking boss,' who is a member of the Screwmen's Association, and he does the hiring, puts men to work, has full charge over every gang" in the ship, and takes orders from superintendent. Testimony of James Byrnes, Minutes of Session [of the Port Investigation Commission] Held in the City of New Orleans, Tuesday, March 10th, 1908, 86–87, in Samuel L. Gilmore Papers.

29. *Daily Picayune*, September 9, 1903; *Union Advocate*, September 14, 1903; *Sunday States*, September 13, 1903.

30. *Daily Picayune*, August 3, 31, September 1, 3–7, 9, 11, 13, 14, 17, 18, 1903; *Daily States*, September 3–6, 8–10, 12–14, 1903; *Times-Democrat*, September 3–6, 8–

11, 13, 14, 1903. Also see: *Report of the Bureau of . . . Labor . . . 1902–1903,* 47; *Galveston Daily News,* September 4, 5, 7, 9–18, 1903; *Mobile Register,* September 5, 11, 12, 1903; *New York Tribune,* September 10, 11, 1903; *Atlanta Constitution,* September 13, 14, 19, 1903; *Savannah Tribune,* September 26, 1903. Stevedore W. Kearney admitted that Chalmette and Westwego were "mighty difficult places to get to and it is not right to order a man over there and not pay him. "See Minutes of the Port Investigation Commission, in Samuel Gilmore Papers, 100–102.

31. *Report of the Bureau of . . . Labor . . . 1902–1903,* 47; *Daily Picayune,* September 16, 18–20, 1903; *Times-Democrat,* September 15–19, 1903; *Daily States,* September 15, 16, 1903; *Atlanta Constitution,* September 19, 1903; *Mobile Register,* September 17, 19, 20, 22, 23, 1903; *Savannah Tribune,* September 26, 1903; *Galveston Daily News,* September 15–20, 22, 23, 1903.

32. *Times-Democrat,* September 1, 1906; *Daily Picayune,* August 26, 1905; August 16, 24, 27, 30, 31, 1906. This power to change work rules unilaterally, employers recognized, could result in the addition of men; in their eyes, failure to incorporate an anti-rule change clause would "leave the men free to make rules at any time which will increase the cost of handling freight"—that is, if the longshoremen adopted a rule adding an extra man or two, ship owners feared their labor expenses might rise between 30 and 60 percent. The agents suspected that the unions planned to spring some new secret rules on them as soon as the contract was signed.

33. Municipal political officials offered the large steamship agents little assistance in their fight against union workers. Like its nineteenth-century predecessor, the Choctaw Club remained heavily dependent upon a white working-class electorate and hence refused to alienate its social base by intervening decisively on capital's behalf. Martin Behrman was elected mayor as the machine candidate in 1904 with the support of many of the city's white union members. Behrman resembled his nineteenth-century Ring counterparts in his indebtedness to and sympathy with labor, yet he also differed from them in his strong solicitations and commitment to the port's men of commerce. See Chapter 6 for further discussion of this theme.

34. *Daily Picayune,* September 1–3, 5, 1906; *Times-Democrat,* August 31, September 1–3, 1906. Longshore unions also addressed the problem of the availability of work at the out-of-town docks of Chalmette, Southport, and Westwego. Stevedores and agents frequently ordered gangs of men to work on ships docked there, but too often longshoremen discovered upon their arrival little work to perform. Under the 1903 contract, they received two-and-a-half hours' pay if there was no work; now they demanded five hours' pay if there was no work and a guarantee of four hours' pay for any work up to four hours. In the 1906 contract settlement, agents reluctantly increased the pay to four hours for no work; any longshoreman who was put to work, for any length of time, would receive no less than five hours' pay.

35. *Daily Picayune,* September 5, 11, 12, 17, 21, 23, 1906; *Times-Democrat,* September 12, 13, 1906.

36. *Daily Picayune,* September 11, 12, 1906.

37. *Daily Picayune,* September 12, 14, 15, 17, 1906.

38. *Daily Picayune,* September 22, 27, 29, 1906; *Times-Democrat,* September 25–27, 1906.

39. In the aftermath of the 1934 general strike, San Francisco's rank-and-file longshore workers similarly relied upon direct action and, later, contractual provisions, to shift a substantial degree of power away from waterfront employers to themselves. See Herb Mills and David Wellman, "Contractually Sanctioned Job Action and Workers' Control: The Case of San Francisco Longshoremen," *Labor History* 28, No.

2 (Spring 1987), 167–95; Bruce Nelson, *Workers on the Waterfront: Seamen, Long-shoremen, and Unionism in the 1930s* (Urbana: University of Illinois Press, 1988), 157–74; Kimeldorf, *Reds or Rackets?*, 110–15.

40. *Times-Democrat*, September 6, 7, 1907; *Daily Picayune*, September 7, 1907; Testimony of J.B. Honor, Minutes of the Port Investigation Commission, 48, in Samuel L. Gilmore Papers.

41. *Daily Picayune*, September 12, 1907; *Times-Democrat*, September 7, 1907; Testimony of W. Kearney, Minutes of the Port Investigation Commission, 118–20, in Samuel L. Gilmore Papers.

42. *Daily Picayune*, August 25, 27, September 5, 7, 1907; *Times-Democrat*, September 7, 1907; Pearce, "The Rise and Decline of Labor," 63–64. Also see testimony before the Port Investigation Commission in *Daily Picayune*, March 28, 1908; *Times-Democrat*, March 28, 1908. Employers also objected to rates of pay and rules governing work at the outlying Chalmette, Southport, and Westwego docks. While accepting in principle the compensating of men sent to these docks, but for whom there was no work, employers demanded a reduction in that payment and an increased flexibility in shifting gangs between ships. The black and white longshore unions rejected the Steamship Conference proposals, demanded a renewal of the old contract, and issued a new ultimatum: all cargo repair and the sewing of torn sacks—work previously done by an outside contractor—would now be done only by longshoremen at longshoremen's wages. *Times-Democrat*, Septemer 7, 1907; *Daily Picayune*, September 5, 7, 1907; *Daily News*, September 7, 1907.

43. *Times-Democrat*, September 10–14, 1907; *Daily News*, September 10, 11, 1907; *Daily Picayune*, September 10–13, 1907; *Galveston Daily News*, September 10, 1907.

44. *Times-Democrat*, September 15, 16, 1907; *Daily Picayune*, September 15, 16, 1907; *Galveston Daily News*, September 15, 16, 1907; *Houston Post*, September 16, 1907; Pearce, "The Rise and Decline of Labor," 65. Also see *The Board of Trade Labour Gazette* XV, No. 10 (October 1907), 297; Minutes of the Meeting of the Board of Directors, September 10, 1907, New Orleans Cotton Exchange Minute Book, L, 63–64. The longshoremen divided along racial lines on the sack sewing demand, with white workers favoring the clause and black workers opposing it. At a "stormy" joint session, whites accused their black co-workers of attempting to deliver the associations to the bosses, and of "having itching palms to be healed by a liberal application of United States currency salve." In a vote that was generally split along black-white lines, the whites upheld the inclusion of the clause. Although Swann agreed to stand by the majority and carry out the longshoremen's law, black unionists were hardly satisfied. "Not a third of our strength . . . was present at the meeting," one black worker complained, "and if we all had been there, there'd be no strike on the levee today." They had more success when they raised the issue at another joint meeting and the unions dropped the clause. But they rejected the employers' proposals as well, and negotiations broke off. *Times-Democrat*, September 7, 8, 11, 1907; *Daily Picayune*, September 7, 9, 11, 1907.

45. Montgomery, "The 'New Unionism' and the Transformation of Workers' Consciousness in America, 1909–22," in *Workers' Control in America* (New York: Cambridge Univ. Press, 1979), 91–112; Nelson, *Workers on the Waterfront*, 6–7.

46. *Daily Picayune*, December 10, 1902; May 14, 1903. Railroads also paid wages substantially lower than those in the longshore trade. While longshoremen earned 40 cents an hour, Illinois Central employees received 15 cents an hour and Morgan dock workers loading the ships of the Cromwell and Southern Pacific lines made 30 cents.

47. *Daily Picayune*, December 10, 19, 20, 30, 1902; *Times-Democrat*, November 9, 19, 30, 1902; *Report of the Bureau of . . . Labor . . . 1902–1903*, 33–34.

48. *Daily Picayune*, December 9, 11–13, 15–18, 20, 1902; *Times-Democrat*, December 9, 1902; *L'Abeille de la Nouvelle Orleans*, December 16, 1902; *Report of the Bureau of . . . Labor . . . 1902–1903*, 34.

49. *Daily Picayune*, December 24, 25, 1902; *Times-Democrat*, December 23 and 24, 1902; *Report of the Bureau of . . . Labor . . . 1902–1903*, 34, 50. Also see *Daily Picayune*, July 7, 1904. The Illinois Central strike serves as a model of what freight handlers were up against in the early twentieth century. Southern Pacific freight handlers also worked without limits (union members claimed a gang of handlers stowed 700 bales of cotton per day) and the protection of union work rules, but their wages of 30 cents an hour were double that of the Illinois Central men. The Universal Freight Handlers Union, ILA Local 402, which represented the Southern Pacific men, had a membership that was roughly half black and half white, and offices were divided between the two. In September 1903, the Universal Freight Handlers demanded that their wages be raised to 40 cents an hour—that is, that they receive the same compensation as the port's regular longshoremen. The freight handlers' union, however, was not a member of the Dock and Cotton Council, which had rejected its application earlier that summer. The Southern Pacific ignored the demands and when the union struck in early September, it imported roughly 300 strikebreakers and about 40 guards from its own secret service department and the Thiels detective agency. Within days the Southern Pacific had resumed full operation. The strike collapsed in less than two weeks. See *Times-Democrat*, August 26, September 3–6, 9–14, 1903; *States*, September 1–6, 8–14, 1903; *Daily Picayune*, December 20, 1901; May 14, 29, September 1–5, 14, 1903; *Union Advocate*, September 14, 1903; *Galveston Daily News*, August 27, September 1, 2, 4, 5, 7, 12–18, 1903; *Mobile Register*, September 12, 1903; *Report of the Bureau of . . . Labor . . . 1902–1903*, 46. The following year, Southern Pacific freight handlers established new unions. Although the *Picayune* initially reported that white Cargo Workers' Union Local 626 would "keep free of the colored brother," it quickly retracted the statement, observing that the union would "work in strict harmony with the colored men's organization and that the best of feelings prevail." *Daily Picayune*, July 8, 9, 1904.

50. *Daily Picayune*, September 3, 4, 5, 1904; September 2–4, 1905; *Times-Democrat*, September 7, 1904; also see: *Labour Gazette* XI, No. 1 (January 1903), 14; *States*, September 8, 1903.

51. *Daily Picayune*, September 8–12, 1905; *Times-Democrat*, September 7–12, 1905; *States*, September 11, 12, 1905.

52. *Daily Picayune*, September 13–23, 1905; *States*, September 13, 15, 16, 18, 19, 21–23, 1905; *Times-Democrat*, September 13–24, 1905.

53. Shortly after the longshore grain trimmers' successful strike in September 1901, but before the Dock and Cotton Council's formation, 100 black round freight teamsters, affiliated with the Team Drivers' International Union, struck to increase their low wages and correct on-the-job abuses. They complained that they often received far less than a full day's pay for working from early morning until late at night; the inadequate facilities for receiving freight of the Illinois Central, Louisville and Nashville, and the Southern Pacific railroads produced delays of up to two hours before the teamsters could unload their drays, time for which the black workers received no remuneration. Boss draymen quickly acquired teams of strikebreakers who worked under the protection of local police. The drivers returned to work under conditions barely different from those existing before they struck. See *Daily Picayune*, September 9, 17, 18, 20, 22, 1901.

54. *Daily Picayune*, October 2, 8, 9, December 2, 8, 9, 19, 1902; *Times-Democrat*,

November 7, 11, 22, December 2, 19, 1902. The only outside assistance came from the teamsters' international union. Following the annual convention of the American Federation of Labor in New Orleans that December, C.F. O'Neill, general organizer for the International Team Drivers' Union in Buffalo, New York, remained in the city to assist the local men. Denounced by Mayor Paul Capdevielle as a northern labor agitator who did not understand that "the temper of the Southern people was different from that of Buffalo," he took charge of the strike and made "each of the members . . . take a pledge that he would not become intoxicated" for the duration.

55. *Daily Picayune,* December 14–19, 1902; *Times-Democrat,* December 14, 19, 1902; *New York Tribune,* December 19, 1902.

56. Similarly, black coal wheelers who coaled ships and unloaded coal barges secured little support from the Dock and Cotton Council. Without the Council's endorsement, roughly 800 members of Coal Wheelers Local Union 45, ILA, struck without success in September 1903 to increase their 35 cents an hour wage to 40 cents. Two years later, 150 wheelers again failed to raise their hourly wages, which had dropped to 30 cents. The Cotton Council appointed a committee to assist with negotiations but went little further. Employers experimenting with mechanical devices and floating loaders easily defeated the strikers, resulting in a breakup of the union. A new Coal Handlers' Union, made up of "conservative members of the old organization," received preferential treatment as "individuals" in the strike's aftermath. In 1907, during the general dock strike in support of the screwmen, wheelers again failed to make headway. See *Union Advocate,* December 28, 1903; February 1, 1904; *Daily Picayune,* September 3, 4, 1903; October 5, 6, November 4, 11, 13–15, 17, 1905; *Times-Democrat,* September 5, 8, October 2, 1903; *States,* September 4–6, 1903; *Report of the Bureau of . . . Labor . . . 1902–1903,* 46.

Despite the tendency of waterfront laborers to support workers in allied trades, jurisdictional divisions remained pronounced. For example, during the summer of 1903, a serious rift between screwmen and longshoremen threatened to break the Dock and Cotton Council apart and shatter the year-and-a half-old carefully fostered solidarity. The issue involved tradition and jurisdiction. While the clearly unskilled work of unloading cotton remained the domain of the screwmen, the screwmen traditionally permitted a certain number of longshoremen near them on board the ship to unload some cotton. The screwmen invoked their technical jurisdiction, barring longshoremen from a single ship and precipitating a major crisis in longshore ranks. When the Dock and Cotton Council sided with the screwmen, longshoremen vowed to defy it. The crisis raised many questions of procedure and principle. Under its rules, the Council could have expelled the longshoremen, but such action would have deprived it of its largest organizations, thereby undermining its strength. This the longshore unions clearly understood. "Let it come," dared one longshoreman. "They can't do it too quick. There are 2,000 longshoremen to 700 screwmen. If they want us out of the Council, we will go, but watch what becomes of the Council." Instead, the Council took no action, and the crisis was narrowly averted when the black and white screwmen backed down. "It is a serious matter for labor unions to begin fighting amongst ourselves," white screwmen's president Robert Trainor conceded. "We have enough to contend with when we combine against the common enemy without spending our forces against each other." Given their precarious position vis-à-vis the steamship agents and the simple fact that the Council could not function without the longshoremen, the screwmen had little choice but to sacrifice principle for a more pragmatic unity. During the crisis, the Elder-Demster Steamship Line repeatedly offered the dissatisfied longshoremen all of the screwmen's work of unloading cotton. Approximately 16 days after the jurisdictional controversy first arose, this unique conflict came to an undramatic end. Satisfied by the screwmen's concession, the longshoremen steadfastly refused to take their work, and agents had to accept the status quo. "We're paying screwmen's wages for longshoremen's work," the Elder-Demster agent conceded. But his workers unloaded his remaining 2700 bales of cotton without incident. See *Daily Picayune*, July 9–17, 23, 1903.

57. *Daily Picayune*, December 18, 1902.

58. For example, stevedore J.B. Honor successfully sued the black and white long-shoremen's unions in 1904. *Daily Picayune*, April 4, 11, 12, 20, 22, 26, May 25, 1904; *Times-Democrat*, October 13, 1903; *Galveston Daily News*, September 9, 1903. The president of the International Longshoremen's Association, Daniel Keefe, praised contracts and condemned sympathy strikes even more vigorously. "Sympathetic strikes can find no suitable apology for their existence during the life of any trade agreement or contract. It certainly means dishonor to any organization to engage in sympathetic strikes when bound by an annual agreement or contract, it is nothing less than suicide. We must, under no circumstances . . . no matter how great the provocation may appear, think of repudiating our contract or agreement. . . . The whole life and progress of organized labor today hinges upon their loyalty to their contracts." See *Proceedings of the Fourteenth Annual Convention of the International Longshoremen, Marine and Transportworkers' Association, 1905*, 26.

In 1908, employers brought suit against James Byrnes, Dock and Cotton Council president and later state Labor Commissioner, Philip Pearsaw, president of the black Coal Wheelers' Union, and E.S. Swann of the black longshoremen, for ordering the Coal Wheelers' Union to refuse to load a steamship on behalf of striking longshoremen in 1907. Indicted in early 1908, the three were convicted of violations of the Sherman Anti-Trust Act in January 1911. They received only a nominal fine. *States*, January 25, 1911. The most compelling analysis of the impact of law and the courts on the strategies and outlooks of the labor movement is William Forbath's "The Shaping of the American Labor Movement," *Harvard Law Review* 102, No. 6 (April 1989), 1111–1256. On earlier efforts of New Orleans employers to utilize the court system to restrain longshore unions, see Christopher L. Tomlins, *The State and the Unions: Labor Relations, Law, and the Organized Labor Movement in America, 1880–1960* (New York: Cambridge University Press, 1985), 61; *Birmingham Labor Advocate*, November 19, 1892, April 22, 1893.

59. *Daily Picayune*, December 11, 13, 1902; *Times-Democrat*, December 12, 13, 15, 1902.

60. *Daily Picayune*, December 15, 1902.

61. *Daily Picayune*, December 15–17, 1902; *Times-Democrat*, December 15, 1902.

62. *Times-Democrat*, March 12, 1908.

63. International Longshoremen's Association, *Proceedings of the Twenty-first Convention* (Port Huron: Riverside Printing Co., 1913), 173. Longshore laborers were not the only group to cross racial lines. In the summer of 1903, black and white lumber mill hands in New Orleans, aided by white longshoreman and AFL organizer Rufus Ruiz, formed an interracial, industrial union (embracing skilled and unskilled men). The organization elected two sets of delegates to represent them before the city's two labor federations—black unskilled delegates attended the Central Labor Union, while skilled whites attended the Central Trades and Labor Union. See *Times-Democrat*, July 20, 1903. On the lumber workers' strike, see *Daily Picayune*, June 14, 16, 23-28, 30, July 7, 10, 1903. Other groups of workers that adopted biracial structures to govern relations between black and white locals included funeral carriage drivers, cigar mak-

ers, horseshoers, and hod carriers. See *Daily Picayune,* January 13, 1902; November 14, 1905; October 16, 1911; *Times-Democrat,* September 2, 1903.

64. *Times-Democrat,* May 6, 15, 1903. The *States* and the *Herald* are quoted in the *Times-Democrat.*

65. *Official Journal of the Proceedings of House of Representatives of the State of Louisiana at the First Regular Session of the Third General Assembly,* May 28, 1908, 200.

66. *Daily Picayune,* March 28, April 25, 1908. Also see: March 24, 1908.

67. *Report of the Bureau . . . of Labor . . . 1902–1903,* 39–40; Porter, "Report of Screwmen's Lockout," 86–87; *Times-Democrat,* April 27, 28, 1903; *Daily Picayune,* April 25, 26, 28, 1903; *Times-Democrat,* April 26–28, 1903. On Keefe and the National Civic Federation, see Tomlins, *The State and the Unions,* 73.

68. *Times-Democrat,* April 28, 1903.

69. *Daily Picayune,* May 3, 1903.

70. *Daily Picayune,* May 5, 9, June 13, 15, 28, September 7, 1903; also see *Times-Democrat,* September 2, 1902; Foster, *Pops Foster: The Autobiography of a New Orleans Jazzman,* 55; "New Orleans Letter," *The Journal of Labor* (Atlanta), September 16, 1904; *Southwestern Christian Advocate,* September 10, 1903; September 6, 1906; There were small cracks in the white union movement's racial front. In 1904, for example, black and white building trades workers not only marched together on Labor Day, but attended the same celebration dinner; the following year, black and white carpenters again marched together apart from the CTLC (with the black marchers behind the whites), but they did not hold a joint dinner. In 1911, some 10,000 union members and their allies marched in a "monster mass meeting" jointly sponsored by the labor movement and the Socialist party to protest the arrest and kidnapping of the McNamara brothers, accused of dynamiting the Los Angles *Times* building. The parade consisted of delegations from the CTLC, the Building Trades Council, the Structural Iron Workers, the Street Carmen, followed by the Teamsters and Carriage Drivers' Union, and black Freight Handlers. See *Southwestern Christian Advocate,* September 8, 1904; *Daily Picayune,* September 3, 5, 1905; *Times-Democrat,* October 5, 1911. Black unionists in other southern cities, including Jacksonville, Mobile, Savannah, and Galveston, marched independently of white workers on Labor Day. See *Florida Times-Union,* September 7, 8, 1903; *Mobile Register,* September 8, 1903; September 2, 3, 1907; *Savannah Tribune,* September 12, 1903; *Galveston Daily News,* September 6, 1898, September 7, 1903. In contrast, black unionists in Atlanta marched in the white union parade as a "fifth division." *Atlanta Constitution,* September 7, 8, 1903. On parades and the representation and creation of cultural meaning, see Mary Ryan, "The American Parade: Representations of the Nineteenth-Century Social Order," *The New Cultural History,* ed. Lynn Hunt (Berkeley: University of California Press, 1989), 131–52.

71. *Daily Picayune,* May 25–27, 30, June 3, 1904.

72. *Daily News,* August 28, 1907; *Times-Democrat,* August 28, 1907; *Daily Picayune,* August 28, 1907; Pearce, "The Rise and Decline of Labor in New Orleans," 64.

73. *Daily Picayune,* August 29, 30, 1907; *Daily News,* August 29, 30, 1907.

74. *Daily News,* August 31, 1907; *Times-Democrat,* August 30, 1907; *Daily Picayune,* August 30, 31, September 1, 1907; *Galveston Daily News,* August 30, 1907; Pearce, "The Rise and Decline of Labor," 63–64; George Carroll Miller, "A Study of the New Orleans Longshoremen's Unions from 1850 to 1962" (unpublished M.A. thesis, Louisiana State University and Agriculture and Mechanical College, 1962), 21.

75. For a discussion of southern Republicans and blacks, see: Hirshson, *Farewell to the Bloody Shirt;* Vincent P. DeSantis, *Republicans Face the Southern Question: The*

New Departure Years, 1877–1897 (New York: Greenwood Press, 1959); Paul D. Casdorph, *Republicans, Negroes, and Progressives in the South, 1912–1916* (Alabama: University of Alabama Press, 1981), 57–60; George Brown Tindall, *The Emergence of the New South 1913–1945* (Baton Rouge: Louisiana State University Press, 1967), 167–69; Londa Lavonne Davis, "After Reconstruction: Black Politics in New Orleans, 1876–1900" (unpublished M.A. thesis, University of New Orleans, 1981). On persistent divisions between "black and tans" and "lily-whites," see *Times-Picayune*, October 5, 1915. *The Crisis* 4, No. 1 (May 1912), 7–8; Walter L. Cohen, "The Situation in Louisiana," *The Colored American Magazine* IX, No. 2 (August 1905), 455–57. There is no in-depth study of New Orleans' black community in the twentieth century. For a fine introduction to the subject, see: Arnold Hirsch, "Race, Rights, and the Reemergence of Black Politics in Twentieth Century New Orleans" (unpub.paper, Univ. of New Orleans, 1988).

76. *Daily Picayune*, February 19–21, March 11, May 4, 7, June 18, 26, September 20, 23, 1904; *Times-Democrat*, February 21, 1907; *Southwestern Christian Advocate*, March 18, 1908; Casdorph, *Republicans, Negroes, and Progressives*, 58–60; Donald E. DeVore, "The Rise from the Nadir: Black New Orleans Between the Wars, 1920–1940" (unpublished M.A. thesis, University of New Orleans, 1983). For a brief discussion of the Equal Rights League, also see Rosenberg, *New Orleans Dockworkers*, 28–31; *The Republican Liberator*, September 9, 1905.

77. *Southwestern Christian Advocate*, December 11, 1902; December 2, 1909; *Daily Picayune*, November 27, 1901; December 3, 1902; *Times-Picayune*, January 1, 1914; November 13, 29, 1918; December 3, 1919; August 5, 1932; *New Orleans Item*, August 4, 1932; "The Poll Tax and the Negro," *The Colored American Magazine* X, No. 3 (March 1906), 155. Some white trade unions required their members to pay poll taxes. See *Daily Picayune*, November 23, 27, 30, December 2, 1901.

78. *Southwestern Christian Advocate*, April 23, 1903; May 7, 1903; *Times-Democrat*, May 3, 1903; *Daily Picayune*, July 31, October 18, November 4, December 3, 1902; *Times-Democrat*, November 4, 1902; *Galveston Daily News*, August 7, 1903. For a discussion of black protests against streetcar segregation in this era, see August Meier and Elliott Rudwick, "The Boycott Movement Against Jim Crow Streetcars in the South, 1900–1906," *The Journal of American History* 55, No. 4 (March 1969), 756–75; Jennifer Roback, "The Political Economy of Segregation: The Case of Segregated Streetcars," *Journal of Economic History* 46, No. 4 (December 1986), 893–917; "The Negro and Justice," *The Independent* LXIII, No. 3072 (October 17, 1907), 923; "On Boycotting Street Cars," *The Colored American Magazine* IX, No. 1 (July 1905), 398–99; *Galveston Daily News*, September 4, 9, 1903. Mobile's streetcar law, enacted by the city's general council and which generated black protest, went into effect in the same week. *Mobile Register*, November 5, 1902.

79. *Times-Picayune*, June 8, 1914.

80. *Daily Picayune*, December 7, 1912; *Times-Picayune*, June 1, 1914. On Providence Hospital, see: *Times-Picayune*, November 2, 30, December 3, 1917; July 6, 1919. For additional information on black fraternal and other organizations, see the special section on "Progressive Negroes" in the *Daily Picayune*, December 30, 1913; "A New Orleans Baseball Park," *The Crisis* 23, No. 1 (November 1921), 20. Vance was also involved in Booker T. Washington's National Afro-American Council. *Daily Picayune*, July 7, 1903.

81. On the ILA resolution, see International Longshoremen's Association, *Proceedings of the Twenty-Fifth Convention, Galveston, Texas, July Fourteenth to Twenty-second, 1919*, 514–15. On the challenge to residential segregation law, see: "Cases

Supported—New Orleans Residential Segregation" (1924–1927), in August Meier, ed., *Papers of the NAACP, Part 5. Campaign against Residential Segregation, 1914–1955* (Frederick, Md.: University Publications of America, 1986), Frames 623 to 783. On black protests during World War One, see Chapter 6.

82. First-and second-generation German immigrants established the United Brewery Workmen in the 1880s. Union members were often uncomfortable with what they saw as the reformist populism and utopianism of the Knights of Labor; during the 1890s, the International supported the Socialist Labor party (SLP). But the SLP's advocacy of dual unionism in opposition to the AFL precipitated a break within the party's ranks and clashed with the Brewery Workmen's own commitment to working with the AFL. As a result, the Brewery Workmen reluctantly abandoned the SLP. Following the split with the SLP, brewery workers aligned themselves with Eugene Debs' Social Democracy and its successor, the Socialist party, in 1901. See John H.M. Laslett, "Marxian Socialism and the German Brewery Workers of the Midwest," in *Labor and the Left: A Study of Socialist and Radical Influences in the American Labor Movement, 1881–1924* (New York: Basic Books, 1970), 14–15, 18–21; Oscar Ameringer, "Unionism: Present and Future," in The Oscar Ameringer Collection, Papers and Publications, Box No. 1, Folder 1–27, undated, in Archives of Labor History and Urban Affairs, University Archives, Wayne State University; "Report of the International Secretaries: Our Political Position," *Proceedings of the Seventeenth Convention of the International Union of United Brewery Workmen of America held in New York City, September 13th to 27th inclusive, 1908,* 132; Philip Foner, *History of the Labor Movement,* Vol. 3, *The Policies and Practices of the American Federation of Labor 1900–1909* (New York: International Publishers, 1964), 211; James O. Morris, *Conflict Within the AFL: A Study of Craft Versus Industrial Unionism, 1901–1938* (1957, rpt. Westport: Greenwood Press, 1974), 16, 20; Paul F. Brissenden, *The I.W.W.: A Study of American Syndicalism* (1919, rpt. New York: Russell & Russell, 1957), 38, 55, 58–59. On the AFL and radical currents, see: J.F. Finn, "AF of L Leaders and the Question of Politics in the Early 1890s," *American Studies* 7, No. 3 (1973), 243–65; Montgomery, "The 'New Unionism,' and the Transformation of Workers' Consciousness," 91.

83. *First Annual Report of the Bureau of Statistics of Labor for the State of Louisiana 1901* (Baton Rouge: *Advocate,* 1902), 190; *Report of the Bureau of Statistics of Labor for the State of Louisiana 1904–1905* (New Orleans: Miller & Brandao, 1906), 61–62, 76–77, 113–15; *Daily Picayune,* September 4, 9, 5, October 30, November 2, 24, December 5, 9, 1901; January 27, November 13, 1902; July 5, 1903; June 18, 1904; November 4, 1905; August 18, September 18, October 6, 12, 1905; *Union Advocate,* August 24, October 5, 19, 1903; January 8, 11, 1904.

84. Patrick McGill before the AFL Executive Board, "The AFL Executive Council Minutes of Meetings, January 20–25, 1908," 43, in *The American Federation of Labor Papers: The Samuel Gompers Era* (Reel 3), 1204 (hereafter referred to as the *AFL Papers);* McGill to C.P. Shea, *The Teamsters* III, No. 2 (December 1905), 20–21; "Report Organizer P. McGill," *The Teamsters* III, No. 10 (August 1906), 40a–40d; *Report of the Bureau of Statistics of Labor for the State of Louisiana, 1904–05* (New Orleans: Miller and Brandao, 1906), 62–63, 76–77, 113–15, 123–24; *Daily Picayune,* September 18, October 6, 12, November 4, 1905; May 24, 1907; *Times-Democrat,* September 27, 1905; June 3, 1907; Covington Hall, "Labor Struggles in the Deep South" (unpublished manuscript in Archives of Labor History and Urban Affairs, University Archives, Wayne State University, and Special Collections, Tulane University), Part III, 1, 4–6. There are several paginations in the manuscript; I have followed

the original typescript pagination. For a short biography of McGill, see *Daily Pica-yune*, December 30, 1913. For a history of the New Orleans dispute, also see *Galveston Journal*, October 11, 25, 1907.

85. "Official: Brewery Workers' Charter Revoked," *American Federationist* XIV (July 1907), 483; "Official Resolutions and Decisions Adopted by the Executive Council of the AFL at Its Meeting in Washington, D.C., March 18–23, 1907," *American Federationist* XIV (May 1907), 332–35; Hermann Schlüter, *The Brewing Industry and the Brewery Workers' Movement in America* (Cincinnati: The International Union of United Brewery Workmen of America, 1910), 219–28; *Daily Picayune*, May 24, 25, 1907. The Brewery Workmen's International secretary Joseph Proebstle repeated an often-made argument when he explained upon his arrival in New Orleans in early June that the union was "only asking the same privilege that is accorded to the Mine Workers, the Longshoremen, the Seamen and other unions, which have jurisdiction over all the men employed in their respective industries." The "men of Local 215," he contended, "were sold in the deal like chattel slaves." *Times-Democrat*, June 4, 1907; *Daily Picayune*, June 4, 1907.

86. "Report of the International Secretaries," *Proceedings of the Seventeenth Con-vention of the International Union of United Brewery Workmen of America . . . 1908,* 128, 131; Leonard to Morrison, June 2, 9, 1907, *The John Mitchell Papers* (Reel 12); Oscar Ameringer, *If You Don't Weaken: The Autobiography of Oscar Ameringer* (New York: Henry Holt and Company, 1940), 194–95; "Report of P. McGill," *The Teamsters* IV, No. 10 (August 1907), 31–32; "Editorial," *The Teamsters* IV, No. 8 (June 1907), 3; Hall, "Labor Struggles in the Deep South," Part III, 5; Schlüter, *The Brewing Indus-try,* 228; Proebstle to Gompers, September 27, 1907, *The AFL Papers* (Reel 35), 1402; *Daily Picayune,* May 24, 25, 30, June 1, 1907; *Daily News,* May 25, 29–31, June 1, 1907; *Times-Democrat,* May 25, June 1, 2, 4, 1907; St. Louis *Post-Dispatch,* June 1, 1907.

87. The total number of members of the Brewery Workmen's unions in New Or-leans remains uncertain, but the figure probably did not exceed 200. Beer drivers composed a smaller but unspecified number; there is no information on the racial breakdown of either the brewery workers or the beer drivers. The numbers could not have been very large.

88. Leonard to Morrison, July 7, 1907, *The John Mitchell Papers* (Reel 12); Tele-gram from Byrnes, Scully, Swann, Wynn, Graff, Lob, Duffy and Ruiz to Samuel Gompers, July 4, 1907, *The AFL Papers* (Reel 35), 1401; "Report of the International Secretaries," *Proceedings of the Seventeenth Convention of the . . . United Brewery Workmen . . . 1908,* 128; Hall, "Labor Struggles in the Deep South," Part III, 12–13; *Daily News,* July 5, 13, 1907; *Daily Picayune,* May 25, 30, June 4, July 6, 1907; *Times-Democrat,* May 30, 31, 1907; *The Weekly People,* September 6, 1907.

89. Leonard to Morrison, July 7, 1907, *The John Mitchell Papers* (Reel 12). Oscar Ameringer offered a similar explanation: Many dock workers "found employment in breweries during the summer, while many brewery workers were to be found at the docks in the winter. Out of this situation developed the exchange of union cards. By simply depositing his dock worker's card with the Brewery Workers' Union, the dock worker became a brewery worker in good standing and entitled to all the rights. By the same token, brewery workers who deposited their cards with dock unions became dock workers in good standing." Ameringer, *If You Don't Weaken,* 195–96. This close relationship did not prevent Brewery Workers' Local 161 from criticizing dock workers for failing to observe its 1903 boycott of the American Brewery. The Screwmen's and Longshoremen's Union Saloon, located on the corner of St. Thomas and Chartres

streets, sold scab beer to union members. Local 161 similarly placed the International Freight Handlers, Local 394, and the white screwmen's union on its list of unions whose members patronized saloons and groceries that sold the American Brewery Company's beer. See *Union Advocate,* August 24, October 19, 1903; *Daily Picayune,* October 20, 1903.

90. Ameringer, *If You Don't Weaken,* 195–96; Leonard to Morrison, June 2, 9, July 7, September 29, October 17, 1907, *The John Mitchell Papers* (Reel 12). The dock and black unions were not alone in their support for the industrial unionists. AFL organizer James Leonard reported that although a larger number of locals had been advised by their international unions to "keep hands off" the jurisdictional conflict, they had in fact displayed "such a lukewarmness [to the AFL position] as to lead one to believe that they are favoring the Brewery Workers." Unions of tailors, elevator constructors, electrical workers, retail clerks, bricklayers and masons, as well as the longshoremen, screwmen, and black interior freight handlers all endorsed the brewery workers and condemned the white AFL Council. CTLC president Lee demanded that the unions endorsing the Brewery Workers immediately retract their comments and support, but his declarations and threats carried little weight with the dissidents. See *Daily Picayune,* May 28, 30, June 5, 6, 7, 8, 1907; *Times-Democrat,* June 5, 1907; *Daily News,* June 2, 5, 14, 1907; Leonard to Morrison, June 9, September 29, 1907, *The John Mitchell Papers* (Reel 12).

91. *Daily Picayune,* June 1, 8, 1907; Laslett, "Marxian Socialism and the German Brewery Workers," 17; Ameringer, *If You Don't Weaken,* 196.

92. *Daily Picayune,* July 6, 11, 13, 19, 21, 1907; Leonard to Morrison, July 14, 1907, *The John Mitchell Papers* (Reel 12). An interracial meeting held July 10 dispatched a committee to meet with the CTLC. The Council, however, objected not to the presence of blacks but rather to Louis Kemper of the Brewery Workers, on the grounds that his union had been expelled from the AFL. Committee members James Hughes and T.R. LeBlanc then refused on principle to address the meeting. A second interracial meeting, on July 18, established another committee of five—three whites (James Hughes of the longshoremen, Thomas Harrison of the screwmen, and Rufus Ruiz, a longshoreman and AFL organizer) and two blacks (T.R. LeBlanc of the CLU and black cotton teamsters' president Dave Norcum)—to work with the Brewery Workers. *Daily Picayune,* July 10, 11, 13, 21, 1907; Leonard to Morrison, July 14, 1907, *The John Mitchell Papers* (Reel 12).

93. *Daily News,* July 25–31, 1907; *Daily Picayune,* July 26–28, 30, 1907; *Item,* July 27, 1907; Hall, "Labor Struggles in the Deep South," Part III, 21–22.

94. Leonard to Morrison, June 2, 9, July 14, September 1, 1907, *The John Mitchell Papers* (Reel 12); "Report of P. McGill," *The Teamsters* IV, No. 10 (August 1907), 32.

95. Hall, "Labor Struggles in the Deep South," Part III, 12, 47; *Daily Picayune,* June 7, 1907; *Daily News,* June 7, 1907; *The Weekly People,* June 22, July 6, September 14, November 2, 1907. At the outset of the crisis in May 1907, Socialist Labor party leader Daniel DeLeon, upon the invitation of the "red faction" of the city's Socialist party, lectured crowds of New Orleans workers on the "Burning Question of Trade Unionism" and "Socialism, the Goal of Trade Unionism," pumped "hot shot into the craft unionism" of the AFL, and praised the recently formed IWW. *Daily News,* May 10, 12, 13, 1907; *Daily Picayune,* May 11, 12, 1907; Hall, "Labor Struggles in the Deep South," Part III, 9–12.

96. Years later Covington Hall related that Tom Gannon, former screwman and then a leader of the brewery workers, explained to him why Ewing did not interfere with Molyneaux's radical journalism. "We Irish screwmen and longshoremen con-

trolled the Eleventh Ward," Gannon told Hall. "We said who should be its boss. Ewing knew this—he also knew that losing the boss-ship of the ward meant losing that of the city." A pro-Brewery Workers committee told Ewing that it wanted "Ameringer, Molyneaux and Hall to use [his *News*] as they please, put anything in it they think best; and if you don't agree to this, it will just be too bad for you in the next election. Ewing agreed, and we literally expropriated the 'official organ' of the Central Trades and Labor Council." Hall, "Labor Struggles in the Deep South," Part III, 49. A "man of refined manners," Molyneaux became secretary of the local Socialist party at age 22 in 1904. The *Picayune* reported that he was a student of both economics and psychic phenomena, being a "hypnotist of some ability." *Daily Picayune,* July 15, 1904.

97. Hall, "Labor Struggles in the Deep South," Part III, 47; Ameringer, *If You Don't Weaken,* 195. For discussions of southern socialists, see McAlister Coleman, "Oscar Ameringer Never Weakened," *The Nation* (n.d.), in the Oscar Ameringer Collection, Wayne State University; Grady McWhiney, "Louisiana Socialists in the Early Twentieth Century: A Study of Rustic Radicalism," *Journal of Southern History* XX, No. 3 (August 1954), 315–36; Thomas Becnell, "Louisiana Senator Allen J. Ellender and IWW Leader Covington Hall: An Agrarian Dichotomy," *Louisiana History* XXIII, No. 3 (Summer 1982), 259–75; James R. Green, *Grass-Roots Socialism: Radical Movements in the Southwest 1895–1943* (Baton Rouge: Louisiana State University Press, 1978), 17, 35–37. For Ameringer's activities, also see *States,* March 25, 1908; *Daily Picayune,* March 24, 1908. On Hall's activities and philosophy, see his letters in *Union Advocate,* February 16, March 16, August 17, 1903; David R. Roediger, "Covington Hall: The Poetry and Politics of Southern Nationalism and Labour Radicalism," *History Workshop Journal,* Issue 19 (Spring 1985), 162–68; Nicholas Lemann, "Revolutionary Patriots: Covington Hall and the Southern Anti-Capitalist Tradition" (unpublished B.A. honors thesis, Harvard University, 1976), 33–48.

98. *Daily News,* October 4, 1907; *The Teamsters* V, No. 8 (June 1908), 10–11. For examples, see *Daily Picayune,* June 20, 30, September 25, 1907; *Daily News,* June 6, July 31, August 1–3, 7, 8, 15, 21, 23, 25, 27, September 4, 6, 7, 10-17, 20, 21, 25, 26, 1907.

99. *Daily News,* August 16–18, 20, 24, 1907; *Times-Democrat,* August 16, 17, 1907; *Daily Picayune,* August 11, 16–18, 1907; *United Labor Journal* X, No. 46 (August 31, 1907), 22a, 33. Dock union officials assumed positions of leadership in the new body. Thomas Harrison, secretary of the white screwmen and former Commissioner of the state's Bureau of Statistics of Labor, was unanimously elected president of the new central body; other elected officers included I.G. Wynn of the black cotton yardmen as vice president; former screwman Thomas Gannon of Brewery Workers' Local 215 as recording secretary; Harry Baptiste of the black Universal Freight Handlers as financial secretary; and James Byrnes, president of the Dock and Cotton Council and the white screwmen as treasurer. Hall, "Labor Struggles in the Deep South," Part III, 22–27; *Daily Picayune,* August 24, 31, 1907; *Galveston Journal,* October 11, 25, 1907.

100. Hall, "Labor Struggles in the Deep South," Part III, 22.

101. In 1915, Thomas P. Woodland, a black labor leader and president of the Central Labor Union, issued an open letter in the pages of the ILA's journal, *The Longshoreman,* that illustrated the complexity of ideology and organizational affiliation. In his page-long appeal, Woodland drew heavily from the 1907 United Labor Council's preamble and constitution. "The capitalist class owns the factories, land, ships, railroads, in fact all the means by which wealth is produced and distributed," Woodland wrote. "Every new invention in machinery, every new discovery of natural forces, has worked to the benefit of the propertied class alone, which has been further

enriched thereby." He argued that "in order to emancipate ourselves from the influence of the class that is hostilely arrayed against the wage-working class, the wage-working class must organize and oppose the power of capital with the power of organized labor and must champion their own interests on the docks, in the cotton presses, the team drivers, in fact, every class of labor that is connected with the shipping interest of the port of New Orleans. . . . The class-conscious power of capital with all its camp-followers, is confronted with the class-conscious power of labor. . . . There is no power on earth strong enough to thwart the will of such a majority conscious of itself. The earth and all its wealth belong to all." Such beliefs, however, did not lead Woodland to embrace either the Socialist party or the IWW. Rather, he admitted that "I am not one of those who believe that capital and labor are natural enemies. On the contrary, the welfare of both is best promoted by a harmonious understanding between the two. . . . Let us ask nothing more than justice from the employing class, but accept nothing less. I believe that 'Labor and Capital should always go hand in hand in the God-like work of promoting the greatest good of the greatest number.' " The solution to labor's plight, Woodland concluded, lay in membership in the ILA and the AFL. As the "channel of expression and activity for the united labor hosts of DOCK WORKERS," the ILA has "succeeded in establishing better wages, shorter hours, better conditions, protection of your job, pensions, and such just demands as every dock worker wants." Thomas P. Woodland, "An Open Letter on Affiliation. 'Why We Should Organize and Affiliate with the International Longshoremen's Association.' " *The Longshoreman* 6, No. 7 (May 1915), 5.

102. Minutes of the Executive Council of the AFL, August 21, 23, 1907, *The AFL Papers* (Reel 3), 1204; also *The John Mitchell Papers* (Reel 43); Leonard to Morrison, September 1, 1907, *The John Mitchell Papers* (Reel 12); *Daily Picayune,* August 24, 1907; *Times-Democrat,* August 17, September 14, 1907; *Daily News,* August 27, September 24, October 11, 1907; Brissenden, *The I.W.W.,* 215.

103. *Report of Proceedings of the Twenty-Eighth Annual Convention of the American Federation of Labor,* November 9–21, 1908, 71–75; Minutes of the Executive Council of the AFL, January 20–25, 1908, *The John Mitchell Papers* (Reel 43); *States,* April 19, 1908; *Brauer Zeitung,* July 4, 11, 1908; Leonard to Morrison, January 12, 1908, *The John Mitchell Papers* (Reel 13); *Daily Picayune,* December 5, 1907, June 5, 24, July 2, September 1, 30, 1908; *Times-Democrat,* June 5, 1908; Schlüter, *The Brewing Industry,* 228; "Brewery Workers' Troubles," *The Teamsters* V, No. 4 (February 1908), 3–8; McGill to The Executive Board, IBT, *The Teamsters* V, No. 8 (June 1908), 10–11.

104. *Daily Picayune,* September 1, October 1, 1909; "Report of Pat McGill," *The Teamsters* V, No. 10 (August 1908), 36–37; *The Teamsters* VI, No. 10 (August 1909), 22–23; "Report of Pat McGill," *The Teamsters* VI, No. 6 (April 1909), 2–4; McGill to Tobin, *The Teamsters* VII, No. 10 (August 1910), 28–31.

105. *Daily Picayune,* August 23, 25, September 4, 7, 10, 1907; *Times-Democrat,* September 10, 1907. Discussion or mention of the fall 1907 strikes can be found in David Montgomery, "The 'New Unionism' and the Transformation of Workers' Consciousness," 93; Jim Stodder and Dave Wells, "A Short History of New Orleans Dockworkers," *Radical America* 10, No. 1 (January–February 1976), 51–55; Rosenberg, *New Orleans Dockworkers,* 115–41; Pearce, "The Rise and Decline of Labor," 63–70; Ameringer, *If You Don't Weaken,* 196–202; Foner, *History of the Labor Movement,* Vol. 3, 250–53; Bernard A. Cook and James R. Watson, *Louisiana Labor: From Slavery to "Right-to-Work"* (Lanham: University Press of America, 1985), 105–8. The experience of Galveston served as the employers' key point of reference—they never tired of describing the differences in the organization of labor between the two ports.

Claiming that screwmen in the Texas port stowed between 300 and 360 bales of cotton per day for wages of $31 per gang of five, agents and stevedores made it clear that their own demand for 200 bales did not even "equalize conditions." Distrustful of their employers' figures, screwmen conducted their own investigations into the labor conditions at New Orleans' rival ports. James Byrnes, president of the white screwmen and the Dock and Cotton Council, and Thomas Harrison, secretary of the white screwmen and former Commissioner of the state's Bureau of Statistics of Labor, visited Port Arthur and Galveston to ascertain for themselves the exact pay, level of output, and other conditions pertaining to the screwmen of these ports. They found the New Orleans agents' assertions about Texas stowing were seriously inflated. With detailed information concerning wages, work loads, and conditions in these cities, the screwmen's committee—ten whites and ten blacks—confronted the stevedores and agents before the commercial exchange committee. Unimpressed, the commercial exchange committee sided with the agents, concluding that the 200-bale a day limit—not the 160-bale limit—constituted a "fair day's work." But workers rejected the non-binding ruling and a joint session of black and white screwmen demanded a wage increase from $5 to $6 a day for regular work and from $6 to $7 a day for foremen for handling the same number of bales. See *Daily News,* September 1, 17, 20–22, 26, 27, 1907; *Times-Democrat,* September 27, 1907; *Daily Picayune,* September 10, 24, 1907; *Galveston Daily News,* September 17, 21, 24, 1907; *Houston Post,* September 18, 19, 25, 1907; Pearce, "Rise and Decline of Labor," 65.

106. *Times-Democrat,* September 27, 29, 30, 1907; *Daily News,* September 27, 1907; *Daily Picayune,* September 18, 19, 20, 22, 23, 27, 29, 30, 1907; *Houston Post,* September 30, 1907; in Cotton Exchange Minute Book, L (Special Collections, Howard-Tilton Memorial Library, Tulane), see Minutes of the Meeting of the Board of Directors, September 29, 1907, 68; Minutes of Monthly Meeting of the Board of Directors, 75; Meeting of Board of Directors, October 12, 1907, 76; Meeting of Board of Directors, October 14, 1907, 77.

107. *Daily Picayune,* September 11, 12, 16, 18–20, 1907; *Times-Democrat,* October 3, 6, 7, 9–11, 15, 16, 1907. Also see W.W. Howe to Attorney General, October 5, 1907, and Attorney General to W.W. Howe, October 15, 1907, in File No. 16, Sub 8, 1907—Cotton—New Orleans (Justice Department Files, National Archives).

108. *Daily Picayune,* September 13, 1907; *Times-Democrat,* October 10, 16, 1907.

109. *Times-Democrat,* October 7, 11, 1907; "The One Hundred and the Modern Mr. Carlyle," *Harlequin* IX, No. 14 (October 12, 1907), 4; Hall, "Labor Struggles in the Deep South," Part III, 36–38.

110. *Times-Democrat,* October 1, 2, 5, 1907; *Daily Picayune,* October 1, 2, 5, 6, 1970; *L'Abeille de la Nouvelle Orleans,* October 1, 2, 1907; *Galveston Daily News,* October 4, 5, 1907; *Houston Post,* October 2, 7, 1907; *New York Daily Tribune,* October 1, 1907; Philip S. Foner, *Organized Labor and the Black Worker 1619–1973* (New York: International Publishers, 1974), 90–92; Foner, *History of the Labor Movement,* Vol. 3, 251–53. Estimated figures for the number of dock workers are drawn from the *Picayune.*

111. *Daily Picayune,* October 6, 7, 11–13, 1907; *Times-Democrat,* October 11, 12, 1907; *L'Abeille de la Nouvelle Orleans,* October 11–13, 1907; "Thunder and Peacock Parade. Ineffective on Labor Troubles," *Harlequin* IX, No. 14 (October 17, 1907), 4–5; *Houston Post,* October 12, 1907; Pearce, "The Rise and Decline of Labor," 63, 67. The unprecedented three-week-long general levee strike tested the unions' alliance and endurance. Journalists reported rumors of serious divisions within the dock workers' ranks as the strike wore on. The longshoremen and several black unions, they

stated, were pressuring the screwmen to settle, even at the expense of surrendering. "There is one thing certain," the *Times-Democrat* reported one unnamed union leader as stating, "the screwmen must settle this matter . . . one way or the other. The other laborers are tired of pulling their chestnuts out of the fire for them. They are the aristocrats of the levee and they have never done anything but harm. . . ." But the anticipated division never erupted. Defying predictions that they would bolt, black longshoremen voted unanimously to endorse the general strike stance. "There was," Oscar Ameringer remembered, "considerably less danger of the Negroes deserting the whites than of the whites deserting the blacks." The Dock and Cotton Council similarly stood firm. Delegates voted to put the compromise proposal of the commerical exchanges—the original 200-bale limit—directly to the striking screwmen. When the men again rejected it, the Council voted unanimously to stand firmly behind them. "There seems to be absolutely no grounds for the published accounts of dissention among the other unions," the *Daily News* reported. "It's no longer the screwmen's strike . . . [F]rom all that can be learned the entire rank and file of the levee unions are determined to fight the matter to the end." After almost 45 days of fruitless negotiations, the *Times-Democrat* noted, the prospects for a settlement of the levee labor problem were as bleak as ever. *Times-Democrat,* October 9, 14, 1907; *Daily Picayune,* October 14, 1907; *Daily News,* October 14, 1907; *Galveston Daily News,* October 14, 1907; *Houston Post,* October 14, 1907.

112. *Times-Democrat,* October 12, 14, 15, 18, 20, 1907; *Daily Picayune,* October 10, 1907; *Harlequin,* October 17, 1907.

113. *Daily News,* October 21, 1907.

114. *Daily News,* October 19, 20–26, 1907; *Times-Democrat,* October 24, 25, 1907; *Daily Picayune,* October 24, 25, 1907.

115. *Daily Picayune,* October 29, 30, November 1–3, 1907; Ameringer, *If You Don't Weaken,* 215; Pearce, "The Rise and Decline of Labor," 68–69; Herbert Northrup, "The New Orleans Longshoremen," *Political Science Quarterly* LVII, No. 4 (December 1942), 529–30.

116. *Daily Picayune,* November 1, 1907; *Daily News,* November 1, 1907; *L'Abeille de la Nouvelle Orleans,* October 30, 1907; Ameringer, *If You Don't Weaken,* 216–17.

117. *Daily Picayune,* November 8, 9, 12–16, 20, 1907; *L'Abeille de la Nouvelle Orleans,* November 3, 7, 9, 13, 1907; *Daily News,* November 8–10, 17, 1907; Pearce, "The Rise and Decline of Labor," 69.

118. *Official Journal of the Proceedings of the House of Representatives of the State of Louisiana at the General Session of the Second General Assembly,* November 11, 1907, 14–15, and November 12, 1907, 25–27; *Official Journal of the Proceedings of the House of Representatives of the State of Louisiana at the First Regular Session of the Third General Assembly,* May 28, 1908, 107–9, 195–202; *Daily Picayune,* October 27, December 18, 1907; February 10, 20–22, March 10–12, 14, 17–19, 24, 25, 29, April 23–26, May 1–4, 1908; *Times-Democrat,* February 5, 6, 8, 9, 18–23, March 10–12, 16, 18, 19–21, 26–28, 30, 1908; *Mobile Register,* February 9, 1908; Savannah *Morning News,* February 5, 6, 1908.

119. *Official Journal of the Proceedings of the House of Representatives of the State of Louisiana,* May 28, 1908, 198–99; *Daily Picayune,* March 11, 1908; *States,* April 29, 1908.

120. *Daily Picayune,* March 11, 24, 28, 1908; *Daily News,* March 11, 14, 1908; Ameringer, *If You Don't Weaken,* 217–19; *Times-Democrat,* April 30, 1908.

121. *Daily Picayune,* March 24, 25, 30, 1908; *Daily News,* March 14, 1908; *States,* March 27, 1908.

122. *Official Journal of the Proceedings of the House of Representatives,* May 28, 1908, 107, 198–200; *Daily Picayune,* May 2, 3, 6, 7, 12, 13, 1908; *Times-Democrat,* May 29, 1908; W.R. Parkerson, "Memoirs, 1857–1912" (in Manuscripts, Rare Books and University Archives, Howard-Tilton Memorial Library, Tulane University), 7–8.

123. *Daily Picayune,* July 1, 2, 7–10, 26, 27, August 2, 1908; *Times-Democrat,* July 2, 3, 5, 16, 1908. In exchange for the employers' concession to employ an extra roller, the longshore unions agreed to sign a five-year contract and to take orders not just from their immediate foreman but from any one of a number of (union) foremen. On August 1, 1908, employers and longshoremen finally signed their new contract.

124. *Daily Picayune,* June 4, July 1–3, 6–26, 1908; *Times-Democrat,* July 1–11, 13–16, 18–21, 25, 1908. Although the Dock and Cotton Council refused to sanction the freight handlers' strike, once again individual longshoremen and even screwmen acted according to their consciences, refusing to work with non-union men replacing the strikers, or to touch cargo that had been "polluted . . . by contact with nonunion handlers." In one case, white longshore president Christy Scully ordered ten gangs of longshoremen to quit unloading a cargo, on the grounds that "union and nonunion men could not work together;" black teamsters' president Dave Norcum similarly ordered his members to not deliver cotton for a particular drayman associated with the railroad. When a ship that had taken on cargo by non-union men at Stuyvesant Docks redocked at Eighth Street, 200 union longshoremen, without authorization, struck the ship. Union longshoremen and screwmen also resisted the entreaties of their officers to load freight that had been placed on the wharves before the strike (and hence not "polluted"). *Times-Democrat,* July 4, 9, 10, 12, 14, 15, 18, 25, 1908.

125. *Daily News,* March 14, 1908; *Daily Picayune,* March 14, 1908; *Times-Democrat,* March 14, 1908.

Chapter 6

1. International Longshoremen's Association, *Proceedings of the Twenty-first Convention Including Proceedings of Executive Council Meeting, Boston, Massachusetts, July 14th to 19th inclusive, 1913,* 174.

2. International Longshoremen's Association, *Proceedings of the Twenty-third Convention, San Francisco, California, November 8, 1915,* 93–94.

3. *Daily Picayune,* May 31, 1913.

4. That same collaborative spirit also stood in sharp contrast with the experiences of Galveston and Mobile, where divisions between white and black dock workers that year resulted in strikes. In Galveston, the conflict centered on the extension of the New Orleans formula of the equal division of work between black and white screwmen, adopted in 1912. In 1913, white workers repudiated their "amalgamation contract" by signing with the largest stevedoring firms providing for the employment of only members of the white ILA local. The International backed up the black unionists, charging the white local with unfair behavior and violations of the International's orders, and ordering a strike of all other union longshore workers, black and white, in support of the black screwmen. In Mobile, the ILA sided with whites opposed to black efforts to challenge a traditional division of labor that reserved timber work for whites, lumber work for blacks. A strike of white and some black unions forced a small black local to accept the status quo. The *Mobile Register* found the New Orleans experience of contract renewal an "education" and advised its readers to "learn by others' experience." *Mobile Register,* September 24, 1913. On the Galveston and Mobile strikes, see ILA, *Proceedings of the Twenty-first Convention . . . 1913,* 98–100, 171, 178–79; *The*

Longshoreman 5, No. 2 (December 1913); *Galveston Daily News,* August 20, 23–28, 31, September 2–5, 7–9, 15–18, 20, 21, 23, 1913; *Houston Post,* September 3–7, 9, 16–21, 1913; *Mobile Register,* September 2–7, 9, 12–17, 19–26, 28–30, October 1, 2, 1913; *The Nation,* May 7, 1914.

5. *Daily Picayune,* September 2, 1910; May 13, 1912; August 1, 1913. Undermining the teamsters' bargaining position were new techniques and plans for reorganization of the cotton transportation process. The Cotton Exchange sought an agreement with the Illinois Central and Mississippi Valley Railroad for the "free in-belting" of cotton for export; that is, the rail lines would bring the cotton directly to the ships' sides, doing away entirely with unloading and draying of bales. Teamsters objected to the change, correctly noting that it would decrease their numbers substantially. In the short run, they forced the Cotton Exchange to abandon its plan. But the potential to bypass the drayage stage may have induced the teamsters to make other compromises when pressured by employers. *Item,* August 31, 1911.

6. *Voice of the People,* September 18, 25, November 20, 1913. On Hall's activities with the Brotherhood of Timber Workers, see James Green, "The Brotherhood of Timber Workers 1910–1913: A Radical Response to Industrial Capitalism in the Southern U.S.A.," *Past and Present,* No. 60 (August 1973), 161–200; Jeff Ferrell and Kevin Ryan, "The Brotherhood of Timber Workers and the Southern Lumber Trust: Legal Repression and Worker Response," *Radical America* 19, No. 4 (1985), 55–74.

7. *Voice of the People,* September 4, 18, 25, October 2, 9, 23, November 13, 20, 27, 1913. On the IWW in New Orleans, also see Bernard A. Cook and James R. Watson, "The Sailors and Marine Transport Workers' 1913 Strike in New Orleans: The AFL and the IWW," *Southern Studies* XVIII, No. 1 (Spring 1979), 120; W.J. Parks to W.B. Wilson, June 20, July 10, 22, 1913, in Subject—Strike—Transport Workers—New Orleans, No. 33-8, Federal Mediation Conc. Serv., RG 280, Nat'l Records Ctr., Suitland Md.

8. "Longshoremen Draw No Line," *New York Age,* August 14, 1913; International Longshoremen's Association, *Proceedings of the Twenty-Second Convention, Including Proceedings of Executive Committee Meeting, Milwaukee, Wisconsin, July 13th to 18th inclusive, 1914,* 11; ILA, *Proceedings of the Twenty-first Convention . . . 1913,* 173–74; ILA, *Proceedings of the Twenty-Third Convention . . . 1915,* 130. Booker T. Washington, "The Negro and the Labor Unions," *Atlantic Monthly,* June 1913. On the formation of the Gulf Coast and South Atlantic District of the ILA, see *Daily Picayune,* September 8, 11–14, 1911; James Maroney, "The International Longshoremen's Association in the Gulf States During the Progressive Era," *Southern Studies* XVI, No. 2 (Summer 1977), 225–32.

9. In 1905, for example, delegate Henry Baptiste, Local 402, ILA, representing the freight handlers of the Southern Pacific, denounced "sister locals" of the International for taking advantage of Local 402's poor position. Apparently some union longshoremen who found temporary employment on the Southern Pacific docks failed to join Local 402 or pay its membership fee. Moreover, the "men do not assist us in unionizing the non-union men," complained Baptiste. Nor did they help the local win recognition from the company. ILA, *Proceedings of the Fourteenth Annual Convention of the International Longshoremen, Marine and Transportworkers' Association. Detroit, Michigan, July 10th to 15th inclusive, 1905,* 191.

10. ILA, *Proceedings of the Twentieth Convention Including Proceedings of Executive Council Meeting. Port Huron, Michigan. July 8th to 13th, inclusive, 1912,* 72.

11. *Daily Picayune,* August 31, 1911; *Item,* August 31, 1911; *Times-Democrat,* August 31, 1911.

12. *Daily Picayune,* August 15, 29, 31, 1911. For a general account of the 1911 Illinois Central strike, see Graham Adams, Jr., *Age of Industrial Violence 1910–15: The Activities and Findings of the United States Commission on Industrial Relations* (New York: Cambridge University Press, 1966), 127–45.

13. *Daily Picayune,* September 28–30, October 1–3, 1911; *Item,* September 29, 30, October 1, 2, 1911; *Times-Democrat,* September 28–30, October 1, 3–6, 1911.

14. *Daily Picayune,* October 3, 9, 14, 1911; July 30, 1910; *Times-Democrat,* September 30, October 3, 4, 1911. Racial tensions may have been present in the failed 1908 Illinois Central strike as well. According to the white press, black freight handlers advocated an early settlement of the conflict, then favored both the Dock and Cotton Council's order to return to work and the signing of a five-year contract; white strikers apparently opposed all these moves. Despite the *Picayune's* and *Times-Democrat's* predictions of an imminent split between the black and white freight handlers, the alliance apparently held firm. There is a lack of existing evidence to substantiate the point fully. *Daily Picayune,* July 6, 1908; *Times-Democrat,* July 21, 1908.

15. *Daily Picayune,* August 11, December 16, 17, 1912; Adams, *Age of Industrial Violence,* 140–45.

16. Frank Putnam, "New Orleans in Transition," *New England Magazine* XXXVI, No. 2 (April 1907), 228–29; "New Orleans Second Largest Port in the United States," *The Longshoreman* (November 1917), 2, reprinted from *Railway and Marine News;* Frank G. Carpenter, "Industrial New Orleans," *Birmingham Age-Herald,* May 20, 1917.

17. Figures from the *Census of the United States,* 1900, 1910, 1920, cited in Blaine A. Brownell, *The Urban Ethos in the South 1920–1930* (Baton Rouge: Louisiana State University Press, 1975), 12.

18. John Smith Kendall, *History of New Orleans,* III (New York: The Lewis Publishing Co., 1922), 614–20; "New Orleans Second Largest Port in the United States," *The Longshoreman* (November 1917), 2.

19. Robert W. Williams, Jr., "Martin Behrman and New Orleans Civic Development, 1904–1920," *Louisiana History* II, No. 4 (Fall 1961), 375–400; Dewey W. Grantham, *Southern Progressivism: The Reconcilation of Progress and Tradition* (Knoxville: The University of Tennessee Press, 1983), 95–97; John R. Kemp, ed., *Martin Behrman of New Orleans: Memoirs of a City Boss* (Baton Rouge: Louisiana State University Press, 1977); Matthew J. Schott, "John M. Parker of Louisiana and the Varieties of American Progressivism" (unpublished Ph.D. dissertation, Vanderbilt University, 1969), 101–4; George M. Reynolds, *Machine Politics in New Orleans, 1897–1926* (1936; rpt. New York: AMS Press, 1968). On the nature of Behrman's support (and lack of support) from labor unions, see Section V below.

20. Julia Truitt Bishop, "The Romance of New Orleans' Public Belt Railway," *Times-Picayune* (Magazine Section), August 1, 1915; Board of Trade, *Cost of Living in American Towns. Report of an Enquiry by the Board of Trade into Working Class Rents, Housing and Retail Prices, Together with the Rates of Wages in Certain Occupations in the Principal Industrial Towns of the United States of America* (London: His Majesty's Stationery Office, 1911), 291; "New Orleans Public Belt Railroad: One of the World's Unique Institutions," *New Orleans Port Record* 1, No. 5 (January 1943), 17–20.

21. Kendall, *History of New Orleans,* III, 599–613; *First Annual Report of the New Orleans Maritime Association for the Year Ending August 31, 1901* (New Orleans: George Muller & Co., 1901), 50–51; H.S. Herring, *History of the New Orleans Board of Trade, Ltd, 1880–1930* (1930), 56–58; Clem G. Hearsey, "History of the Port of

New Orleans: Terms of Lease to Louisiana Construction and Improvement Company, Etc.," in *Martin Behrman Administration Biography, 1904–1916* (New Orleans: John J. Weihing Printing Co., 1917).

22. Flo Field, "New Conditions on New Orleans' Water Front," *Times-Picayune* (Magazine Section), April 8, 1917; Flo Field, New Orleans' Cotton Trade and Facilities," *Times-Picayune* (Magazine Section), April 22, 1917; Julia Truitt Bishop, "Personal Visit to the Great Port of New Orleans," *Times-Picayune* (Magazine Section), July 4, 1915; *The Longshoreman,* November 1917; Carpenter, "Industrial New Orleans," *Birmingham Age-Herald,* May 20, 1917; F.J. Springer, "Development of the Port of New Orleans," *Scientific American* CXV, No. 6 (August 5, 1916); Kendall, *History of New Orleans,* III, 599–613, 625–26; *New Orleans: The Nation's Second Port, the South's Greatest City* (New Orleans: Board of Commissioners, 1925), 15–25; "The Public Belt Railroad," in *Martin Behrman Administration Biography;* Frederic J. Haskin, "Great Warehouse to Revolutionize Trade in Cotton," *Times-Picayune,* November 18, 1915; Mildred Cram, *Old Seaport Towns of the South* (New York: Dodd, Mead & Co., 1917), 273–74; *Times-Picayune,* April 27, November 29, 1919; March 28, 1920; *Times-Democrat,* May 2–4, 1908.

23. *Times-Picayune,* May 1, July 16, 1914; *Daily Picayune,* May 3, 1908. For discussions of the process of mechanization on U.S. docks, also see James A. Jackson and Robert H. Rogers, "The Status of Cargo Handling in American Marine Terminals," *General Electric Review* XIX, No. 2 (February 1916), 127–32; Barnes, *The Longshoremen,* 39–40; U.S. Congress, Senate, Commission on Industrial Relations, *Final Report and Testimony,* 2 vols., 64th Cong., 1st sess., 1916, Doc. 415, III; *Galveston Daily News,* September 17, 26, 1913.

24. Stewart Vanderveer, "On the Fruit Wharf of New Orleans," *Times-Picayune,* July 11, 1920; *Times-Democrat,* August 30, 1905; *Times-Picayune,* August 26, 1919; *Galveston Daily News,* September 28, 1913. For the 1913 strike of United Fruit laborers and seamen, see Merl E. Reed, "Lumberjacks and Longshoremen: The I.W.W. in Louisiana," *Labor History* 13, No. 1 (Winter 1972), 53–57; Cook and Watson, "The Sailors and Marine Transport Workers' 1913 Strike in New Orleans," 111–22.

25. *Times-Democrat,* September 12, 14, 16, 1909; *Daily Picayune,* September 12–16, 1909; September 2, 1910; "Loading a Liner," *Daily Picayune,* December 15, 1912; Thomas Ewing Dabney, "S.P. Docks Here Most Efficient Line Operates," *Item,* April 4, 1920.

26. *Times-Picayune,* August 27, 1915.

27. *Times-Democrat,* June 19–21, 25, 26, 1913; *Daily Picayune,* June 19–21, 24, 26, 1913.

28. *Times-Picayune,* August 27, 1915. Also see: *Times-Picayune,* August 6, 8, 26, September 2, 4, 5, 1915.

29. *Times-Picayune,* March 26, April 30, May 5, July 8, 1914; "Report of President on Conference with Steamship Agents, Stevedores and Longshoremen," June 10, 1914, in Board of Commissioners of the Port of New Orleans, Minute Book, Volume 7 (Reel 3), 1386, in New Orleans Public Library.

30. *Voice of the People,* July 14, 1914; also see May 14, 28, 1914.

31. *Times-Picayune,* July 25–27, 1914; Minutes of Meeting on June 29, July 8, 1914, in Board of Commissioners of the Port of New Orleans, Minute Book, Volume 8 (Reel 3), 1403–11.

32. *Times-Picayune,* July 9, 15, 23, 1914; Minutes of Meetings on July 8, 14, 24, August 3, 12, 1914 in Board of Commissioners of the Port of New Orleans, Minute Book, Volume 8 (Reel 3), 1403–41.

33. *Times-Picayune,* July 26–31, August 7, 13, September 11, 12, 14–16, 22, December 13, 1914; *Voice of the People,* May 7, 14, 28, July 14, 1914; Minutes of Meetings on August 3, 12, 1914 in Board of Commissioners of the Port of New Orleans, Minute Book, Volume 8 (Reel 3), 1430–41. For other examples of the introduction of labor-saving machinery, see *Times-Picayune,* July 21, 1915; *Item,* April 17, 18, 1920. On the initial impact of the war, see James L. McCorkle, Jr., "Louisiana and the Cotton Crisis, 1914," *Louisiana History* XVIII, No. 3 (Summer 1977), 303–21.

34. *Times-Picayune,* March 29, April 9, 10, June 3, 13, 1917; January 8, 20, 23, March 3, 1918; *States,* April 1, 1917. For impact of war on the South, see George B. Tindall, *The Emergence of the New South 1913–1945* (Baton Rouge: Louisiana State University Press, 1967), 49; Jacqueline Hall et al., *Like a Family: The Making of a Southern Cotton Mill World,* 184–190; Gary Fink, *Labor's Search for Political Order: The Political Behavior of the Missouri Labor Movement, 1890–1940* (Columbia: University of Missouri Press, 1973), 61–81. Also see David Kennedy, *Over Here: The First World War and American Society* (New York: Oxford University Press, 1980), 280–81. For activities against alleged subversives by intelligence agents working out of New Orleans, see documents contained in U.S. Military Intelligence Reports, *Surveillance of Radicals in the United States 1917–1941* (Frederick, Md.: University Publications of America, 1984), Reel 1, Frames 230–92; Gulfport and Biloxi *Daily Herald,* September 7, 1917; *New Orleans Item,* September 7, 1917; *Times-Picayune,* October 6, 1917.

35. *Times-Picayune,* April 8, 12, 18, May 26, June 5–7, 19, July 11, September 7, 11, October 8, 1917; April 5, September 12, 1918; *New Orleans Item,* September 4, 25, 1917; *New Orleans States,* April 8, 9, 12–14, September 4, 9, 25, 1917. Also see William J. Breen, "Sectional Influences on National Policy: The South, the Labor Department, and the Wartime Labor Mobilization, 1917–1918," *The South Is Another Land: Essays on the Twentieth-Century South,* ed. Bruce Clayton and John A. Salmond (Westport: Greenwood Press, 1987), 69–83; Jane Lang Scheiber and Harry N. Scheiber, "The Wilson Administration and the Wartime Mobilization of Black Americans, 1917–18," *Labor History* 10, No. 3 (Summer 1969), 433–58.

36. *Times-Picayune,* May 2, 1917. The Building Trades Council adopted similar resolutions.

37. *New Orleans States,* April 28, 1917; *Times-Picayune,* April 28, May 13, July 27, 1917; January 20, July 26, 1918.

38. *Times-Picayune,* April 2, 3, July 17, September 4, 5, 1917; January 8, 20, May 11, 18, June 29, October 29, 30, November 6, 7, 1918. On labor's stance toward wartime agencies, also see *Times-Picayune,* March 1, 1919.

39. *Times-Picayune,* April 30, June 4, August 15, September 4, 27, October 2–4, 1918.

40. *Times-Picayune,* October 5, 1918.

41. *Times-Picayune,* May 4, September 23, 1918; *Ninth Biennial Report of the Department of Commissioner of Labor and Industrial Statistics of the State of Louisiana 1916–1918* (New Orleans: Hauser Printing Co., 1918), 15–16. On black migration in this era, see Emmet J. Scott, *Negro Migration During the War* (1919; rpt. New York: Arno Press, 1969); James R. Grossman, *Land of Hope: Chicago, Black Southerners, and the Great Migration* (Chicago: University of Chicago Press, 1989); Gavin Wright, *Old South, New South: Revolutions in the Southern Economy Since the Civil War* (New York: Basic Books, 1986), 198–206; Peter Gottlieb, *Making Their Own Way: Southern Blacks' Migration to Pittsburgh, 1916–30* (Urbana: University of Illinois Press, 1987), 12–38; Hall et al, *Like a Family,* 183–90; Florette Henri, *Black Migration: Movement North, 1900–1920* (New York: Anchor Press/Doubleday, 1975), 49–80; Jacqueline

Jones, *Labor of Love, Labor of Sorrow: Black Women, Work, and the Family from Slavery to the Present* (New York: Basic Books, 1985), 152–95; William Tuttle, *Race Riot: Chicago in the Red Summer of 1919* (New York: Atheneum, 1970); William H. Harris, *The Harder We Run: Black Workers Since the Civil War* (New York: Oxford University Press, 1982), 51–76; Carole Marks, "Black Workers and the Great Migration North," *Phylon* XLVI, No. 3 (June 1985), 148–61.

42. *Times-Picayune,* April 16, 28, July 27, 1917; January 20, 1918; June 11, 1919; *Journal and Guide,* July 7, 1917 (Item 299, 1 of 2, frame 33. Labor: War: Negro Labor and World War I), in *Hampton University Newspaper Clipping File.*

43. *New Orleans Item,* October 12, 1916; *Times-Picayune,* February 4, 1918.

44. Testimony of Harry Keegan, *Proceedings of National Adjustment Commission held at New Orleans, Louisiana, October 30, 1918. In the matter of Wages of Deep Sea Longshoremen,* 19–20, Tray 54, South Atlantic, September 1917–February 1919, United States Shipping Board, RG 32.

45. Alexander M. Bing, *War-time Strikes and Their Adjustment* (New York: E.P. Dutton & Company, 1921), 1; Melvyn Dubofsky, "Abortive Reform: The Wilson Administration and Organized Labor, 1913–1920," *Work, Community and Power: The Experience of Labor in Europe and America, 1900–1925,* ed. James E. Cronin and Carmen Sirianni (Philadelphia: Temple University Press, 1983), 197–220; also see Kennedy, *Over Here;* David Montgomery, 'Immigrants, Industrial Unions, and Social Reconstruction in the United States, 1916–1923," *Labour/Le Travail* 13 (Spring 1984), 101–13; Valerie Jean Conner, *The National War Labor Board: Stability, Social Justice, and the Voluntary State in World War I* (Chapel Hill: The University of North Carolina Press, 1983).

46. *Sixth Annual Report of the United States Shipping Board, June 30, 1922* (Washington: Government Printing Office, 1922), 20; Gordon S. Watkins, "Labor Problems and Labor Administration in the United States During the World War, Part II: The Development of War Labor Administration," *University of Illinois Studies in the Social Sciences* VIII, No. 4 (September 1919), 139. Also see Jonathan Schneer, "The War, the State and the Workplace: British Dockers During 1914–1918," *Social Conflict and the Political Order in Modern Britain,* ed. James E. Cronin and Jonathan Schneer (London: Croom Helm, 1982), 96–112.

47. B.M. Squires, "The Strike of the Longshoremen at the Port of New York," *Monthly Labor Review* IX, No. 6 (December 1919), 100–101; *Second Annual Report of the United States Shipping Board, December 1, 1918* (Washington: Government Printing Office, 1918), 86–87; War Department, Office of the Secretary, *A Report of the Activities of the War Department in the Field of Industrial Relations During the War* (Washington: Government Printing Office, 1919), 56–57; Roy S. MacElwee and Thomas R. Taylor, *Wharf Management: Stevedoring and Storage* (New York: D. Appleton, 1921), 64; *The Longshoreman,* September 1917, October 1917, July 1918, and August 1918; "No Longshoremen Strikes Expected," *The Survey* 38, No. 23 (September 8, 1917); Charles P. Larrowe, *Maritime Labor Relations on the Great Lakes* (East Lansing: Michigan State University, Labor and Industrial Relations Center, 1959), 41–46.

48. David Montgomery, "The 'New Unionism' and the Transformation of Workers' Consciousness in America, 1909–22," in *Workers' Control in America: Studies in the History of Work, Technology, and Labor Struggles* (New York: Cambridge University Press, 1979), 93–101; Bing, *War-time Strikes,* 1–10.

49. *New Orleans States,* October 11, 1919; *Tenth Biennial Report for the Depart-*

ment of Commissioner of Labor and Industrial Statistics of the State of Louisiana 1919–1920 (New Orleans: Hauser Printing Co., 1920).

50. *Times-Picayune,* September 1–5, 7, 8, 11, 1917; *New Orleans Item,* September 4, 5, 11, 1917; *States,* September 1–4, 7, 9, 11, 1917; Little Rock *Union Labor Bulletin,* September 8, 1917. New Orleans employers recognized that the high cost of living made substantial wage increases imperative, but the boss draymen contested the amount demanded by the teamsters. The strike surprised some observers, who feared that the Dock and Cotton Council was unable to control its black teamsters. This small black union, supported by the black screwmen, ignored the advice of white screwmen president Thomas Harrison against any walkout, and the black strikers' opinion of the white labor leader "weakened in consequence" of his cautiousness. Yet the teamsters' membership in the Council afforded them a basic protection: since the Council earlier had endorsed the teamsters' demands, the boss draymen could not employ strikebreakers without activating the circuit of waterfront solidarity.

51. *New Orleans Daily States,* September 19, 20, 1917; *New Orleans Item,* September 6, 17, 19–21, 1917; *Times-Picayune,* September 13, 17–21, 1917. An additional issue was overwork. New kinds of freight, a heightened premium on schedule, and labor shortages had induced agents and stevedores to attempt to push their workers harder. Complaining of overwork, longshoremen renewed their old demand for more men and reduced hours.

52. *New Orleans States,* September 20, 1917; *New Orleans Item,* September 24, 1917; *Times-Picayune,* September 21–23, 1917. The car freight handlers received a flat rate of 20 cents an hour for unloading rail cars near the river. In September they demanded 40 cents an hour for day work, 60 cents an hour for night work, and a piece scale of different sums for carrying different kinds of freight. Rail contractors rejected the proposal outright, offering instead a new flat rate of 25 cents an hour for day work and 35 cents for night work.

53. *Daily States,* September 22, 24, 25, 1917. The establishment of a "labor treaty" between employers and longshore unions with the formation of the National Adjustment Commission did not stop 30,000 shipyard workers in San Francisco or 6500 longshoremen in New York from striking in September 1917. "Settling Strikes in the Shipping Trades," *The Survey* 38, No. 26 (September 29, 1917), 578.

54. *Daily Item,* September 24–26, 1917; *Daily States,* September 25, 30, 1917; *Times-Picayune,* September 24–26, 1917.

55. E.A. Kelly to W.Z. Ripley, January 4, 1919, in "Folder: Gulf Ports Coastwise Award," U.S. Shipping Board, RG32; *Times-Picayune,* September 11, 1917; January 23, August 7, 8, 15, September 18, October 23, 30, November 6, 8, December 19, 1918; January 21, February 25, 1919; January 5, 6, 1920; *New Orleans States,* November 6, 1918; "Recent Wage Awards by National Adjustment Commission," *Monthly Labor Review* VIII, No. 2 (February 1919), 465–68. In Box 54: South Atlantic, September 1917–February 1919, U.S. Shipping Board, RG32, see "Proceedings of National Adjustment Commission Held at New Orleans, Louisiana, October 31, 1918. In the Matter of Coastwise Longshoremen" (hereafter referred to as "Proceedings . . . 1918. In the Matter of Coastwise Longshoremen"); "Proceedings of Local Adjustment Commission Held at New Orleans, Louisiana, November 7, 1918. In the Matter of Wages of Freight Handlers" (hereafter referred to as "Proceedings . . . 1918. In the Matter of Wages of Freight Handlers"); "Proceedings of National Adjustment Commission Held at New Orleans, Louisiana, October 30, 1918. In the Matter of Wages of Deep Sea Longshoremen" (hereafter referred to as "Proceedings . . . 1918. In the Matter of Wages of Deep

Sea Longshoremen"). Also see "In the Matter of Coal Wheelers and Trimmers' Local No. 45, I.L.A. . . . January 8, 1919, before the Local Wage Adjustment Commission," Tray 14: Deluth-Gulf Coast, Folder: Coal Trimmers—New Orleans; *New Orleans States,* November 6, 1918. In some cases, companies acted voluntarily to raise wages. With no strike imminent, the Morgan Line took preemptive action, offering their dock workers a 10 percent wage increase to 40 cents an hour in September 1917. Upon occasion, lower ranking workers struck to capture the Board's attention. In August 1918, a one-day strike by 700 Southern Pacific dock workers resulted in an order that the company recognize its unions. Frustrated over a delay in a Board ruling, 200 to 300 Illinois Central men struck for one day in October of that year.

56. *New Orleans States,* September 27, 1917; Murphy to Stevens, September 27, 1917; Tray 14: Deluth-Gulf Coast, Folder L-Gulf Coast, U.S. Shipping Board, RG32; Telegram from Murphy to Chairman, NAC, July 15, 1918, in Box: New Orleans-New York 28, Folder, Southern Pacific Docks—New Orleans, U.S. Shipping Board, RG32.

57. "Proceedings . . . 1918. In the matter of Wages of Deep Sea Longshoremen," 104–7; also see *New Orleans Item,* September 21, 1917; *Times-Picayune,* September 20, 1917.

58. Hendren to Bass, October 18, 1918, in Folder: 2 Gulf Conference, Tray 14: Deluth-Gulf Coast, U.S. Shipping Board, RG 32. On the night work issue, also see testimony of stevedore Kearney in "Proceedings of Hearings before the National Adjustment Commission held at New Orleans, Louisiana, beginning October 28, 1919," 118–19, in Tray 55: South Atlantic, February-November 1919, U.S. Shipping Board, RG 32 (hereafter referred to as "Proceedings . . . 1919").

59. S.T. DeMilt to H. Keegan and A. Workman, January 9, 1919, in "New Orleans-New York 28 Folder: New Orleans Complaints," U.S. Shipping Board, RG 32.

60. "Proceedings . . . 1918. In the matter of Wages of Deep Sea Longshoremen," 102, 108, 134.

61. In "Load of Flour Controversy" (New Orleans), Box: Buffalo-New Orleans 27, U.S. Shipping Board, RG 32, see P.T. Murphy to T.V. O'Connor, March 15, 1918; Murphy to H. Keegan and A. Workman, November 10, 1917; S.T. DeMilt to H. Keegan and A. Workman, February 7, 1918; Henren to J. Palfrey, November 9, 1918. (Hereafter referred to as "Load of Flour Controversy.")

62. "Proceedings . . . 1918. In the matter of Wages of Deep Sea Longshoremen," 121, 103, 123.

63. "Proceedings . . . 1918. In the matter of Wages of Deep Sea Longshoremen," 102–3; In "Load of Flour Controversy" file, see: Murphy to O'Connor, March 14, 1918; P.T. Murphy to H. Keegan and A. Workman, November 10, 1917; H. Keegan and A. Workman to P.T. Murphy, P.T. Murphy to Stevens, February 8, 1918.

64. P.T. Murphy to R.B. Stevens, February 8, 1918, in "Load of Flour Controversy."

65. *The Longshoreman* 9, No. 11 (September 1917), 1; "Proceedings . . . 1918. In the matter of Wages of Deep Sea Longshoremen," 171; *New Orleans Item,* September 21, 1917; *New Orleans States,* September 19, 1917.

66. "Proceedings . . . 1918. In the matter of Wages of Deep Sea Longshoremen," 11.

67. "Proceedings . . . 1919," 171.

68. P.T. Murphy to Bass, March 9, 1918, in "Load of Flour Controversy"; "Proceedings . . . 1919," 160; also see "Proceedings . . . 1918. In the Matter of Wages of Deep Sea Longshoremen," 10.

69. "Proceedings . . . 1919," 78–79, 118, 125, 171, 182–83, 191; "Proceedings . . . 1918. In the Matter of Wages of Deep Sea Longshoremen," 116.

70. P.T. Murphy to R.B. Stevens, February 8, 1918, in "Load of Flour Controversy"; "Proceedings . . . 1918. In the Matter of Wages of Deep Sea Longshoremen," 104–7, 181.

71. W. Elliot Brownlee, *Dynamics of Ascent: A History of the American Economy* (Chicago: The Dorsey Press, 1988, 2nd ed.), 377–80; Robert H. Ferrell, *Woodrow Wilson and World War I, 1917–1921* (New York: Harper and Row, Publishers, 1985), 196–99; *Times-Picayune,* January 28, 1920.

72. *Times-Picayune,* January 29, February 12, May 3, June 2, 11, September 6, 20, 28, November 20, 25, 1918; March 1, May 3, 1919; "Report of District President J.H. Fricke to Convention, Gulf Coast District, ILA," *The Longshoreman* (September 1918); "Correspondence: A.T. Golden, Recording Secretary of Local 972 to T.V. O'Connor, January 1, 1918," *The Longshoreman* (March 1918); *American Federationist* XXV, No. 1 (January 1918), 66; *American Federationist* XXV, No. 5 (May 1918), 399; *American Federationist* XXV, No. 8 (August 1918), 704; *American Federationist* XXVI, No. 6 (June 1919), 524; *American Federationist* XXVI, No. 11 (November 1919), 1067; *American Federationist* XXVI, No. 12 (December 1919), 1155; *American Federationist* XXVII, No. 3 (March 1920), 270. For a discussion of the wave of organizing in the South during this era, see Tindall, *Emergence of the New South,* 332–38; Earl Lewis, "At Work and at Home: Blacks in Norfolk, Virginia, 1910–1945" (unpublished Ph.D. dissertation, University of Minnesota, 1984), 88–106; Montgomery, *Fall of the House of Labor,* 380–82.

73. *Christian Science Monitor,* May 24, 1918 (Item 299, 2 of 2, frame 126, Labor: War: Negro Labor and World War I), *Hampton University Newspaper Clipping File; Times-Picayune,* October 3, 1918.

74. *Times-Picayune,* November 20, 25, 1918.

75. *Times-Picayune,* February 14, 15, 27, 1920; *New Orleans Item,* February 14–17, 22, 26, 27, 28, 1920; in File 170–1012, RG 280, see J.S. Meyers, Commissioner of Conciliation, to H.L. Kerwin, March 1, 1920; Meyers to Cunningham, June 9, 1920; Kerwin to Meyers, February 13, 1920.

76. *Times-Picayune,* August 28, 31, September 1–3, 9, 1919; *New Orleans States,* October 11, 1919; U.S. Shipping Board Weekly Report Ending September 16, 1919, p. 2, in Records of T.V. O'Connor, 1921–33. Correspondence about Labor Matters, 1919–23. Manning Scale, Weekly Reports, U.S. Shipping Board, RG 32.

77. *New Orleans States,* October 6, 15, 17, 25, 26, 1919; *Times-Picayune,* June 4, 5, 7, 8, 11, October 8, 17, 26, November 12, 13, 1919; in File No. 170–813, Subject: Fruit Dispatch Co., New Orleans, Louisiana, RG280, see "Prelminary Report," September 22, 1919; J.W. Bridwell, Commissioner of Labor, to Kerwin, October 3, 1919; United Fruit Memorandum, November 15, 1919; J.W. Bridwell to E.J. Cunningham, December 30, 1919; Meyers to Kerwin, December 29, 1919. On the larger strike wave, see Lieut. Col. George R. Goethals to Adjutant General Southeastern Department, December 31, 1919, in Military Intelligence Division Correspondence, 1917–41, 10634–691 to 106334–753 Declassified NND 740058 Box No. 3648, RG165, Records of the WFGS.

78. *Times-Picayune,* September 18, 27–29, 1919; Robert P. Bass, *Marine and Dock Labor: Work, Wages, and Industrial Relations During the Period of the War. Report of the Director of the Marine and Dock Industrial Relations Division, United States Shipping Board* (Washington: Government Printing Office, 1919); B.M. Squires, "The Strike of the Longshoremen at the Port of New York," *Monthly Labor Review* IX, No. 6 (December 1919), 102–4; B.M. Squires, "Peace along Shore: How the Longshore-

men Settle Differences with Employers," *The Survey* LXIV, No. 15 (August 2, 1920), 573–74; *Times-Picayune,* April 21, 1919.

79. *New Orleans States,* October 13, 14, 1919; Squires, "The Strike of the Longshoremen at the Port of New York," 95–115; Squires, "Peace along Shore," 574–75.

80. *Times-Picayune,* October 26, 1919.

81. See: *New Orleans States,* October 11–16, 18–24, 26, 28, 29, 31, November 1, 2, 6–10, 1919; *Times-Picayune,* October 12–14, 16, 17, 22, 24, 26, 30, November 1, 6–10, 12, 1919.

82. *New Orleans States,* October 13, 17, 21, 31, November 1, 2, 6, 1919.

83. *New Orleans States,* November 7–10, 1919; *Times-Picayune,* November 24, 1919; also see "Proceedings . . . 1919"; *The Union Review* (Galveston) 1, No. 32 (November 28, 1919).

84. *Times-Picayune,* December 24, 28–31, 1919.

85. *Times-Picayune,* March 31, April 8, 9, 1920; *New Orleans Item,* March 20, 1920; *New Orleans States,* April 7, 1920.

86. *Times-Picayune,* January 17, 18, 1920.

87. *New Orleans States,* January 5–8, 16, 17, February 15, 1920; *New Orleans Item,* January 5–7, 9–11, 17, 1920; *Times-Picayune,* November 13, 1919; January 6, 7, 9, 16–18, 1920; ILA, *Proceedings . . . 1921,* 45–46. The problem extended well beyond New Orleans, reflecting Congressional failure to extend protection to the coastwise steamship companies. New York and Galveston both suffered long and costly strikes of coastwise longshoremen as a result. See "The Coastwise Longshoremen," *The Survey* XLIV, No. 10 (June 5, 1920), 48; *Galveston Daily News,* March 30, 1920.

88. *New Orleans States,* February 5, 7, 8, 1920; *New Orleans Item,* February 5–8, 1920; *Times-Picayune,* February 6–8, 1920.

89. *New Orleans States,* February 8, 9, 12–15, 18, 1920; *New Orleans Item,* February 7, 9–16, 19, 1920; *Times-Picayune,* February 9, 11, 13, 14, 17, 1920.

90. *New Orleans States,* February 28, 29, March 2, 1920; *New Orleans Item,* February 20, 24, 28, 29, March 1, 2, 1920; *Times-Picayune,* February 20, 24, 29, March 1, 2, 1920.

91. James C. Maroney, "The Galveston Longshoremen's Strike of 1920," *East Texas Historical Journal* XVI, No. 1 (1978), 34–38; William D. Angel, Jr., "Controlling the Workers: The Galveston Dock Workers' Strike of 1920 and Its Impact on Labor Relations in Texas," *East Texas Historical Journal* XXIII, No. 2 (1985), 14–27; "The Harbor Strikes," *The Survey* XLIV, No. 12 (June 19, 1920), 396; *Galveston Daily News,* March 20–22, 26, 30, April 9, 16, 19, 20, 23, May 6, 7, 11, 12, 14–19, 22–24, 27–31, June 2–18, 20–22, 24, 29, 30, July 4, 9–12, 15–23, 25, 29, 31, August 1, 3, 4–7, 9, 10, 12–17, 19–21, 1920; Little Rock *Union Labor Bulletin,* September 10, 1920.

92. *New Orleans States,* March 2, 5, 1920; *New Orleans Item,* March 3–6, 8, 1920; *Times-Picayune,* March 3–7, 23, 1920.

93. *New Orleans States,* April 29, 30, May 1, 5, 1920; *New Orleans Item,* April 29, 30, 1920; *Times-Picayune,* May 1, 2, 5, 7, 1920; *Galveston Daily News,* May 7, 1920; "Memorandum from Edith Foote to Chairman, Subject: New Orleans Longshore Negotiations, September 29, 1920," in File No. 621–3–2, Part 1, To December 31, 1920. Subject: Labor and Labor Conditions—Longshoremen and Stevedores, New Orleans, LA, U.S. Shipping Board, RG 32.

94. *Times-Picayune,* March 17, April 10, May 5, 8, 14, 18, 20, 22, June 4, August 2, 13–16, 1920; *New Orleans Item,* March 17, 1920; *Galveston Daily News,* May 22, 29, 1920.

95. *Times-Picayune,* October 12, 27, 1920.

96. *Times-Picayune,* May 28, August 14, 15, 25, September 22, 23, 29, 1920; also see telegrams and memorandą in File No. 621-3-2, Part 1, To December 31, 1920. Subject: Labor and Labor Conditions—Longshoremen and Stevedores, New Orleans, LA, U.S. Shipping Board, RG 32.

97. Williams, "Martin Behrman and New Orleans Civic Development," 375–400; Grantham, *Southern Progressivism,* 95–97; John R. Kemp, ed., *Martin Behrman of New Orleans: Memoirs of a City Boss* (Baton Rouge: Louisiana State University Press, 1977); Schott, "John M. Parker of Louisiana," 101–4; George M. Reynolds, *Machine Politics in New Orleans, 1897–1926* (1936; rpt. New York: AMS Press, 1968); Harold Zink, *City Bosses in the United States: A Study of Twenty Municipal Bosses* (Durham: Duke University Press, 1930), 317–33.

98. On machine patronage and the labor movement, see *Times-Picayune,* August 15, 1920. Also see *Daily Picayune,* October 14, 1904; *Times-Picayune,* October 20, 1921; on the Democratic Labor League, see *Times-Picayune,* May 6, September 16, October 12, 15, 16, November 4, 10, 1904; Zink, *City Bosses,* 323.

99. J.W. Leigh, "Labor Hells in Dixie," *Industrial Pioneer* 1, No. 6 (October 1923), 11–12.

100. Kendall, *History of New Orleans,* II, 548; Schott, "John M. Parker of Louisiana," 111; Grantham, *Southern Progresivism,* 97–98; "New Orleans, A Commission City," *The Literary Digest* XLV, No. 11 (September 14, 1912), 406.

101. On Parker's election, see Schott, "John M. Parker of Louisiana," 341–46; T. Harry Williams, *Huey Long* (1969; rpt. New York: Vintage, 1981), 131–36; Kemp, ed., *Martin Behrman,* 289–95; Reynolds, *Machine Politics,* 208–10; Philip Reilly Collins, "The Old Regular Democratic Organization of New Orleans" (unpublished M.A. thesis, Georgetown University, 1947), 103, 107, 111; Tindall, *Emergence of the New South,* 236–37; *New Orleans States,* January 9, 1920; *The Labor Record,* December 19, 1919 (in Louisiana Collection, Howard-Tilton Memorial Library, Tulane University).

102. *New Orleans Item,* March 21, 26, April 4, 6, 1920; *New Orleans States,* March 26, May 6, 1920; *Times-Picayune,* March 26, April 5, 7, May 22, 23, 26, 30, June 3, 5, July 11, 1920.

103. *New Orleans Item,* April 26, 1920; *New Orleans States,* May 4, 11, 18, 28, 1920; *Times-Picayune,* May 20, June 3, 4, 1920.

104. *New Orleans States,* June 4, 1920; *Times-Picayune,* June 5–7, 12, 13, 16, 1920; *Galveston Daily News,* June 5, 10, 13, 16, 18, 1920.

105. Reynolds, *Machine Politics in New Orleans,* 215.

106. *New Orleans Item,* February 2, 1920. The boilermakers' leader, and other union ODA backers, drew up a long list of complaints against the mayor. In one instance, several years earlier, Behrman employed only one boilermaker with a number of black helpers to install the syphon at the Melponene Drainage station, despite protests by the state labor commissioner and the boilermakers' union. In another case, the union's secretary claimed that Behrman refused to assign a union member to the Public Belt Railroad shops, informing the men that he "did not owe any allegiance to organized labor, and particularly to the boilermakers' union, and that he would place men to work in the Public Belt only when he saw fit." Union supporters of the ODA now repeatedly condemned the city's past stance toward unions of its municipal employees, charging that Behrman had held down the wages of garbage collectors as well as city engineers and other workers of the Sewerage and Water Board, and other city departments. Moreover, they claimed that when Public Belt switchmen struck against "deplorable conditions" several years earlier, Behrman instructed the police to "subdue them" while he put strikebreakers to work; striking carmen of the New Orleans

Railway and Light Company found Behrman actively opposed to their wage demands before a government mediator. Lastly, the ODA supporters blamed Behrman for the firm, anti-labor position adopted by the Dock Board against piledrivers on the Industrial Canal during their recent strike. *New Orleans Item,* September 5, 1920; *Times-Picayune,* August 5, 15, 21, 27, 1920.

107. *New Orleans Item,* February 2, 1920. *Times-Picayune,* August 5, 15, 17, 1920. The boilermakers explicitly denounced Behrman for failing to place a union boilermaker in the shops of the Public Belt Railroad. The union's recording secretary T.P. Martin claimed that Behrman told the union that he did "not owe any allegiance to organized labor . . . and that he would place men to work in the Public Belt only when he saw fit. This fact alone will demonstrate what an enemy to organized labor Martin Behrman has been." *Times-Picayune,* August 17, 1920.

108. *New Orleans Item,* September 2, 1920. For other charges against the machine's labor record, see *Times-Picayune,* August 21, 27, September 5, 1920; *New Orleans Item,* September 5, 1920; Reynolds, *Machine Politics in New Orleans,* 212–15.

109. Schott, "John M. Parker of Louisiana," 106–7; *Times-Picayune,* August 12, 1920.

110. *New Orleans Item,* February 2, 1920.

111. *New Orleans Item,* February 2, March 3, April 16, 1920; *Times-Picayune,* April 16, 30, May 27, August 5, 12, 17, September 4, 5, 7, 1920. Divisiveness in organized labor's ranks over partisan political endorsements was hardly new. A similar controversy over the "labor vote" broke out some sixteen years earlier, during Martin Behrman's first campaign for mayor. Since formal rules barred the Central Trades and Labor Council from endorsing political candidates, a number of labor leaders formed the Union Labor League, or Democratic Labor League, in 1903, adopting a platform calling for municipal ownership of public utilities, the eight-hour day law, and woman's suffrage. The League split in September 1904, however, when its executive board voted overwhelmingly to endorse machine candidate Martin Behrman for mayor (to replace the outgoing Paul Capdevielle) and the rest of the regular Democratic ticket. A storm of protest over the endorsement immediately ensued. Supporting the move were Central Trades and Labor Council president Robert E. Lee, longshoreman and former Dock and Cotton Council president Harry Keegan, longshore leader, and First Recorders' Court Judge James Hughes, and screwman and state Labor Commissioner Thomas Harrison. Opposing the endorsement were the head of the Democratic Labor League, cotton yardman and Dock and Cotton Council president Frederick Grosz, and William McGilvray of the plumbers' union. Before the election in November, locals of street carmen, retail clothing clerks, plumbers, printing pressmen, broommakers, and carpenters and joiners passed resolutions repudiating the League's alleged promise to "deliver the union vote," upholding their political "independence and freedom." Grosz, who ran as a Home Rule (reform) candidate for commissioner of public works, lost. The issue went beyond the record of the Democratic party regulars. According to the Central Trades and Labor Council newspaper, *Union Advocate,* "the lesson we should learn . . . is to keep 'labor' and all other 'class' distinctions out of politics, especially out of Democracy, which is broad enough for all classes and distinctions of our people to meet within its ranks upon political equality." And, on another occasion, it advised: "Let 'union labor' and all others keep out of 'class' politics,—be citizens." *Union Advocate,* January 25, February 1, 1904; *Daily Picayune,* December 4, 1902; September 30, 1903; April 4, 11, May 6, 14, 27, June 4, 17, September 16, October 12–17, 20–22, 24–27, 30, November 4, 10, 1904; *Times-Democrat,* October 24, 1903.

The following year, recriminations from the political controversy broke out in the Central Trades and Labor Council's ranks. McGilvray, delegate of the plumbers' union, successfully sponsored a resolution requiring that all Council delegates be union members actively employed in the trades they represented—effectively barring from the Council all "politicians" such as the machinists' Robert E. Lee (then Commissioner of Labor Statistics), the longshoremen's Judge James Hughes (Behrman's chief clerk), the screwmen's J. Jamison (employed by the Commissioner of Police and Public Buildings), the steam fitters' Edward Gleason (General Superintendent for the Department of Public Works and a member of the legislature), the stage hands' Andy Hamilton (deputy clerk in the civil district court), as well as former Labor Commissioner Thomas Harrison of the screwmen, who was then not working on the docks. The white longshoremen's and screwmen's unions denounced the move, declared that individual labor organizations alone were the proper bodies to decide who represented them, and threatened to withdraw from the Council, taking other unions with them, if it failed to repeal the McGilvray resolution. Shortly thereafter, Council delegates acquiesced, and the plumbers recalled delegate McGilvray. *Times-Democrat,* September 26, 27, 1905; *Daily Picayune,* September 24, October 14, 15, 24, 29, November 11, 1905.

112. The *Item* noted that voting was light in riverfront neighborhoods, "while the uptown and back of the ward vote, whence comes the independent strength, ran unusually high. . . . These heavy slumps along the river front are doubtless accounted for in large part by the thousands of votes taken off the registration books by Registrar William Bell. . . . The steam has all oozed out of [the machine's] . . . rank-and-file . . . [and by mid-day there had occurred] not a single disturbance—not even from the happy-hunting grounds of the thugs and plug-uglies that always theretofore carried elections for the Machine by club, knife and pistol, in the riverfront precincts of the downtown wards." *New Orleans Item,* September 14, 1920; also see Reynolds, *Machine Politics in New Orleans,* 215–16.

113. *Times-Picayune,* May 4, June 6, 12, 17, 1920; Little Rock *Union Labor Bulletin,* May 21, 1920. For a discussion of the open shop drive and the American Plan, see Montgomery, *Fall of the House of Labor,* 438–39; David Brody, "The American Worker in the Progressive Era," in *Workers in Industrial America: Essays on the Twentieth Century Struggle* (New York: Oxford University Press, 1980), 44–46; Allen M. Wakstein, "The National Association of Manufacturers and Labor Relations in the 1920s," *Labor History* 10, No. 2 (Spring 1969), 163–76; Tindall, *Emergence of the New South,* 338–39.

114. *Tenth Biennial Report of the Department of Commissioner of Labor and Industrial Statistics of the State of Louisiana, 1919–1920,* 75; *Thirteenth Biennial Report of the Department of Commissioner of Labor and Industrial Statistics of the State of Louisiana, 1925–1926,* 83; *Times-Picayune,* March 16, May 28, 31, June 5, 9, 24, July 2, 3, 8, September 1–4, 10, 12, 15, 17, 18, 21–24, 1921.

115. For discussion of the U.S. Shipping Board and the 1921 seamen's strike, see Bruce Nelson, *Workers on the Waterfront: Seamen, Longshoremen, and Unionism in the 1930s* (Urbana: University of Illinois Press, 1988), 54–55, 67; Montgomery, *Fall of the House of Labor,* 403; Joseph P. Goldberg, *The Maritime Story: A Study in Labor-Management Relations* (Cambridge: Harvard University Press, 1958), 93–104; Hyman Weintraub, *Andrew Furuseth: Emancipator of the Seamen* (Berkeley: University of California Press, 1959), 156–59. Also see *Times-Picayune,* March 29, April 4, 20, 26, May 1, 4, 5, 7, 8, 16, 17, 20, 25, 28, June 19, September 10, 22, 1921.

116. Untitled report, September 9, 1921; Weems to Ringwood, September 8, 1921; O'Connor to Santa Cruz, September 21, 1921; Santa Cruz to O'Connor, September

24, 1921; in File 621–3–2, Part 3: From September 1, 1921 to December 31, 1921. Subject: Labor and Conditions . . . Box 1404, U.S. Shipping Board, RG 32.

117. ILA, *Proceedings of the Twenty-sixth Convention. Buffalo, New York, July Eleventh to Eighteenth, 1921,* 71; *Times-Picayune,* October 2, 4, November 2–5, 1921.

118. *New Orleans States,* November 19, 21, 1921; *New Orleans Item,* November 21, 1921; *Times-Picayune,* November 22, 1921; Arthur Raymond Pearce, "The Rise and Decline of Labor in New Orleans" (unpublished M.A. thesis, Tulane University, 1938), 74–75; *Eleventh Biennial Report of the Department of Commissioner of Labor of the State of Louisiana 1921–1922* (New Orleans: Standard Printing Company, 1922), 64.

119. *New Orleans States,* November 27, 1921; *New Orleans Item,* November 27, 1921.

120. *New Orleans Item,* November 30, 1921; also November 21, 25, 27, 1921; *New Orleans States,* November 21, 23, 24, 26, 1921; *Times-Picayune,* November 22, 23, 25, 27, 1921; United States Shipping Board and Emergency Fleet Corporation Consolidated Report, Week Ending September 24, 1921, p.5, R 632.

121. *New Orleans Item,* November 21, 25, 26, 1921; *New Orleans States,* November 23, 1921; *Times-Picayune,* November 23, 25, 26, 1921.

122. *New Orleans Item,* November 21, 25, 27, 1921; *Times-Picayune,* November 22, 23, 25–27, 1921; *New Orleans States,* November 21, 23, 24, 26, 1921.

123. *New Orleans States,* November 24, 1921; *Times-Picayune,* November 24–26, 1921; *New Orleans Item,* November 26, 1921; Pearce, "The Rise and Decline of Labor," 75.

124. *New Orleans States,* November 24, 26, 29, December 2, 1921; *Times-Picayune,* November 23, 26–30, December 2–4, 1921; *New Orleans Item,* November 25–27, 30, December 2, 1921.

125. *New Orleans Item,* November 21, 1921; Jenkins to O'Connor, October 21, 1921, File 621-3-2, Box 1404, U.S. Shipping Board, RG32. The U.S. Shipping Board's moderate approach to the New Orleans strike in part reflected a change in federal position toward labor disturbances. Calvin Coolidge became president in August 1923, following the death of President Warren Harding. One of his first tests on the industrial relations front centered on a strike by the United Mine Workers in the Central Competitive Field. Despite a reputation as an opponent of unions, Coolidge barely intervened in the walkout, establishing, in historian Robert Zieger's words, "a pattern of his response to labor issues during his years in office. Far from encouraging antilabor extremists, Coolidge largely let labor alone . . . [avoiding] action wherever possible, viewing each situation with canny aloofness, and shunning rash and dramatic acts and pronouncements." Robert Zieger, *Republican and Labor 1919–1929* (Lexington: University of Kentucky Press, 1969), 156–57.

126. *Times-Picayune,* November 29, 30, December 2, 1921; *New Orleans States,* November 29, December 2, 1921; *New Orleans Item,* December 2, 1921; Pearce, "The Rise and Decline of Labor," 75; Report of President Anthony J. Chlopek, "South Atlantic and Gulf Coast District, Conditions and Wages, 1921–1922," in ILA, *Proceedings of the Twenty-seventh Convention, Boston, Mass., May fourteenth to May twenty-first, 1923,* 50; Christy to Kerwin, December 2, 1921, File: Subject—Longshoremen, New Orleans, Louisiana, No. 170–1577, RG280.

127. *Times-Picayune,* September 14–16, 20, 21, 27, 30, 1923; *New Orleans Item,* October 15, 16, 19, 20, 25, 30, 1923; "South Atlantic and Gulf Coast District, 'Wage Negotiations with Employers,' " ILA, *Proceedings of the Twenty-eighth Convention, Montreal, Quebec, August Tenth to Fifteenth, 1925,* 89–90; Pearce, "The Rise and

Decline of Labor," 80–85; *Twelfth Biennial Report of the Department of Commissioner of Labor of the State of Louisiana 1923–1924* (New Orleans: American Printing Co., 1924), 37–38; Sidney Terry, "The Great New Orleans Strike," *The Industrial Pioneer* 1, No. 7 (November 1923), 5–6. Also see Herbert R. Northrup, "The New Orleans Longshoremen," *Political Science Quarterly* LVII, No. 4 (December 1942), 531–33.

128. *Times-Picayune,* September 15, 1923; *New Orleans Item,* September 15, 24, 27, 1923.

129. *New Orleans Item,* September 21, 1923; *Times-Picayune,* September 17–20, 1923.

130. J.W. Leigh, "Social Conditions in South," *The Industrial Pioneer,* 1, No. 7 (November 1923), 9; "South Atlantic and Gulf Coast District: Wage Negotiations with Employers," ILA, *Proceedings of the . . . 1925 Convention,* 89.

131. *New Orleans Item,* September 21, 23, 26, 28, 30, October 4, 5, 7, 9–11, 13, 1923; *Times-Picayune,* September 21, 22, 24, 25, 27, 29, October 1–12, 14–16, November 1, 1923; Gulfport and Biloxi *Daily Herald,* September 24, 25, 27, October 3–5, 13, 15, November 1, 1923.

132. In No. 170–2268, Subject: Longshoremen, New Orleans, RG280, see E.H. Dunnigan, Commissioner of Conciliation, to H.L. Kerwin, Director of Conciliation, U.S. Department of Labor, October 27, 1923; Dunnigan to Kerwin, October 25, 1923. In File No. 621-3-2, Part 6, From October 16, 1923 to November 14, 1923, Subject: Labor and Labor Conditions—Longshoremen and Stevedores, Box 1405, RG32, U.S. Shipping Board, see "Immediate Release" (Press Release, #800), October 19, 1923; *Birmingham Age-Herald,* October 19, 1923.

133. In Norfolk, 2500 longshoremen struck to raise wages from 65 cents an hour to 80, rejecting their employers offer of 75 cents. Stevedores in the Hampton Roads area took quick advantage of the region's "superabundance of labor," putting large numbers of strikebreakers to work under police protection. One black strikebreaker was fired upon and killed, apparently by black strikers, shortly after the agents and stevedores announced their intention to resume permanent non-union operations. The U.S. Shipping Board declared its willingness to settle on the employers' original offer of 75 cents, although it refused to sign any contract with the union. By the middle of the month, private employers followed suit, resuming operations with their old workers on an open shop basis. *The Virginia-Pilot and The Norfolk Landmark,* October 7–14, 16, 1923; Gulfport and Biloxi *Daily Herald,* October 15, 17, 19, 20, 23, 1923; *Birmingham Age-Herald,* October 20, 22, 1923.

134. "Resolution Unanimously Adopted at Meeting of New Orleans Steamship Association Held Wednesday October 17, 1923," in File No. 170–2268, Subject: Longshoremen, New Orleans, RG 280; in same file, also see: Santa Cruz to Jenkins, October 19, 1923; Winthrop L. Marvin to T.V. O'Connor, October 19, 1923; O'Connor to Marvin, October 20, 1923; Santa Cruz to Jenkins, October 25, 1923; *Twelfth Biennial Report . . . 1923–1924,* 37–38; *New Orleans Item,* October 11, 18–22; *Times-Picayune,* October 18–24, 27, 29, 1923; Gulfport and Biloxi *Daily Herald,* October 19, 20, 22, 23, 1923; Carroll George Miller, "A Study of the New Orleans Longshoremen's Unions from 1850 to 1962" (unpublished M.A. thesis, Louisiana State University and Agriculture and Mechanical College, 1962), 27–29.

135. On unsuccessful strikes in Mobile, Pensacola, Gulfport, Galveston, Houston, Texas City, and the Hampton Roads district, see: Gilbert Mers, *Working the Waterfront: The Ups and Downs of a Rebel Longshoreman* (Austin: University of Texas Press, 1988), xiv; "South Atlantic and Gulf Coast District: 'Wage Negotiations with Employers,' " ILA, *Proceedings of the Twenty-Eighth Convention, Montreal, Quebec,*

August Tenth to Fifteenth, 1925, 89–94; Gulfport and Biloxi *Daily Herald,* September 26–29, October 1–6, 15, 17, 19, 20, 23, November 2, 4, 1923; *The Virginia-Pilot and The Norfolk Landmark,* October 7–14, 16, 1923.

136. Sterling D. Spero and Abram L. Harris, *The Black Worker: The Negro and the Labor Movement* (1931; rpt. New York: Atheneum, 1969), 186. While this pathbreaking book remains indispensable reading on African–American and labor history, there are simply too many factual inaccuracies in the authors' account of the New Orleans experience to treat it as reliable or definitive. Some contemporary black radicals were more impressed with the Dock and Cotton Council—indeed, with the Gulf Coast ILA—than subsequent historians have been. In a 1921 article praising the IWW-affiliated Marine Transport Workers Industrial Union of Philadelphia for its racial outlook, *The Messenger,* the journal of A. Philip Randolph and Chandler Owen, offered this impression of southern longshore workers: "Negro and white longshoremen of the Gulf ports have participated in a strike shoulder to shoulder on the docks, conducted by the International Longshoremen's Association, which is a comparatively conservative labor organization, [and have] contributed a stirring example of interracial solidarity to the history of race relations in America." *The Messenger 3* (July 1921), 214–15, quoted in Philip S. Foner and Ronald L. Lewis, *The Black Worker: A Documentary History from Colonial Times to the Present,* Vol. 6: *The Era of Post-War Prosperity and the Great Depression 1920–1936* (Philadelphia: Temple University Press, 1981), 413.

137. Daniel Rosenberg, *New Orleans Dockworkers: Race, Labor, and Unionism 1892–1923* (Albany: State University of New York Press, 1988), 169. Also see Robert C. Francis, "Longshoremen in New Orleans: The Fight Against 'Nigger' Ships," *Opportunity* 14 (March 1936), 82–85, 93; Northrup, "The New Orleans Longshoremen," 532–33; Herbert R. Northrup, *Organized Labor and the Negro* (New York: Harper & Brothers, 1944), 149–50; Dave Wells and Jim Stodder, "A Short History of New Orleans Dockworkers," *Radical America* 10, No. 1 (January–February 1976), 55; Miller, "A Study of the New Orleans Longshoremen's Unions," 24–25.

138. *Proceedings of the Twenty-seventh Convention, Boston ,Mass., May fourteenth to May twenty-first, 1923,* 248–55. In an article entitled "Longshoremen's Protective Union Benevolent Association Was Organized January 1872," in a special section on black New Orleans, the *Picayune* stated that the black longshoremen's union had 967 members, and 3,500 "passive members" in 1911. *Daily Picayune,* December 30, 1913 (Part 6).

139. "*Special Report:* Investigator #36 Reports, New Orleans, September 13, 1923," File 621-3-2, Part Five, From July 1, 1923 to October 15, 1923. Subject: Labor and Labor Conditions. . . , RG32, U.S. Shipping Board. In same file, see: Santa Cruz to Jenkins, Subject: Labor Situation, September 17, 1923.

Epilogue

1. E. J. McGuirk, Chairman, Labor Committee, New Orleans Steamship Association, to T. Seemes Walmseley, Mayor of New Orleans, July 15, 1930, in File 170–5859, Subject: Longshoremen. New Orleans, Louisiana. RG 280. In same file, also see: "To the Public of New Orleans" from "Organization Committee of Longshoremen," November 7, 1930; *Fourteenth Biennial Report of the Department of Commissioner of Labor and Industrial Statistics of the State of Louisiana 1927–28* (New Orleans: Wetzel Printing, 1928), 23; *Times-Picayune,* October 20, 1927.

2. Sterling D. Spero and Abram L. Harris, *The Black Worker: The Negro and the*

Labor Movement (1931; rpt. New York: Atheneum, 1969), 186–87; Robert C. Francis, "Longshoremen in New Orleans: The Fight Against 'Nigger' Ships," *Opportunity* (March 1936), 82–83; McGuirk to Walmsley, July 15, 1930, p. 5, in File 170–5359, RG 280.

3. "Longshore Labor Conditions in the United States—Part I," *Monthly Labor Review* 31, No. 4 (October 1930); "Longshore Labor Conditions in the United States—Part II," *Monthly Labor Review* 31, No. 5 (November 1930); also published as Boris Stern, *Cargo Handling and Longshore Labor Conditions* (U.S. Bureau of Labor Statistics Bulletin No. 550, Washington, D.C., 1932), 88–89. In File No. 170–5859–A, Subject: Longshoremen, New Orleans, RG280, see Howard Calvin, Commissioner of Conciliation, to H.L. Kerwin, Director, U.S. Conciliation Service, March 12, 1932; T.J. Darcy to John E. Jackson, IN RE: Stevedores' & Longshoremen's Lock-out, April 7, 1931. In File No. 170–5859, Subject: Longshoremen: New Orleans, Louisiana, RG 280, see Organization Committee of Local 1226 and Local 231, ILA, to U.S. Labor Commissioner, October 16, 1930; Organization Committee of Longshoremen To the Public of New Orleans, "Our Success Means Your Success," November 7, 1930; W. H. Rogers to H.L. Kerwin, March 13, 1931; "Reaction and Semi Slavery," (joint union flyer), no date; Canfield to Kerwin, September 30, 1930; McGuirk to Walmsley, July 15, 1930.

4. "Longshore Labor Conditions—Part II," 11–13; Miller, "A Study of the New Orleans Longshoremen's Unions," 27–31; David Lee Wells, "The ILWU in New Orleans: CIO Radicalism in the Crescent City, 1927–1957" (unpublished M.A. thesis, University of New Orleans, 1979); Paul Peters and George Sklar, *Stevedore: A Play in Three Acts* (New York: Covici Friede, 1934); Gilbert Mers, *Working the Waterfront: The Ups and Downs of a Rebel Longshoremen* (Austin: University of Texas Press, 1988); "Strike of Longshoremen on the Gulf Coast," *Monthly Labor Review* 42, No.2 (February 1936), 392–95; Robert C. Francis, "Dock Trouble in New Orleans," *The Crisis* 42 (December 1935), 372; Bruce Nelson, "The Triumph and Travail of a Left-led Union: The ILWU, from San Francisco to New Orleans," paper delivered at the Organization of American Historians meeting, March 1990, Washington, D.C.; Douglas L. Smith, *The New Deal in the Urban South* (Baton Rouge: Louisiana State University Press, 1988), 17, 21–22, 191–92.

INDEX

African Methodist Ministers Alliance, 219, 220
Agnew, Thomas, 79, 82
Airey, Thomas L., 66–67, 72
American Federation of Labor (AFL), 121, 166, 189–91, 194–95, 221, 229; and black workers, 151, 152, 153–54, 191–92; in New Orleans, 114, 118, 150–52, 153–54, 189–91, 141–42 (*see also* Central Labor Union (CLU); Central Trades and Labor Council (CTLC))
American Sugar Refinery, 230
Ameringer, Oscar, 193, 194, 199, 201
Anti-Lottery League, 146
arbitration, 98, 223, 225–26, 231, 238; and longshore disputes of 1906–07, 71, 72, 73, 74, 75
artisanal republicanism, 6, 7, 32
Association of Commerce, 211, 233, 235, 236, 238, 242

Badger, A. S., 55, 56, 57–58
Baltimore *Sun,* 143
Banks, Nathaniel P., 5, 15, 258 n.12, 259 n.15
Banville, John, 242
Barnes, Joseph W., 54–55
Behan, William J., 71, 81
Behrman, Martin, 183, 199–200, 213, 214, 239–42; anti-union policies of, on public works, 214, 240–41; cooperation of, with commercial elite in port modernization, 211, 240–41, 313 n.33; efforts of, to settle strikes in private sector, 171, 175, 178, 183, 198, 199–200, 224, 235, 237–38, 241; as leader of Choctaw Club, 211, 237; union leaders' divided support for, 237, 240, 241–42, 337–38 n. 106, 338–39 n. 111
Benson, William, 243
Bessant, Alcide, 125, 128
biracial unionism, 22, 56, 120–21, 192, 195–96, 254–55; attacks on, from outside, 155, 160, 169, 181–82, 199–200, 201–2; black unionists' support for, 52, 155, 163, 203; conditions favorable to, 74–75, 93, 119, 120, 158–59; elsewhere in U.S., 43, 91–92, 287 n.63, 307 n.121; freight handlers and, 100–101, 102–3, 105; as important reason for workers' power on New Orleans waterfront, 35, 64–65, 134, 159; Knights of Labor and, 43, 89, 91–92; limits of, 154–55, 183–85; longshoremen and, 95–98, 112,

134, 156–58, 184–85, 254–55, 314 n.44; in other New Orleans industries, 94–95, 317–18 n.63, 318 n.70; screwmen and, 158, 163–64, 298 n.40, 312 n.25; structure of, 94, 95, 149 (*see also* Cotton Men's Executive Council; Dock and Cotton Council); temporary repudiation of, by whites, in mid-1890s, 118, 122–23, 134–40, 143–45; white unionists' support for, 183, 188–89, 199–200, 207, 312 n.25; *see also* work-sharing
Bishop, Julia Truitt, 212
Black, Wilbert, 242
Black Codes, 18
Black Republican, 8
black community of New Orleans: before and during Civil War, 13–15, 16, 19–20, 262 n.33; class divisions among, 13, 14–18, 26, 32, 50–52; community organizations of, 13–14, 50–51, 84–86, 218–19; challenges by, to discrimination, 28, 88–89, 185–87, 218–19, 267 n.75; jobs held by, 14–15, 126, 154; jobs held on the waterfront, 38–39, 250; newspapers of, *see Black Republican; New Orleans Tribune; Southwestern Christian Advocate; Weekly Louisianian; Weekly Pelican;* and politics, 8, 25, 26–27, 145, 151 (*see also* Republican party, blacks as primary base of support for); population of, 13, 14, 126, 211, 275 n.51; prominence of black union leaders in, 84–86, 87, 186, 187, 188, 218, 285 n.41; violence against, 18–19 (*see also* riots)
Black Worker, The (Spero and Harris), 92, 249
Blanchard, Newton, 200
Blassingame, John, 50, 83
Board of Trade, 115, 211
Bonacich, Edna, 124
Breen, James F., 127
Breen, John, 115, 116, 127, 139
breweries, 189–95, 321 n.87, 321–22 n.89
brokers, 107
Brotherhood of Timber Workers, 205
Brower, M. E., 101
Brown, S. P., 90
Building Trades Council, 219, 234
Burke, E. A., 45, 46, 47
Butler, Benjamin, 4, 5, 14
Byrnes, James, 201, 234, 241, 242

Callahan, John, 115, 116, 152, 213–14, 215
Capdevielle, Paul, 147, 157, 168, 183

345